Medicinal Compounds and Nanotechnology in Therapeutics and Precision Healthcare

This book reviews the evolution, development, and application of medicinal compounds, beginning with their historical context and advancing to modern innovations. It describes the therapeutic potential of bioactive compounds from microorganisms, the role of antioxidants and anti-inflammatory agents in managing chronic diseases, and the use of small molecules in pain management and mental healthcare. It further explores immunomodulatory compounds, epigenetic modifiers, and the latest advancements in drug development, including integrative medicine approaches and the applications of nanotechnology in traditional medicine. Additionally, it addresses the use of immunomodulatory agents such as methotrexate and paclitaxel in cancer treatment, regulatory considerations for ensuring safety and efficacy in medicinal compounds, and the growing field of nutraceuticals. Toward the end, it covers the extraction, profiling, and absorption, distribution, metabolism, excretion, and toxicity (ADMET) evaluation of medicinal compounds; intellectual property challenges; and the significance of traditional medicine in modern healthcare.

Key Features:

- Reviews the development of medicinal compounds from historical remedies to modern innovations.
- Presents the therapeutic potential of bioactive compounds from microorganisms and the crucial roles of antioxidants and anti-inflammatory agents in disease management.
- Describes the innovations in drug delivery systems, epigenetic modifiers, and immunomodulatory agents.
- Provides guidance on regulatory considerations and intellectual property in medicinal compound development.
- Examines the role of nanotechnology and the integration of traditional and modern therapies.

This book is intended for researchers of ethnopharmacology, pharmaceutical sciences, and medicinal chemistry.

Medicinal Compounds and Nanotechnology in Therapeutics and Precision Healthcare

Edited by
Vibhav Gautam,
Hemant Kumar Gautam,
Rajiv Pathak, and Lalit Kumar

CRC Press
Taylor & Francis Group
Boca Raton London New York

CRC Press is an imprint of the
Taylor & Francis Group, an **informa** business

Designed cover image: Swati Singh, Vibhav Gautam, Hemant Kumar Gautam, Rajiv Pathak, and Lalit Kumar

First edition published 2027
by CRC Press
2385 NW Executive Center Drive, Suit 320, Boca Raton FL 33431

and by CRC Press
4 Park Square, Milton Park, Abingdon, Oxon, OX14 4RN

CRC Press is an imprint of Taylor & Francis Group, LLC

ISBN: 9781041225843 (hbk)
ISBN: 9781041225867 (pbk)
ISBN: 9781003733010 (ebk)

DOI: 10.1201/9781003733010

Typeset in Sabon
by codeMantra

This book is dedicated to all researchers and scholars who are committed to advancing science for the betterment of human health.

We also dedicate this work to our mentors, whose guidance has shaped our academic and professional journeys, and to our families for their unwavering support, patience, and encouragement.

Contents

Foreword

Medicinal Compounds and Nanotechnology in Therapeutics and Precision Healthcare represents a significant and timely contribution to the rapidly evolving field of biomedical and pharmaceutical sciences. By integrating diverse perspectives on medicinal compounds with advances in nanotechnology and precision medicine, this volume addresses some of the most pressing challenges in modern healthcare.

The editors have successfully compiled a collection of chapters that span foundational concepts and advanced innovations, including small molecule therapeutics, epigenetic regulation, natural product-based drug discovery, computational approaches, and theranostic strategies. The inclusion of emerging topics such as targeting cancer stem cells and nanotechnology-driven interventions highlights the forward-looking nature of this work.

One of the key strengths of this book lies in its interdisciplinary approach, which reflects the convergence of multiple scientific domains in the pursuit of more effective, targeted, and personalized therapies. The content is presented in a clear and structured manner, making it accessible to students while also providing valuable insights for researchers and professionals.

This volume is expected to serve as a useful reference for academic and research communities and to contribute meaningfully to ongoing advancements in therapeutic science and precision healthcare.

Prof. Satya Narayan Sankhwar
Director, Institute of Medical Sciences
Banaras Hindu University, Varanasi, India

Preface

The field of medicinal science is undergoing a transformative shift driven by advances in molecular biology, nanotechnology, and computational approaches. *Medicinal Compounds and Nanotechnology in Therapeutics and Precision Healthcare* aims to capture these developments and present a comprehensive overview of emerging strategies in drug discovery and targeted therapy.

This edited volume brings together contributions from experts across diverse domains, highlighting the critical roles of small molecules, natural products, and epigenetic modulators in modern therapeutics. This book also explores the growing impact of nanotechnology in enhancing drug delivery, improving bioavailability, and enabling precision targeting in complex diseases such as cancer and neurological disorders.

The chapters are thoughtfully organized to provide a logical progression from the historical evolution of medicinal compounds to advanced developments in biomedical nanotechnology, computational drug discovery, and theranostics. Particular emphasis has been placed on interdisciplinary approaches, reflecting the integration of chemistry, biology, pharmacology, and computational sciences in addressing current healthcare challenges.

This book is intended to serve as a valuable resource for students, researchers, academicians, and professionals in pharmaceutical sciences, biotechnology, and related fields. It is our hope that this work will contribute to a deeper understanding of the subject and inspire further innovation in the development of effective and personalized therapeutic strategies.

Despite careful efforts to ensure accuracy and completeness, any inadvertent errors remain our responsibility. We welcome constructive feedback from readers for future improvements.

Acknowledgments

The editors express their sincere gratitude to all contributing authors for their valuable chapters and for sharing their expertise across a wide range of emerging topics in medicinal science and nanotechnology. Their scholarly contributions have been instrumental in shaping the quality and scope of this volume.

We are deeply thankful to our mentors, colleagues, and collaborators for their insightful discussions, constructive feedback, and continuous encouragement throughout the preparation of this book.

We also acknowledge the support provided by our respective institutions, which have fostered an environment conducive to academic excellence and research innovation.

Special thanks are extended to the editorial and production team at Taylor & Francis for their professional guidance, timely support, and assistance in bringing this book to fruition.

Finally, we express our heartfelt gratitude to our families for their patience, understanding, and constant encouragement during the course of this work.

Dr. Vibhav Gautam also acknowledges the efforts of PhD student Ms. Swati Singh in generating the cover image for this book using a photoacoustic imaging platform at Banaras Hindu University, Varanasi, India.

About the Editors

Vibhav Gautam received his postgraduate degree in Biotechnology from Savitribai Phule Pune University, Pune, India. He earned his PhD degree in Biotechnology from the National Institute of Plant Genome Research, India (degree awarded by Jawaharlal Nehru University, New Delhi, India). He is currently working as an assistant professor at the Centre of Experimental Medicine and Surgery, Institute of Medical Sciences, Banaras Hindu University, Varanasi, India. Presently, Dr. V. Gautam's work involves the use of various spectroscopy-based techniques to identify the lead compound(s) derived from fungal endophytes that show potential activity against breast cancer. Furthermore, his work also includes the use of various microscopic techniques to study the surface topology and architecture of bioengineered nanoparticles mediated by fungal endophytes.

Hemant Kumar Gautam is a microbiologist with extensive academic and international research experience. He earned his Master's and Ph.D. degrees in Microbiology from the Indian Agricultural Research Institute, New Delhi, followed by postdoctoral research in France and Israel. He is currently a Professor of Biotechnology at Delhi Skill and Entrepreneurship University, Delhi, and formerly served as Chief Scientist at the Institute of Genomics and Integrative Biology, New Delhi. He has made substantial contributions to microbiology, genomics, and biotechnology research and has extensive experience in teaching, research, and scientific leadership. He is affiliated with several universities and is an active member of numerous national and international scientific and academic organizations. Dr. H. K. Gautam is a recipient of the SARC Award, Bharat Excellence Award, Biotechnology Award, Israel Government Fellowship, UNESCO Fellow, and International Project Reviewer RBUCE-UP, UniverSud, Paris. He has authored over 100 academic papers and two books and has submitted over 500 novel sequences to the NCBI database. He has visited several countries, including France, Israel, Australia, Bulgaria, China, Thailand, Germany, the United States, Singapore, Vietnam, Malaysia, and Ukraine. He is a biosafety expert from the Department of Biotechnology, India. Dr. H. K. Gautam is involved in a variety of interdisciplinary projects, including in the areas of microbiology, microbial biotechnology, genomics, and nanobiotics. His goal is to develop techniques for rationally engineering next-generation smart antimicrobials by understanding their fundamental features and exploiting such structures to battle the "antimicrobial resistance" silent pandemic knocking on the door.

Rajiv Pathak began his academic journey with an MSc from the University of Lucknow, India (2007–2009). He went on to pursue a PhD at the CSIR-Institute of Genomics and Integrative Biology, Delhi, India (2010–2016), where his research focused on understanding the role of potential G-quadruplex-forming sequences in microbial genomes. Following this, he undertook a scientifically enriching postdoctoral tenure in the Department of Genetics at the Albert Einstein College of Medicine, New York, USA (2017–2022). Since 2022, he has been serving as a staff scientist in the same department. With over nine years of postdoctoral research experience in virology and related biomedical fields, Dr. Pathak has contributed significantly to multiple interdisciplinary projects aimed at advancing therapeutic strategies and addressing key questions in infectious disease research. His work spans a wide spectrum, including the functional relevance of G-quadruplexes in microorganisms, the development of novel therapeutic approaches against multidrug-resistant bacteria, HIV-1 latency and persistence, early replication kinetics of SARS-CoV-2, and the molecular pathogenesis of SMARCB1-associated atypical teratoid/rhabdoid tumors (a rare pediatric cancer). His scientific expertise encompasses a diverse range of model systems, from microorganisms—including viruses, multidrug-resistant bacteria, and yeast—to human cell lines, iPSC-derived microglial cells, and COVID-19 autopsy samples. He possesses excellent communication and interpersonal skills, reflected in his extensive scholarly output: over 19 original research publications (including nine as first or co-first author) and 15 review articles/book chapters (with two as first/co-first author and six as corresponding author). In total, he has authored approximately 34 peer-reviewed scientific articles published in high-impact international journals such as *Nucleic Acids Research*, *Acta Neuropathologica*, and *Nature Communications*. Dr. Pathak's pioneering contributions have received international recognition, including accolades at prominent scientific forums such as the prestigious Cold Spring Harbor Laboratory in New York. His commitment and dedication to scientific advancement are further demonstrated by his active involvement in the academic community. He has served as an editorial board member for six journals and reviewed over 60 manuscripts for more than 20 respected journals, including top-tier publications such as *Nature Communications* and *Cell Press*. Notably, he was invited to serve as a guest editor for the *Proceedings of the National Academy of Sciences* (PNAS). In addition to his editorial responsibilities, Dr. Pathak has also contributed as an international grant reviewer for the National Science Center (NSC) of Poland under its OPUS-26 funding scheme.

Lalit Kumar received his MBBS and Master of Surgery (MS) degrees from the Institute of Medical Sciences, Banaras Hindu University, Varanasi, India. He subsequently completed his MCh degree in urology from S.P. Medical College, Bikaner, India. He is currently working as an assistant professor in the Department of Urology, Institute of Medical Sciences, Banaras Hindu University, Varanasi, India. Dr. Kumar has undergone advanced training through prestigious fellowships, including a Fellowship in Laparoscopic and Robotic Urology at AIIMS, New Delhi; the ATLAS Internationale Fellowship in Robotic Surgery by the Vattikuti Foundation at Roswell Park Cancer Institute, USA; and observerships at Singapore General Hospital and National University Hospital, Singapore. He is an expert in minimally invasive laparoscopic and robotic surgeries for the management of various urological tract pathologies. Currently, his work involves the integration of high-throughput transcriptomic and metabolomic profiling to uncover molecular signatures associated with various urological cancers. Furthermore, Dr. Kumar's research focuses on biomarker discovery and the development of novel diagnostic and prognostic tools to enhance early detection and precision management of bladder and prostate cancers.

Chapter 1

Small molecules in pain management and mental healthcare

Aditi V. Bagade[1], *Anjali Yadav*[2], *Nityanand Srivastava*[3], *and Ajeet Kumar*[4]

[1]Department of Neurology, Washington University School of Medicine in St. Louis, St. Louis, MO

[2]Department of Genetics, Washington University School of Medicine in St. Louis, St. Louis, MO

[3]Department of Cell Biology, Albert Einstein College of Medicine, New York

[4]Department of Psychiatry, Washington University School of Medicine in St. Louis, St. Louis, MO

1.1 INTRODUCTION

The World Health Organization (WHO) estimated that neurological conditions are the leading cause of disability worldwide based on the Global Burden Of Diseases 2021. Chronic pain and mental illness are highly interconnected. Epidemiological studies indicate that 20%–30% of adults experience chronic pain while 10%–20% suffer from mental health disorders such as depression and anxiety [1]. Notably, up to half of individuals with chronic pain also exhibit symptoms of mental illness, highlighting a bidirectional relationship where each condition exacerbates the other. Both conditions share overlapping neurobiological pathways, including dysregulation in brain regions like the prefrontal cortex and limbic system, and involve common neurotransmitters such as serotonin, norepinephrine, and dopamine. Small-molecule therapeutics play a crucial role in managing both pain and mental illness by targeting specific molecular pathways that underlie these conditions. In this chapter, we discuss clinically used small molecules with targets that are commonly used as therapeutics to manage pain and mental health.

DOI: 10.1201/9781003733010-1

1.2 PAIN

An estimated 1.5 billion individuals worldwide live with chronic pain due to modern lifestyle factors, with its burden steadily increasing as populations around the world continue to age. Prolonged exposure to chronic pain is believed to contribute to additional health complications, further deteriorating individuals' quality of life [1]. As a result, it is essential to utilize effective treatment strategies to manage this prevalent and debilitating condition. The International Association for the Study of Pain (IASP) characterizes pain as an unpleasant sensory and emotional experience associated with tissue injury but can also emerge in some conditions without any identifiable tissue damage. In general, pain-sensing nerve endings are activated at the site of tissue damage (e.g., trauma or inflammation) [2]. Once activated, these sensory nerves transmit the signal to the brain via primary afferent neuron and ascending neuron pathways (e.g., spinothalamic tract) [3]. Therefore, it is pertinent to understand the mechanisms of pain transduction, transmission, and perception, as it is essential for elucidating pain pathways (Figure 1.1). Further, pain without tissue damage arises due to central sensitization of primary sensory neurons, such as fibromyalgia, which is characterized by widespread pain and heightened pain sensitivity without clear peripheral injury, while post-surgical pain may persist beyond normal healing due to maladaptive neural plasticity [4]. Similarly, neuropathic pain represents a multifaceted and highly disabling condition affecting millions of individuals worldwide. Despite the availability of multiple therapeutic approaches, clinical efficacy remains suboptimal, and many treatments are associated with considerable adverse effects. In the context of surgery, acute postoperative pain is frequently insufficiently controlled, with approximately 80% of patients reporting pain of moderate to severe intensity [5]. In response to these challenges, contemporary pain management paradigms have shifted toward multimodal analgesia, an approach that integrates different classes of analgesics across the perioperative period to more effectively target both nociceptive and centrally mediated components of pain.

1.2.1 Management of pain

In the 1600s, European physicians started opium use to relieve pain in patients, and by the 1900s, morphine and heroin were introduced as medication for pain [6,7]. Chloroform and ether were introduced as anesthetics for surgery in the mid-19th century [7]. In 1897, Felix Hoffmann synthesized aspirin from salicylic acid of willow bark [8]. It reduced the salicylic acid side effects and became the first widely used non-opioid analgesic. Later in the 1970s, John Vane and colleagues demonstrated that aspirin and other anti-inflammatory medicines exert their effects by inhibiting the cyclooxygenase (COX) enzyme [9]. These drugs, more commonly

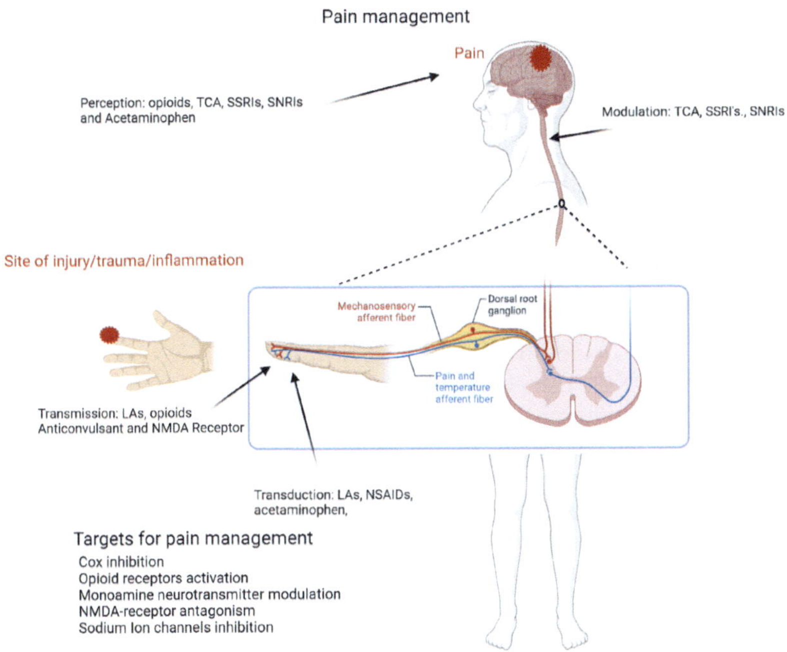

Figure 1.1 Illustration of different pharmacological targets along with various steps of pain pathway for management of pain. Pain transmission occurs across multiple levels, from tissue trauma to brain perception, various pharmacological interventions block the specific steps of pain pathway. At the site of injury, local anesthetics (LAs), anti-inflammatory agents (COX inhibitors) provide direct pain relief. Additionally, opioids and anticonvulsant, and NMDAR antagonist stop the transmission of pain. Monoamine modulators including TCAs, SNRIs and SSRIs stop the modulation of pain signal in spinal cord. Perception of pain in brain can be modulated by monoamine modulators and acetaminophen.

referred to as nonsteroidal anti-inflammatory drugs (NSAIDs), function by inhibiting COX enzymes. Globally, COX inhibitors are among the most frequently used medications for the treatment of acute and chronic nociceptive pain, including inflammatory pain. Following the widespread use of NSAIDs, local anesthetics (LAs) like lidocaine were developed to block sodium channels and suppress nerve signal transmission. In the 1990s, anticonvulsants like gabapentin emerged as effective treatments for neuropathic pain by modulating calcium channels and regulating neurotransmitter release. Beyond their traditional use, tricyclic antidepressants (TCAs) and serotonin-norepinephrine reuptake inhibitors (SNRIs) have found a role in chronic pain therapy by amplifying descending neuron inhibitory pathways, offering an effective and mechanistically distinct strategy to control persistent pain. Some recent targets include calcitonin gene-related peptide (CGRP) inhibitors, such as zavegepant, which have transformed migraine treatment, and suzetrigine, a selective voltage-gated sodium channel (VGSC) inhibitor for pain. These advancements have provided alternatives to opioids and broadened treatment options. Overall, multiple targets participate in the management of several types of pain by modulating specific steps in the pain pathway (Figure 1.1). In the following section, we discuss each class of molecules and its target in the management of pain.

1.2.1.1 Nonsteroidal anti-inflammatory drugs

The action of NSAIDs involves the suppression of COX enzymes, both COX-1 and COX-2, thereby reducing the production of prostaglandins, key drivers for triggering pain, inflammation, and elevated body temperature [10]. Over the past four decades, a wide range of COX inhibitors have been developed. Further, advances in understanding the function of COX enzyme facilitate the development of selective COX-2 inhibitors, or coxibs, which have seen extensive use and promotion over the last two decades. NSAIDs exhibit varying levels of COX-1 selectivity and offer effective pain relief, fever reduction, and anti-inflammatory effects, which make them valuable for managing diverse clinical conditions. However, despite their clinical significance, they also carry notable risk factors such as gastrointestinal (GI) side effects, including ulceration and bleeding [10–12]. Therefore, COX-2 selective inhibitors were designed to be safer alternatives and intended to reduce damage to the gastric lining. In addition to the GI risk, several studies revealed potential cardiovascular risks, emphasizing the need to carefully weigh their therapeutic benefits against safety considerations in clinical practice [13–16]. COX-2 is majorly involved in inflammation, as its expression is typically low but increases in response to inflammation and pain (nociception) [17,18]. Based on COX inhibition selectivity, NSAIDs are listed in Table 1.1 along with their COX selectivity

Table 1.1 Example of NSAIDs with their target selectivity and condition where they use

Drug name	*COX selectivity*	*Use*
Aspirin	COX-1 > COX-2	Pain, inflammation, fever, cardiovascular protection
Ibuprofen	COX-1 > COX-2	Pain, inflammation, fever, arthritis
Naproxen	COX-1 > COX-2	Pain, inflammation, arthritis
Indomethacin	COX-1 > COX-2	Severe inflammation, arthritis, gout
Diclofenac	COX-2 > COX-1	Pain, inflammation, arthritis
Meloxicam	COX-2 > COX-1	Pain, inflammation, arthritis
Nabumetone	COX-2 > COX-1	Pain, inflammation, arthritis

(Continued)

Table 1.1 (Continued) Example of NSAIDs with their target selectivity and condition where they use

Drug name	*COX selectivity*	*Use*
Nimesulide	COX-2>COX-1	Pain, inflammation, osteoarthritis (hepatotoxicity concerns)
Etodolac	COX-2>COX-1	Pain, inflammation, osteoarthritis, rheumatoid arthritis
Etoricoxib	COX-2 Selective	Pain, osteoarthritis, rheumatoid arthritis
Celecoxib	COX-2 Selective	Pain, osteoarthritis, rheumatoid arthritis
Rofecoxib	COX-2 Selective	Pain, osteoarthritis, withdrawn
Valdecoxib	COX-2 Selective	Pain, osteoarthritis, withdrawn
Lumiracoxib	COX-2 Selective	Pain, osteoarthritis, withdrawn in some countries
Piroxicam	COX-1 =COX-2	Arthritis, long-term inflammation
Ketoprofen	COX-1 =COX-2	Pain, inflammation, fever
Sulindac	COX-1 =COX-2	Pain, inflammation, arthritis
Oxaprozin	COX-1 =COX-2	Arthritis, long-term inflammation

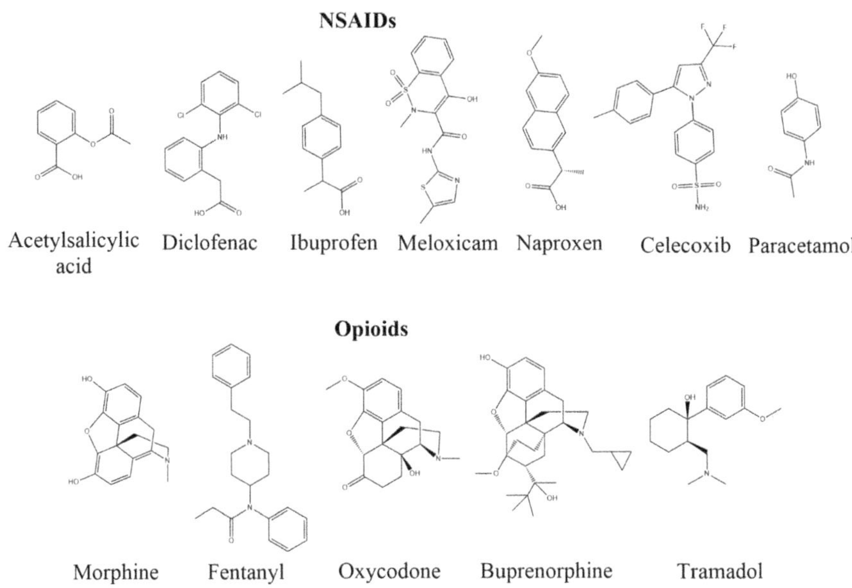

Figure 1.2 Illustration of chemical structure of various NSAIDs and opioids.

and the common conditions they are used for, while the chemical structures of some common NSAIDs are illustrated in Figure 1.2.

Initially developed for chronic pain management, COX-2-selective inhibitors have demonstrated efficacy comparable to that of traditional non-selective NSAIDs. Subsequently, their clinical use has gradually expanded to include the treatment of acute medical and postoperative pain. However, emerging safety concerns have prompted a critical re-evaluation of their risk-management profile in clinical practice. Consequent to emerging safety concerns, several COX-2 inhibitors, such as rofecoxib (Vioxx), valdecoxib (Bextra), and lumiracoxib (Prexige), were discontinued from clinical use [16,19,20]. One common COX-2 inhibitor that has been approved in both the United States and the European Union (EU) is celecoxib. In addition to celecoxib, the EU has also approved etoricoxib, which shares a pharmacological profile with rofecoxib but has not been authorized for clinical use in the U.S. Etoricoxib has demonstrated greater efficacy than nonselective NSAIDs in reducing pain and inflammation, whereas celecoxib provides increased potency at higher doses [21]. NSAIDs are available in various formulations, including oral, topical, and suppository forms, and are widely used for musculoskeletal pain, postoperative pain, arthritis, tendonitis, back pain, toothaches, menstrual cramps, and bursitis.

Topical diclofenac and ketoprofen offer moderate pain relief for osteoarthritis, though evidence for other chronic pain conditions is limited. When used short-term for postoperative pain at the lowest effective dose, NSAIDs are safe and can reduce opioid use and related side effects [22]. However, care must be taken to avoid hepatotoxicity, including the other health risks discussed earlier, such as GI bleeding, cardiovascular events, renal dysfunction, and delayed bone healing, especially at high doses. Paracetamol (acetaminophen) is a commonly used drug for alleviating pain and reducing fever, with only modest anti-inflammatory effects. Although its precise mechanism remains unclear, paracetamol is thought to act primarily within the central nervous system (CNS) by suppressing prostaglandin synthesis through COX enzymes, potentially COX-3, and by modulating both endocannabinoid and serotonergic pathways [23–25]. It is metabolized to *N*-arachidonoylphenolamine (AM404), a compound that may contribute to its analgesic effects by activating transient receptor potential vanilloid type 1 receptors [25–27]. However, acetaminophen is not considered an NSAID, and when combined with NSAIDs, it enhances pain relief by targeting both central and peripheral pain mechanisms, reducing the need for higher doses of either drug.

1.2.1.2 Opioids

Opioids are a class of substances derived from the poppy plant and can be classified into three groups based on their origin, such as natural opioids directly obtained from the poppy plant, which include substances such as codeine, morphine, and opium. They are also called opiates, extracted from the resin surrounding the poppy seeds. Semi-synthetic opioids are created in the laboratory by chemical modification. Natural opioids combined with synthetic elements to create more potent compounds than their natural counterparts, such as heroin, oxycodone, hydrocodone, hydromorphone, and oxymorphone.

Synthetic opioids are entirely synthesized in laboratories and do not derive from the poppy plant. Fentanyl, methadone, and tramadol are examples of synthetic opioids. In addition to these exogenous opioids, the body produces endogenous opioid peptides, including enkephalins, dynorphins, and endorphins, through the nervous system and other cell types. These endogenous opioids have been shown to exert significant analgesic effects. Chemical structures of several opioid types are illustrated in Figure 1.2.

Opioids exert their effects by agonizing specific receptors in the brain and throughout the body, thereby reducing the transmission of pain signals to the CNS [28,29] (Table 1.2). Opioid receptors belong to the Class A (rhodopsin-like) family of G protein-coupled receptors [29]. Four distinct subtypes have been identified—μ, δ, κ, and nociceptin receptors (MOR, DOR, KOR, and NOR, respectively), which share over 60% sequence homology [30,31].

Table 1.2 Opioids along with their primary target and conditions where they are used

Opioids	*Primary targets*	*Mechanism of action/use*
Morphine	MOR agonist	Activates MOR, inhibiting pain signaling in the CNS
Fentanyl	MOR agonist	Highly potent MOR agonist, used for severe pain and anesthesia
Oxycodone	MOR agonist	Moderate-to-strong analgesic, often used for chronic pain
Hydrocodone	MOR agonist	Like oxycodone, used for moderate to severe pain
Methadone	MOR agonist	Long-acting opioid used for pain and opioid dependence
Buprenorphine	Partial MOR agonist, KOR receptor antagonist	Used for pain management and opioid dependence
Tramadol	Weak MOR agonist	Provides pain relief via opioid and non-opioid mechanisms
Tapentadol	MOR agonist	Like tramadol but with stronger MOR activity
Codeine	Prodrug converted to morphine (MOR agonist)	Mild-to-moderate pain relief, often combined with acetaminophen
Hydromorphone	MOR agonist	More potent than morphine, used for severe pain
Oxymorphone	MOR agonist	Strong opioid for severe pain management
Nalbuphine	KOR agonist, partial MOR antagonist	Used for moderate to severe pain with reduced addiction risk
Butorphanol	KOR agonist, partial MOR antagonist	Used for migraine, labor pain, and post-op pain
Pentazocine	KOR agonist, weak MOR antagonist	Less potent analgesic, used for moderate pain

All four receptor subtypes are coupled to inhibitory G proteins (Gαi and Gαo) and, when activated by either endogenous or exogenous ligands, inhibit adenylate cyclase and initiate a range of intracellular signaling cascades that decrease neuronal excitability [30,32]. The anti-nociceptive effects of opioid receptor activation are primarily mediated through modulation of ion channels. Following receptor activation, the Gβγ subunit separates from the G protein complex and suppresses presynaptic calcium channel activity, leading to diminished calcium-dependent vesicle fusion and reduced neurotransmitter release [33–36]. Opioids suppress neurotransmission by modulating calcium and potassium channels, producing rapid effects within seconds to minutes, whereas with chronic exposure, these effects can persist over longer timescales, ranging from hours to days [30,37,38]. Abruptly interrupted exposure to opioid signaling can promote excitatory synaptic plasticity and cause opioid-induced hyperalgesia [39–41]. Most analgesic opioids show a greater affinity for the μ-opioid receptor compared to other opioid receptor subtypes, and their analgesic effects are therefore primarily mediated through MOR activation [28,42].

Opioid receptors are widely expressed across both the CNS and peripheral nervous system (PNS), where they govern a range of physiological functions from pain relief at spinal and supraspinal levels to emotional responses such as euphoria and dysphoria [43–45]. Targeting peripheral opioid receptors in the PNS is considered a potentially safe and more selective approach to pain management, as it can provide analgesia without eliciting euphoria symptoms [46,47]. Opioid receptors represent critical therapeutic targets for treating both acute and chronic pain, with agents such as hydrocodone, oxycodone, morphine, codeine, and fentanyl commonly prescribed to manage severe pain. In intensive care units (ICUs), opioids are crucial for pain management and are among the oldest and most effective medications available for relieving pain [48]. Commonly used opioids in these settings include morphine sulfate, fentanyl, and sufentanil, all known for their high effectiveness in pain relief. However, it is important to note that opioids can accumulate in patients with renal failure, increasing the risk of toxicity. Therefore, careful monitoring and cautious administration are essential for these patients. The efficacy of opioids differs significantly between acute and chronic pain, with stronger evidence supporting their use in acute pain [49–51]. They are considered a cornerstone in perioperative pain management and are highly effective for controlling severe pain following surgical procedures, including major thoracic and abdominal operations, spinal fusion, cesarean section, and coronary artery bypass grafting.

1.2.1.3 Anticonvulsants

Anticonvulsants, also referred to as antiepileptic or anti-seizure medications, are U.S. Food and Drug Administration (FDA)-approved as the first-line treatment in neurointensive care units for the prevention of seizures and status epilepticus by regulating abnormal electrical activity in the brain [52–54]. Beyond their role in treating epilepsy, these medications are utilized for various non-epileptic conditions, including symptomatic pain management, migraines, and certain psychiatric disorders [55–58]. They modulate neuronal excitability through several mechanisms, such as blocking sodium channels, inhibiting voltage-gated calcium channels, enhancing GABAergic activity, and reducing glutamatergic activity [52,59–61]. Common anticonvulsants are listed in Table 1.3 with targets and their modes of action.

Anticonvulsants are considered first-line treatments for a range of conditions such as neuropathic pain, trigeminal neuralgia, migraine headaches, and myofascial pain [61–64]. Among the most commonly used antiepileptic drugs for managing neuropathic pain and fibromyalgia are carbamazepine, gabapentin, lamotrigine, oxcarbazepine, pregabalin, topiramate, and valproic acid [65–67]. Clinical research supports the use of gabapentin and

Table 1.3 List of anticonvulsants along with their primary targets and mechanism of action

Anticonvulsant	*Primary targets*	*Mechanism of action*
Gabapentin	α2δ subunit of voltage-gated calcium channels (VGCCs)	Reduces excitatory neurotransmitter release
Pregabalin	α2δ subunit of VGCCs	Inhibits calcium influx, reducing neurotransmitter release
Carbamazepine	Voltage-gated sodium channels (VGSCs)	Stabilizes neuronal membranes, reducing hyperexcitability
Oxcarbazepine	VGSCs	Like carbamazepine but with fewer side effects
Lamotrigine	VGSCs, glutamate receptors	Inhibits sodium channels and reduces glutamate release
Valproic acid	GABA transaminase, VGSCs	Enhances GABAergic activity, inhibits sodium channels
Topiramate	AMPA/kainate glutamate receptors, VGSCs, GABA-A receptor	Inhibits excitatory neurotransmission and enhances GABA
Levetiracetam	Synaptic vesicle protein 2A (SV2A)	Modulates neurotransmitter release
Zonisamide	VGSCs and VGCCs	Inhibits neuronal excitability
Lacosamide	Slow inactivation of VGSCs	Stabilizes hyperexcitable neuronal membranes
Ethosuximide	T-type calcium channels	Reduces abnormal firing in absence seizures
Felbamate	NMDARs, GABA-A receptors	Inhibits excitatory and enhances inhibitory signaling
Perampanel	AMPA glutamate receptors	Selective non-competitive AMPA receptor antagonist
Rufinamide	VGSCs	Reduces excessive neuronal firing

pregabalin in reducing pain severity in targeted neuropathic conditions, such as fibromyalgia, peripheral diabetic nerve pain, postherpetic nerve pain, and central nervous system-related neuropathic pain [68–71]. These drugs have been FDA-approved, and multiple studies demonstrate their ability to reduce pain intensity by at least 50% [72–74]. Carbamazepine and oxcarbazepine are considered first-line options for trigeminal neuralgia [75,76]. In contrast, evidence for lacosamide and lamotrigine indicates a lack of effectiveness, while other anticonvulsants either lack sufficient evidence or have not been proven effective in managing neuropathic pain.

1.2.1.4 Antidepressants

Antidepressants are primarily recognized for their role in treating major depression; their analgesic properties are often underappreciated. Evidence suggests that certain TCAs and SNRIs are beneficial in controlling neuropathic and chronic pain, independently of their antidepressant effects [77–79]. Figure 1.3 illustrated the synaptic mechanism of action of

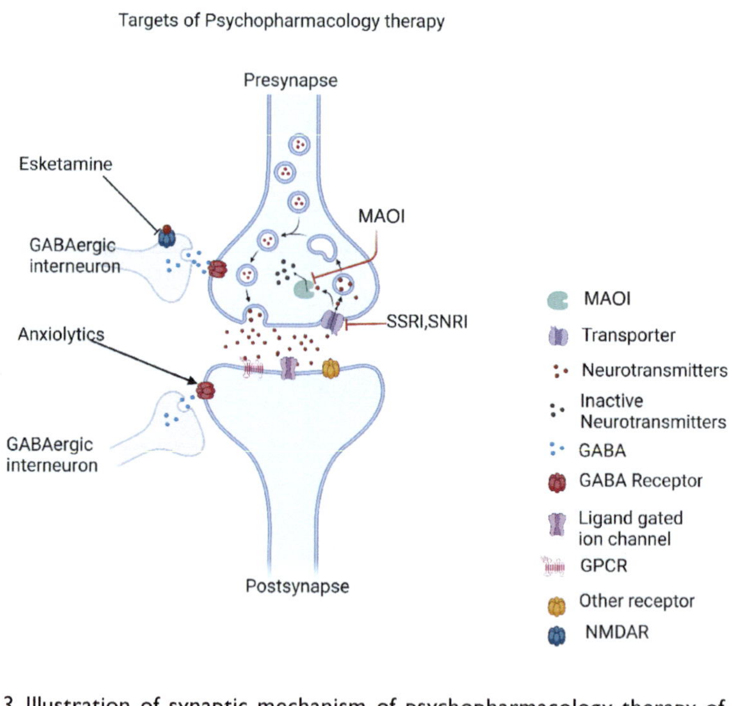

Figure 1.3 Illustration of synaptic mechanism of psychopharmacology therapy of pain and mental illness: Monoamine modulators (SSRI, SNRI, MAOI) enhance neurotransmitters' availability in the synaptic cleft by blocking reuptake transporters and the monoamine oxidase enzyme. Anxiolytics increase the GABA response and cause neuronal inhibition. Presynaptic disinhibition of neurotransmission by NMDAR antagonist esketamine.

common antidepressants used in pain and mental illness. TCAs like amitriptyline, nortriptyline, and imipramine are considered the first choice for neuropathic pain because they work in multiple ways [79]. They block the reuptake of serotonin and norepinephrine, enhance the body's natural pain-blocking pathways, and affect sodium and calcium channels [79,80]. They also have other effects that help reduce pain. These drugs are used for conditions such as diabetic neuropathy, shingles-related pain, fibromyalgia, chronic back pain, and migraines [80,81]. However, the evidence supporting TCAs is mixed; some studies show strong benefits, while others highlight side effects like drowsiness, dry mouth, and dizziness, which can make them hard to tolerate [82,83].

On the other hand, SNRIs like duloxetine, venlafaxine, and milnacipran are also effective for managing neuropathic and musculoskeletal pain [84–86]. These drugs elevate CNS levels of serotonin and norepinephrine, promoting pain suppression and decreasing central sensitization. Among them, duloxetine is FDA-approved for conditions like diabetic neuropathy, fibromyalgia, arthritis, chronic back pain, and chemotherapy-related neuropathy [87,88]. Studies show it provides moderate pain relief, often without relying on its antidepressant effects. Although TCAs tend to be more powerful, their side effects can limit their use [89]. SNRIs, especially duloxetine, are becoming more popular due to their similar pain relief and better tolerability. In contrast, selective serotonin reuptake inhibitors (SSRIs) like fluoxetine and paroxetine have low response to pain and are not usually recommended for pain treatment. Based on mode of action, antidepressants are listed in Table 1.6 along with the common conditions they are used for, while the chemical structures of all classes of antidepressants are illustrated in Figure 1.4. In summary, antidepressants are important for

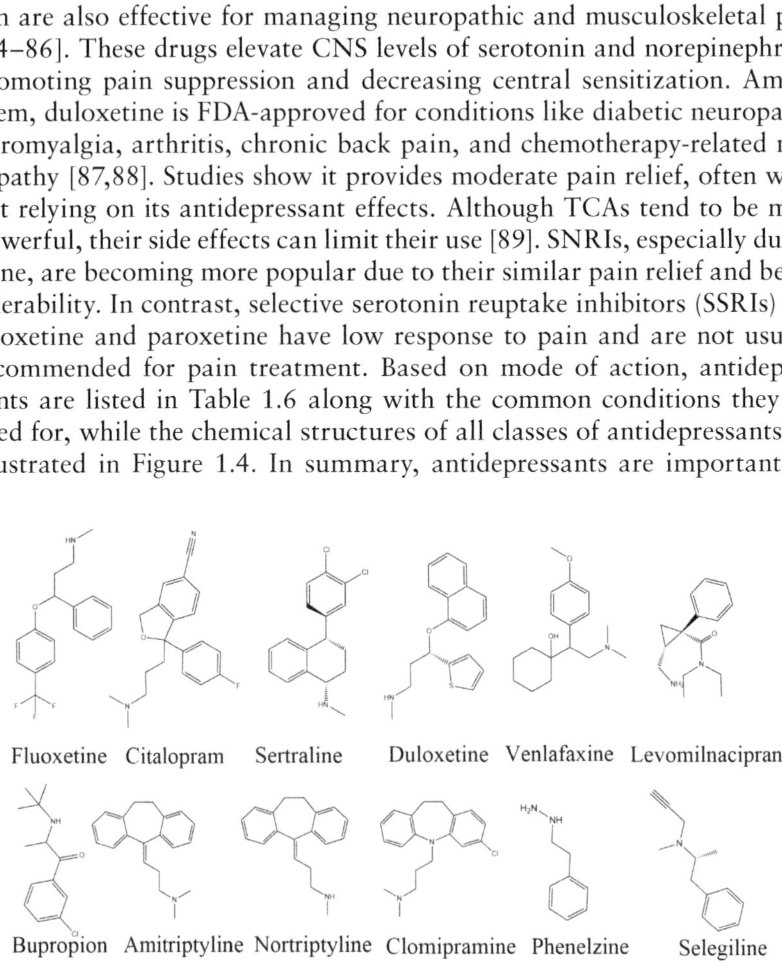

Figure 1.4 Illustration of the chemical structure of a recent emerging molecule to manage depression and some commonly used anticonvulsants, anxiolytics, and psychostimulants.

treating neuropathic pain. While TCAs are still widely used, SNRIs are increasingly preferred because they are safer and easier to tolerate, making them a better option for long-term pain management. We provide a more detailed discussion on antidepressants and their mode of action in the mental health section.

1.2.1.5 Local anesthetics

LAs are extensively employed in surgical anesthesia and pain management. They exert their effects primarily by inhibiting VGSCs, thereby preventing nociceptive afferent neurons from reaching the excitation threshold and blocking pain signal transmission [90,91]. LAs bind to the intracellular region of VGSCs, keeping them in an inactive state and preventing sodium ions from entering the neuron during depolarization, effectively stopping the initiation and transmission of nerve impulses [92–94]. It leads to an increase in the neuronal threshold potential, disruption of action potential conduction, and inhibition of nociceptive signal transmission, resulting in analgesia. LAs predominantly affect sensory nerve fibers, particularly unmyelinated C fibers and thinly myelinated Aδ fibers that mediate pain and temperature sensations [95]. At higher concentrations, LAs can also inhibit the function of larger myelinated motor fibers, leading to temporary muscle paralysis [96,97]. Clinically used LAs share a common chemical structure comprising three components: a hydrophilic nitrogen-containing group that facilitates cytoplasmic diffusion and water solubility; a lipophilic phenyl group that enhances membrane penetration and increases potency; and an intermediate chain that determines whether the drug belongs to the amide or ester class [98]. Based on their chemical structure, LAs are classified as amides or esters, which determines their metabolic pathway. Amide-type agents, including lidocaine, bupivacaine, ropivacaine, and mepivacaine, undergo hepatic metabolism and have a relatively longer duration of action, whereas ester-type drugs, such as procaine, chloroprocaine, and tetracaine, are rapidly hydrolyzed by plasma esterases, resulting in a shorter effect and action [98].

LAs are particularly effective in managing moderate to severe acute and chronic pain, offering the advantage of regional analgesia for localized pain relief without the need for systemic analgesics or general anesthesia, thereby reducing the risk of systemic toxicities [99]. To summarize, common LAs along with their targets and mode of action are enlisted in Table 1.3. However, their clinical use is limited by their relatively short duration of action due to rapid metabolism and redistribution. To extend their effectiveness, various strategies have been developed, including combining different LAs to optimize onset and duration, incorporating adjuncts such as epinephrine to reduce systemic absorption and prolong action, using opioid adjuvants to enhance analgesic effects, and developing advanced

Table 1.4 Local anesthetics with their primary targets and mechanism of action

Local anesthetic	*Target*	*Mechanism of action*
Lidocaine	VGSCs	Blocks sodium ion influx, stabilizes the inactivated state, prevents action potential propagation
Bupivacaine	VGSCs (preferentially Nav1.7, Nav1.8)	Prolonged blockade of sodium channels, strong affinity for cardiac VGSCs
Ropivacaine	VGSCs (Nav1.7, Nav1.8)	Like bupivacaine but with less cardiotoxicity, blocks sensory and motor fibers
Mepivacaine	VGSCs	Intermediate duration of action, used in regional anesthesia
Prilocaine	VGSCs	Less vasodilation than lidocaine
Articaine	VGSCs	Rapid onset, used primarily in dental anesthesia, metabolized by plasma esterases
Procaine	VGSCs	Short acting, metabolized by plasma cholinesterases, low potency
Tetracaine	VGSCs	Potent, long acting, used in spinal and ophthalmic anesthesia
Chloroprocaine	VGSCs	Short acting, rapid metabolism, used in epidural anesthesia
Levobupivacaine	VGSCs (Nav1.7, Nav1.8)	S-enantiomer of bupivacaine, reduced cardiotoxicity, used in regional anesthesia
Etidocaine	VGSCs	Long-acting, more potent motor blockade than sensory blockade

drug delivery systems such as liposomes, hydrogels, and microneedles to ensure sustained release and extended blockade duration [100–102]. These approaches have significantly improved the clinical utility of LAs, enabling more effective pain management with fewer systemic side effects (Table 1.4).

1.2.1.6 NMDAR antagonists

Located at excitatory synapses, *N*-methyl-D-aspartate receptors (NMDARs) are ionotropic glutamate receptors that play a pivotal role in neuronal communication and synaptic plasticity. Dysfunction of these receptors contributes to several neurological conditions, such as neuropathic pain, prompting the exploration of NMDAR antagonists for clinical use [103]. This has led to the exploration of NMDAR antagonists as possible clinical interventions [104]. Several NMDAR antagonists, including ketamine, memantine, methadone, dextromethorphan, riluzole, and levorphanol, are utilized for neuropathic pain management [104,105]. Ketamine, a potent NMDAR antagonist, is particularly effective in blocking receptor activation and reducing neuronal hyperexcitability associated with neuropathic pain [106]. Clinical studies support its efficacy in pain relief, though its potential for abuse and psychotomimetic effects limits widespread use [50,107]. Memantine has demonstrated clinical benefits in conditions such as complex regional pain syndrome and post-mastectomy pain [108]. It has been shown to reverse mechanical sensitivity and lower pro-inflammatory cytokine levels in preclinical models [109]. Unlike ketamine, memantine is generally well tolerated and does not carry a risk of abuse, making it a safer alternative for long-term use [110]. However, while some clinical trials support its effectiveness, broader clinical validation is needed. Other NMDAR antagonists, including methadone, dextromethorphan, riluzole, and levorphanol, have shown potential in preclinical and clinical studies, but strong evidence supporting their standalone use in neuropathic pain management remains limited [111]. Overall, NMDAR antagonists offer promising options for treating neuropathic pain, though their clinical application is often constrained by side effects, abuse potential, and variability in patient response.

1.2.1.7 Recently approved molecules for pain management

Recent advancements in small-molecule therapeutics have significantly transformed the landscape of pain management by offering targeted, non-opioid alternatives. One of the most promising developments in migraine treatment is the emergence of gepants, which act as selective CGRP receptor antagonists [112]. Orally available gepant molecules, including ubrogepant, rimegepant, and atogepant (Figure 1.5), function by blocking CGRP receptors, thereby preventing migraine attacks or alleviating symptoms during an episode [113]. In 2023, nasal spray was also approved as a gepant, known as zavegepant [114]. Unlike monoclonal antibody-based CGRP inhibitors, small-molecule CGRP antagonists offer greater flexibility in dosing and faster onset of action, making them particularly beneficial for acute migraine relief [115]. Additionally, because gepants do not cause vasoconstriction like traditional triptans, they are a safer option for patients with cardiovascular risk factors [116].

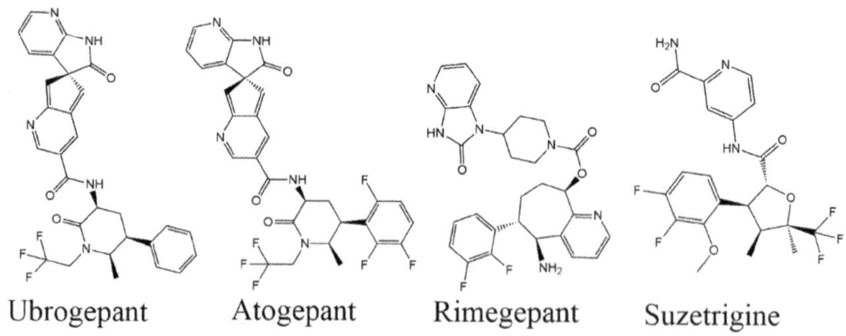

Figure 1.5 Illustration of chemical structure of various recent emerging small molecules of gepants for migraine and suzetrigine (Nav1.8 inhibitor) for pain.

Beyond gepants, other small molecules have also gained attention in pain management, particularly through the development of selective VGSC inhibitors. Suzetrigine (Figure 1.5) is a highly selective VGSC Nav1.8 inhibitor that effectively reduces pain transmission without affecting CNS function [117,118]. Predominantly located in dorsal root ganglion sensory neurons, Nav1.8 sodium channels are a key contributor to the propagation of pain signals [119–123]. By selectively targeting the Nav1.8 channel, suzetrigine provides effective pain relief while minimizing sedation and the potential for addiction, which are major concerns with opioid-based therapies [123,124]. The specificity of small-molecule inhibitors like suzetrigine allows for a more refined approach to pain relief, reducing side effects commonly associated with broad-spectrum sodium channel blockers.

The growing interest in small-molecule drugs for pain management is driven by their oral bioavailability, ease of administration, and rapid action, making them attractive alternatives to biologics. Additionally, their ability to selectively target pain-related pathways offers a safer and more effective strategy compared to traditional analgesics and opioids. With ongoing research and development, small molecules continue to expand the therapeutic options for pain management, addressing the need for non-addictive, well-tolerated, and long-lasting treatments. These advancements highlight the potential of small-molecule drugs to redefine pain relief strategies, providing patients with more accessible and effective options while reducing the reliance on opioids.

1.3 MENTAL HEALTH DISORDERS

Mental health disorders are conditions that lead to alterations in thinking, cognition, emotions, and behavior, which are clinically significant. It is often linked to disability and discomfort in key areas of functioning and might include risk of self-harm. It can be in several different forms such as anxiety disorders, depression, bipolar disorder, post-traumatic stress disorder (PTSD), eating disorders, schizophrenia, disruptive behavior, antisocial disorders, and neurodevelopmental disorders. The Institute for Health Metrics and Evaluation (IHME) 2021 report estimated that 13.9% of the global population was affected by mental disorders. The WHO 2022 Mental Health Report approximated that mental disorders are a leading cause of disability worldwide and account for 1 in 6 years lived with disability [125]. A 2024 study published in *The Lancet Psychiatry* shows that 50% of people worldwide will develop a mental health disorder in their lifetime [126]. This study explains that mental disorders, such as depression, anxiety, schizophrenia, and substance use disorders, impact individuals, families, and societies and significantly contribute to the global disease burden [126]. The COVID-19 pandemic further emphasized vulnerabilities in mental health systems, with reported increases in anxiety and depression cases, as well as concerns about rising PTSD rates in certain populations [127,128]. The COVID-19 pandemic crisis has highlighted the pressing need for robust mental health policies, increased funding, and community-based involvement. Global initiatives, such as the WHO's Mental Health Action Plan, emphasize integrating mental health into primary healthcare, expanding access to both psychological and pharmacological treatments, and reducing stigma through education and advocacy. In the current era, psychopharmacology plays a crucial role in managing mental health conditions by using medication to treat various psychiatric disorders. In the following section, we discuss how psychopharmacology and potential small molecules are evolving in their targeted applications for managing mental health.

1.3.1 Management of mental health

In the earlier days of the 19th century, mental health was unspoken of and poorly understood, and there was stigma around this area of life. Due to lack of awareness, people suffering from mental disorders were often cast aside, ill-treated, and feared. The "mental hygiene movement," in 1908, came from a determined group of individuals who had experienced mental illness and the professionals who treated them [129,130]. This movement transformed asylums, promoted early intervention, and emphasized mental wellness over illness. Hippocrates presented the concept of mental illness as a medical condition during the era of ancient Greece [131]. The 19th and early 20th centuries saw substantial progress, with Emil Kraepelin's classification of mental disorders and Sigmund Freud's psychoanalytic theories [132–134]. In the 19th century, asylum physicians used sedative and hypnotic small molecules, such as bromides, chloral hydrate, hyoscine, paraldehyde, sulfonal, and narcotics, more for sedating and calming instead of treating mental illness [135]. Particularly, psychiatrist Henry Maudsley debated that drugs could not directly eliminate delusions or deeper psychological issues [135]. On ethical grounds, he also objected to chemical restraint and compared it to the previously abandoned practice of mechanical restraint [135,136]. He advocated that pharmaceutical sedation might be just as harmful. The 20th century introduced barbiturates (discovered in 1903), which, like their predecessors, were easy to administer and effective in sedation but not therapeutic in treating mental disorders [135,137]. The beginning of modern psychopharmacology was marked by the synthesis of chlorpromazine in 1950, used in schizophrenia treatment by 1952. Unlike earlier sedatives and hypnotics, chlorpromazine was the first drug believed to actively treat mental illness, rather than simply masking symptoms.

The discovery of chlorpromazine opened a decade of unprecedented progress, culminating in the development of several psychotropic agents that characterized modern psychiatry, including the development of TCAs and monoamine oxidase inhibitors (MAOIs), which laid the groundwork for modern therapies in antidepressants. Since their introduction, small-molecule drugs targeting neurotransmitter systems have become a standard treatment for mental health conditions. Various classes of small-molecule drugs act on specific neural pathways to modulate brain function and alleviate symptoms. There are five major classes of small molecules separated by their mechanism, which are discussed in the following sections with examples.

1.3.2 Antipsychotics

Antipsychotic drugs are used to treat a host of psychotic disorders, including bipolar disorder, schizophrenia, and certain forms of severe depression. These molecules typically modulate neurotransmitter activity of dopamine and serotonin via blocking postsynaptic dopamine D2 receptors or partial agonism of dopamine receptors. In general, antipsychotics are divided into two classes: first-generation (typical) antipsychotics and second-generation (atypical) antipsychotics, which differ in their mechanisms of action [138–140]. Atypical antipsychotics are the second-generation drugs that exhibit improved efficacy and reduced extrapyramidal activity side effects [138,141]. However, a more precise classification of antipsychotics has recently been introduced based on their in vitro binding profiles for a better understanding of their modes of action, as summarized in Table 1.5 [142,143]. The chemical structures of some common antipsychotics are illustrated in Figure 1.6.

Table 1.5 Classification of antipsychotics based on primary target with example molecules

Classes	*Receptor target*	*Example of drugs*
Typical antipsychotic	Antagonist of D2R	Haloperidol, loxapine, fluphenazine
	Antagonist of D2R, 5-HT2A	Chlorpromazine, thioridazine, trifluoperazine
Atypical antipsychotic	Antagonist of D2R	Amisulpride
	Antagonist of D2R, 5-HT2A	Risperidone, olanzapine, sertindole
	Antagonist of D2R, 5-HT2A, and adrenergic receptors	Clozapine, quetiapine
	Partial agonist of D2R, 5-HT1A and antagonist of 5-HT2A	Aripiprazole, cariprazine
	Antagonist of 5-HT2A and 5-HT2C	Pimavanserine

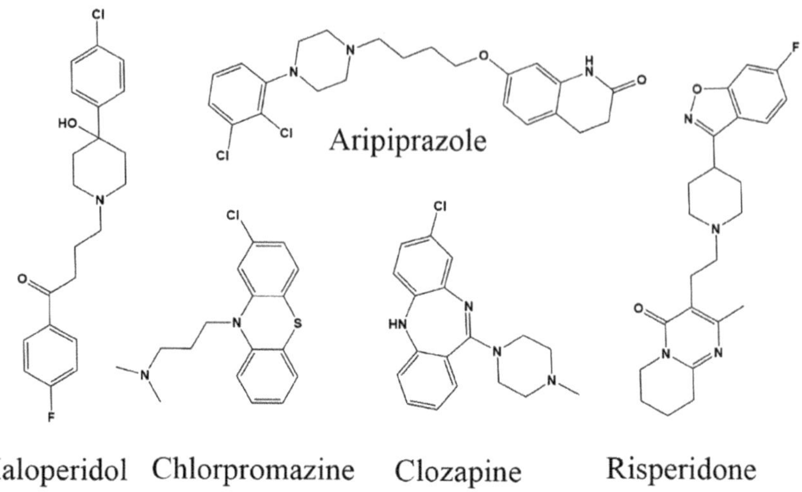

Figure 1.6 Illustration of chemical structures of various antipsychotics, including typical and atypical.

1.3.2.1 Typical antipsychotic (dopamine D2 receptor blocker)

Typical antipsychotic drugs act by blocking dopamine D_2 receptors (D2R) in the brain and are used for the treatment of schizophrenia, acute psychosis, and other psychiatric conditions requiring rapid symptom control [144–146]. Even though these drugs reduce hallucinations and delusions effectively, they have restricted impact on symptoms like social withdrawal and cognitive impairment [146]. For example, haloperidol, a high-potency first-generation antipsychotic, is an emergency drug used against acute psychosis and agitation due to its strong dopamine-blocking property [147–149]. However, it has a substantial risk of extrapyramidal symptoms, such as muscle rigidity and tremors [150,151]. Chlorpromazine, a typical antipsychotic drug used for schizophrenia, is another drug with low potency, and severe nausea causes significant sedation and anticholinergic effects, such as dry mouth and constipation [152]. With the effectiveness of these drugs, they are accompanied by serious side effects, including tardive dyskinesia, a potentially irreversible movement disorder, and neuroleptic malignant syndrome, a life-threatening reaction characterized by high fever, muscle rigidity, and autonomic instability [135,152]. The second-generation antipsychotics overcome these limitations and have a lower risk of motor side effects.

1.3.2.2 Atypical antipsychotic

These are the second-generation antipsychotic drugs that help reduce both positive (hallucinations, delusions) and negative symptoms (social withdrawal, cognitive impairment) [146]. They are also called serotonin-dopamine antagonists and are widely used in the treatment of schizophrenia, bipolar disorder, and other psychiatric conditions. Unlike typical antipsychotics, which primarily block dopamine D_2 receptors, atypical antipsychotics also act on serotonin 5-hydroxytryptamine (5-HT) neurotransmitter receptors and 5-hydroxytryptamine receptor 2A (5-HT2A) (Table 1.5), leading to improved efficacy and fewer extrapyramidal symptoms [135,142,153]. For example, Clozapine, one of the most effective atypical antipsychotics, is used for treatment-resistant schizophrenia but requires regular blood monitoring due to the risk of agranulocytosis, a rare but serious condition characterized by dangerously low white blood cell counts [154,155]. Another commonly used atypical antipsychotic, risperidone, is effective in treating schizophrenia and bipolar disorder with a lower risk of sedation but may cause hyperprolactinemia, leading to side effects such as menstrual irregularities and gynecomastia [156,157]. Olanzapine, known for its strong antipsychotic effects, is beneficial for mood stabilization but has a substantial risk of weight gain and metabolic syndrome due to blocking the histamine H1 receptor in the brain, which can lead to diabetes and cardiovascular complications [158,159]. The antipsychotic drug, aripiprazole, is preferred for patients who need a lower risk of sedation and weight gain. It is a partial dopamine agonist, which offers a different mechanism by modulating dopamine activity rather than fully blocking it [160]. Despite their advantages, atypical antipsychotics also have certain risks, as many can cause metabolic disturbances, sedation, and increased cardiovascular risk. However, their dual dopamine-serotonin action makes them more effective and better tolerated than first-generation antipsychotics, leading to their widespread use in modern psychiatric treatment [161,162]. The most recent atypical drug, pimavanserine, was approved by the FDA in 2016 without dopamine receptor activity to treat psychotic symptoms of Parkinson's disease [163]. Unlike other antipsychotics, pimavanserine only interacts with 5-HT2A and 2C receptors [161].

1.3.3 Antidepressant (monoamine modulator)

Antidepressants work by regulating the levels and activity of key neurotransmitters, including serotonin, norepinephrine, and dopamine, and are classified as monoamine modulators [131]. They reduce symptoms of depression, anxiety disorders, and other mood disorders. These drugs act by either blocking neurotransmitter reuptake transporters, inhibiting their metabolism, or modulating receptor activity, as a result of which increases monoamine availability in the synaptic cleft and neurotransmission enhances [164]. Figure 1.3 illustrates the mechanism of action for all classes of antidepressants at the synaptic level. Besides the neurotransmitter signaling, antidepressants also modulate neurotrophic factors such as brain-derived neurotrophic factor (BDNF) [165,166]. BDNF is considered a surrogate marker of depression; multiple clinical and preclinical studies

Table 1.6 Classes of monoamine modulator antidepressants along with conditions where they are used

Class	*Drug name*	*Common use*
SSRI	Fluoxetine	Depression, anxiety, OCD, panic disorder
	Sertraline	Depression, anxiety, PTSD, OCD
	Paroxetine	Depression, anxiety, OCD, PTSD
	Citalopram	Depression, anxiety, OCD
	Escitalopram	Depression, anxiety
	Fluvoxamine	Depression, OCD, anxiety disorders
SNRI	Venlafaxine	Depression, anxiety, panic disorder
	Duloxetine	Depression, anxiety, neuropathic pain
	Desvenlafaxine	Depression, anxiety
	Levomilnacipran	Depression
	Milnacipran	Fibromyalgia, depression (outside the United States)
NDRI	Bupropion	Depression, smoking cessation, ADHD
TCA	Amitriptyline	Depression, neuropathic pain, migraine prevention
	Nortriptyline	Depression, neuropathic pain
	Imipramine	Depression, enuresis (bedwetting)
	Clomipramine	Depression, OCD
	Doxepin	Depression, anxiety, insomnia, pruritus
MAOI	Phenelzine	Depression (last-line treatment)
	Tranylcypromine	Depression (last-line treatment)
	Selegiline	Depression, Parkinson's disease (low dose)
	Isocarboxazid	Depression (last-line treatment)

suggested lower expression levels in the brain of suicidal victims and chronically stressed rodents that were associated with depression-like symptoms [166–168]. Treatment with antidepressants boosted BDNF brain expression, supporting neurogenesis and synaptic remodeling, particularly in the hippocampus and prefrontal cortex, areas that are often affected in depression [168]. Based on their mechanism of action, antidepressant drugs are classified into several classes, as listed in Table 1.6, along with the disease conditions in which they are used.

1.3.3.1 Selective serotonin reuptake inhibitors

SSRIs are a commonly prescribed class of antidepressants because of their safety, efficacy, and tolerability. Hence, they are the first-line pharmacological treatment for depression and related disorders. As the name suggests, they work by blocking the serotonin transporter (SERT), preventing the reuptake of serotonin into presynaptic neurons and mounting its availability in the synaptic cleft [131,169]. SERT inhibition-enhanced serotonin neurotransmission improves mood and emotional regulation, resulting in SSRIs being effective in the management of major depressive disorder (MDD), generalized anxiety disorder, obsessive-compulsive disorder (OCD), panic disorder, social anxiety disorder, and PTSD [131]. Common SSRIs include fluoxetine, sertraline, paroxetine, citalopram, escitalopram, and fluvoxamine, each with given pharmacokinetic properties that influence their clinical use, as listed in Table 1.6. For example, fluoxetine is known to have a long half-life, which helps in withdrawal symptom reduction upon discontinuation, while escitalopram has reduced side effects because of its highly selective nature [170,171]. Sertraline is often used for both depression and anxiety disorders, whereas paroxetine, despite its strong antidepressant effects, is associated with higher risks of weight gain and withdrawal symptoms [172,173]. Due to the many positive outcomes of SSRIs, they are preferred over older antidepressants like TCAs and MAOIs because they have minimal effect on histamine, adrenergic, or muscarinic receptors, reducing sedation, cardiotoxicity, and anticholinergic side effects. Although SSRIs are well tolerated, they can cause nausea, GI discomfort, sexual dysfunction, weight changes, and insomnia, with some individuals experiencing withdrawal symptoms if treatment is abruptly stopped, particularly with shorter half-life drugs like paroxetine [174]. However, there is a risk of having too much serotonin in the body due to combined effects of SSRIs and other serotonergic agents. It is a rather unusual but potentially lethal state called serotonin syndrome that causes agitation, hyperthermia, and autonomic instability [175]. Despite these risks, SSRIs remain a first-line therapy for mood and anxiety disorders due to their effectiveness, safety, and favorable side effect profile, making them a foundation in modern psychiatric pharmacology.

1.3.3.2 Serotonin-norepinephrine reuptake inhibitors

SNRIs are the FDA-approved medications used for the treatment of multiple disorders, including depression, generalized anxiety disorder, panic disorder, MDD, PTSD, and chronic pain conditions, such as fibromyalgia and neuropathic pain. They work by modulating both serotonin and norepinephrine levels in the brain; dual action makes SNRIs effective in treating a wide range of disorders [131,176]. Commonly prescribed SNRIs are venlafaxine, duloxetine, desvenlafaxine, levomilnacipran, and milnacipran, each differing in their affinity for serotonin and norepinephrine transporters (Table 1.6). Venlafaxine modulates serotonin and norepinephrine, and it is often prescribed when first-line treatments are ineffective [177]. Venlafaxine exhibits side effects like nausea, hot flushes, constipation, dizziness, and increased blood pressure [178]. It exhibits dose-dependent effects on norepinephrine reuptake; at higher doses, it also affects dopamine reuptake, augmenting its healing effect [131]. However, on discontinuing, venlafaxine exhibits withdrawal symptoms due to its short life [179]. To overcome withdrawal symptoms, doctors recommend switching to fluoxetine, which is a long-acting SSRI, and then tapering off fluoxetine, which works well to reduce discontinuation symptoms. While duloxetine is used for diabetic peripheral neuropathy or fibromyalgia, there is major depression and neuropathic pain [180]. In a study to assess the efficacy and safety of antidepressants in treating nonspecific chronic low back pain, it was shown that they significantly reduced pain and improved quality of life, but at higher doses, they showed adverse effects [181]. Overall, SNRIs provide a wider profile of antidepressant activity than SSRIs, particularly for low-energy individuals, painful complaints, or treatment-resistant depression. Their role in mood regulation and pain management renders them a crucial option in today's psychiatric and neurological management.

1.3.3.3 Norepinephrine-dopamine reuptake inhibitors (NDRIs)

NDRI antidepressants are used to treat depression, anxiety, attention deficit hyperactivity disorder (ADHD), and neurodevelopmental conditions. Their mode of action involves increasing the levels of both norepinephrine and dopamine neurotransmitters by blocking their respective reuptake transporters, providing a unique approach to managing depression [182]. Unlike SSRIs or SNRIs, NDRIs do not significantly affect serotonin levels, making them particularly beneficial for individuals whose primary depressive symptoms include fatigue, low motivation, or anhedonia (inability to feel pleasure in pleasurable situations). Instead, NDRIs enhance dopaminergic and noradrenergic neurotransmission without directly affecting serotonergic neurotransmission, making them useful in patients who cannot endure SSRIs or SNRIs [131,183]. The most prescribed NDRI is bupropion, which is used to treat MDD, seasonal affective disorder, and as an adjunct therapy for SSRI-induced sexual dysfunction [184]. Overall, NDRIs like bupropion offer an important substitute to serotonin-based antidepressants, especially for individuals who need a treatment that enhances energy, motivation, and cognitive function without the risk of sexual dysfunction or weight gain. However, susceptible individuals can experience seizure risk and anxiety symptoms due to increased norepinephrine and dopamine activity [185]. A pharmacovigilance study showed that people on NDRIs have a risk of experiencing adverse cardiovascular events.

1.3.3.4 Tricyclic antidepressants

TCAs are an older class of antidepressants that act by blocking the reuptake of two neurotransmitters, namely serotonin and norepinephrine, and non-selectively interact with multiple other receptors, including cholinergic (alpha-1 and alpha-2), muscarinic, and histamine H1 receptors [131,135]. TCAs work like other reuptake inhibitors and similarly raise levels of these neurotransmitters in the synaptic cleft, leading to improved neurotransmission [135]. Based on clinical trials, the FDA has approved various TCAs that are effective and safe to use. However, they are used when SSRIs or other treatments fail for MDD [186,187]. Since they have a higher risk of adverse effects, they are the second-line treatment due to their anticholinergic activity and lower overdose threshold [188]. The side effects include sedation, dry mouth, constipation, blurred vision, weight gain, and cardiovascular effects, making them less favorable as a first-line treatment [89,189]. Additionally, alpha-adrenergic antagonism leads to orthostatic hypotension, increasing the risk of dizziness and falls, especially in elderly patients [89]. TCAs are used for patients with severe or treatment-resistant depression. Beyond depression, TCAs are also prescribed for off-label uses such as chronic pain conditions, migraines, and OCD [190,191]. Some of the most frequently prescribed TCAs include amitriptyline, imipramine, nortriptyline, clomipramine, and doxepin, each with varying selectivity for serotonin and norepinephrine reuptake blockade listed in Table 1.6. Amitriptyline, for example, is widely used for depression and neuropathic pain, whereas imipramine is frequently approved for panic disorder and nocturnal enuresis (bedwetting) in children [192]. Clomipramine with strong serotonergic activity is used for OCD, whereas nortriptyline and desipramine have stronger norepinephrine effects, making them useful in cases of low-energy depression.

1.3.3.5 Monoamine oxidase inhibitors

MAOIs are a class of antidepressants that work by irreversibly inhibiting the enzyme monoamine oxidase (MAO), which is responsible for breaking down serotonin, norepinephrine, and dopamine in the brain. Blocking the MAO enzyme increases the levels of these neurotransmitters, leading to improved mood, motivation, emotional regulation, and cognitive function [131]. MAOIs are beneficial not only for depression but also for neurodegenerative conditions like Parkinson's disease. MAO has two classes, MAO-A and MAO-B; inhibition of MAO-A enhances serotonin and norepinephrine. MAO-B inhibitors block the enzyme that breaks down dopamine and increase its levels, which modestly helps to improve motor symptoms in Parkinson's disease [193]. Due to their irreversible inhibition of MAO, the effects of these drugs persist for weeks after discontinuation, requiring a washout period before switching to other antidepressants. Common MAOIs include phenelzine, tranylcypromine, isocarboxazid, and selegiline, each differing in potency and selectivity (Table 1.6). Phenelzine is frequently used for treatment-resistant depression and social anxiety disorder, while tranylcypromine has stimulant-like properties and is often prescribed for patients with depression characterized by low energy and anhedonia [135]. Isocarboxazid is another option but is less commonly used due to its side effects. Selegiline, a selective MAO-B inhibitor at lower doses, is primarily used in Parkinson's disease but, at higher doses, also inhibits MAO-A, making it useful as an antidepressant [194]. The selegiline transdermal patch provides an advantage over traditional oral MAOIs bypassing the digestive system, reducing the need for strict dietary restrictions. Although MAOIs were among the earliest antidepressants developed, their use has significantly declined due to severe dietary restrictions, drug interactions, and the risk of hypertensive crises [131]. However, they remain effective for treatment-resistant depression and atypical depression, particularly in individuals who do not respond to other antidepressant classes.

1.3.3.6 Recent emerging antidepressant molecules

In recent years, several new antidepressant medications with unique mechanisms of action have been approved, focusing on non-monoamine pathways, specifically targeting NMDAR/glutamate pathways, gamma-aminobutyric acid (GABA)-A receptors, or anti-inflammatory or metabolic targets; for example, some chemical structures of these molecules are illustrated in Figure 1.7 offers new options for individuals with MDD, especially those who have not responded to traditional treatments. Nasal spray of esketamine was approved by the FDA in 2019 for treatment-resistant depression [195]. It is the S-enantiomer of ketamine and functions as an NMDAR antagonist, leading to a rapid increase in glutamate levels and promoting synaptic plasticity [196] (Figure 1.3). Unlike traditional antidepressants that may take weeks to show effects, esketamine can reduce depressive symptoms within hours [197]. Due to potential side effects such as dissociation and increased blood pressure, its administration is restricted to certified medical settings under supervision. In 2022, the FDA approved a combination of dextromethorphan and bupropion oral medication for MDD. Dextromethorphan, commonly known as a cough suppressant, acts on NMDARs and sigma-1 receptors, while bupropion inhibits the reuptake of norepinephrine and dopamine [198]. Together, they modulate glutamatergic and monoaminergic systems, offering a novel approach to depression treatment. Clinical trials have demonstrated that the combination can provide significant symptom improvement, with some patients experiencing relief within a week. Gepirone is a selective 5-HT1A receptor partial agonist approved by the FDA in 2023 as an antidepressant [199]. Unlike traditional SSRIs, it directly modulates serotonin receptors, which may result in a different side effect profile, potentially offering advantages such as a lower incidence of sexual dysfunction and weight gain. The neuroactive steroid zuranolone approved in 2024 for the treatment of postpartum depression acts as a positive allosteric modulator of GABA-A receptors [200,201]. Its unique mechanism offers rapid antidepressant effects, with some patients reporting improvement within days. Notably, zuranolone is administered as a short course, typically over 2 weeks, which may reduce the burden of chronic medication use and associated side effects. These emerging

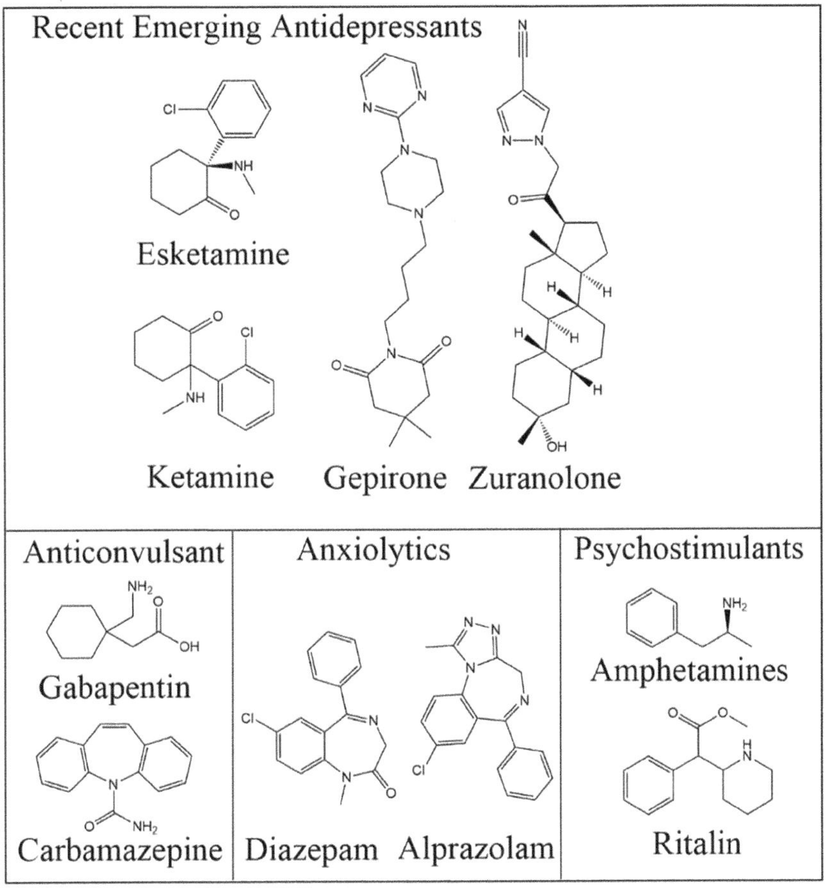

Figure 1.7 Illustration of chemical structure of recent emerging molecules to manage depression and some commonly used anticonvulsants, anxiolytics, and psychostimulants.

therapies represent significant advancements in the treatment of depression, particularly for individuals who have not benefited from traditional antidepressants. Their diverse mechanisms of action provide clinicians with a broader array of tools to tailor treatments to individual patient needs.

1.3.4 Mood stabilizer (ion and neurotransmitter modulator)

Mood stabilizers have a crucial role in managing bipolar disorder and preventing extreme mood swings. These medications modulate neuronal excitability and neurotransmitter release. Lithium carbonate and lithium citrate are the oldest and most well-studied mood stabilizers [135].

1.3.4.1 Lithium

It is a gold standard mood stabilizer for bipolar disorder and elicits a mood-stabilizing effect by affecting several neurotransmitter systems and intracellular signaling pathways [135,202]. Although the mechanism of action of lithium is not completely understood, it is believed to modulate neurotransmitter signaling (serotonin, dopamine, glutamate), enhance neuroprotection, and regulate second messenger systems such as inositol monophosphatase inhibition [203].

1.3.4.2 Anticonvulsants

Anticonvulsants, originally developed to manage epilepsy, are widely used as mood stabilizers in conditions such as bipolar disorder, schizoaffective disorder, and rapid cycling mood disorders because they possess the ability to influence neurotransmitter function, stabilize neuronal excitability, and suppress mood changes [52,53,204,205]. These drugs primarily function by blocking sodium channels, enhancing GABAergic inhibition, and modulating glutamate transmission, all of which result in their mood-stabilizing effect [52, 59]. We summarize anticonvulsants in Table 1.3 along with primary targets and mechanisms of action. Some of the most widespread anticonvulsants used for mood disorders include valproate, carbamazepine, and lamotrigine, each with specific therapeutic effects and side effects. Valproate is indicated for rapid cycling bipolar disorder, providing quicker symptom control compared to lithium in certain instances [206]. It elevates the levels of GABA and suppresses excessive firing of neurons, thereby stabilizing mood. But valproate is also associated with severe side effects, including weight gain, sedation, tremors, liver toxicity, pancreatitis, and teratogenicity. Carbamazepine is also an anticonvulsant used in the treatment of bipolar disorder, particularly in manic or mixed-state patients who are agitated, violent, or cyclothymic [207]. It stabilizes mood changes by blocking VGSCs and inhibiting excitatory neurotransmitter action. Carbamazepine is effective in non-responder patients to lithium or valproate but has the side effects of dizziness, sedation, and nausea [208]. Lamotrigine is best suited for bipolar depression and not mania; therefore, it is ideal for those patients with a high frequency of depressive relapses but not too extreme an elevation of mood [205,208]. It is a mood stabilizer that acts by blocking sodium channels and modulating glutamate transmission to preclude depressive relapses. One of its advantages is that it is better tolerated than valproate and lithium, with fewer side effects of weight gain and sedation. Gabapentin and pregabalin (Figure 1.7) are structural analogs of GABA but do not directly bind to GABA receptors [209]. Instead, they modulate the alpha-2-delta subunit of voltage-gated calcium channels in the CNS [209]. Both reduce the flow of calcium into presynaptic neurons, which in turn decreases the release of excitatory neurotransmitters, such as glutamate, substance P, and norepinephrine, thereby making them effective anticonvulsants and analgesics for several neuropathic pain syndromes [210]. Clinically, gabapentin is used as an adjunctive therapy for partial-onset seizures and has been found to be effective in managing neuropathic pain associated with disorders like postherpetic neuralgia and diabetic neuropathy, while pregabalin is approved for managing neuropathic pain and epilepsy.

1.3.5 Anxiolytics (GABA modulators)

GABA is a primary inhibitory neurotransmitter of the brain, and anxiolytic medications increase the GABA function to reduce neuronal excitability of postsynaptic neurons that produce a calming, sedative, and anxiolytic effect [211] (Figure 1.3). These drugs promote chloride ion influx into neurons, leading to hyperpolarization and reduced neuronal firing, which helps alleviate anxiety and related symptoms [211,212] (Figure 1.3). Anxiolytics are GABA modulators that act by binding to GABA-A receptors, a ligand-gated chloride ion channel, thus increasing inhibitory signaling and thereby stabilizing neural circuits [211] (Figure 1.3). These medications are widely used for generalized anxiety disorder, panic disorder, insomnia, seizure management, and muscle relaxation, with some also playing a role in anesthesia induction. The GABA-A receptor is a pentameric protein complex composed of five subunits arranged around a central chloride ion channel [213,214]. There are around 19 known subunits, in which the most common families are α1–6, β1–3, γ1–3, and δ subunits [213]. These subunits are selected from a diverse pool, with the most common receptor composition being (2α, 2β, 1γ). The subunit composition influences the receptor's pharmacological properties and its sensitivity to different drugs [215,216]. Among the most prescribed GABA-modulating anxiolytics, benzodiazepines are widely used for short-term anxiety relief, panic attacks, and acute stress reactions due to their rapid onset of action [216]. These medications, including diazepam, lorazepam, alprazolam, clonazepam, and temazepam, bind to the benzodiazepine site on GABA-A receptors, increasing the efficiency of GABA neurotransmission [217]. Figure 1.7 illustrates the chemical structure of diazepam and alprazolam, commonly used as anxiolytics. Benzodiazepines act as positive allosteric modulators of GABA-A receptors by binding to the benzodiazepine site, which is located at the interface between the α and γ subunits of the GABA-A receptor [217]. Once benzodiazepines bind to receptor sites, they enhance the affinity of GABA function and produce benefits in certain conditions [217]. While they are effective in reducing anxiety, they have a substantial risk of dependence, withdrawal symptoms, memory impairment, sedation, and respiratory depression, particularly when combined with alcohol or opioids. As a result, benzodiazepines are typically recommended only for short-term or intermittent use to minimize the risk of tolerance and addiction [218]. Before benzodiazepines, widely available barbiturates, also a GABA-A-positive allosteric modulator, bind to a specific site on the transmembrane domain of the β-subunit [217]. Barbiturates were commonly prescribed for anxiety and insomnia, but their use has drastically declined due to their high overdose potential and severe respiratory depression risk [219]. Some common barbiturates such as phenobarbital, secobarbital, and pentobarbital act by prolonging the duration of chloride channel opening at GABA-A receptors, leading to more profound sedation and a greater risk of fatal overdose, particularly when taken in excess or combined with other depressants [219]. Because of their dangers, barbiturates are now primarily reserved for seizure management and anesthesia induction rather than anxiety treatment. A safer alternative for managing sleep-related disorders, particularly insomnia, comes from the non-benzodiazepine GABA-A receptor agonists, commonly referred to as Z-drugs, which include zolpidem, zaleplon, and eszopiclone [216]. Although they share a similar mechanism of action with benzodiazepines, Z-drugs exhibit greater selectivity for GABA-A receptors containing the α1-subunit, which is associated with sedative effects rather than anxiolysis or muscle relaxation [217]. These medications are associated with sleep regulation, producing sedative effects with less potential for dependence and cognitive impairment compared to benzodiazepines [220]. However, Z-drugs are suitable for short-term use only in cases of chronic insomnia because they carry risks of next-day drowsiness, memory disturbances, sleepwalking, and unusual behaviors during sleep [220]. More recently, neurosteroids and novel GABA-A receptor-positive allosteric modulators have emerged as promising alternatives for treating anxiety and depression with fewer side effects [201]. Brexanolone and zuranolone are both synthetic neurosteroids that act as positive allosteric modulators of GABA-A receptors; recently, both were approved for postpartum depression, offering rapid symptom relief with a different mechanism than traditional antidepressants [195,201].

1.3.6 Psychostimulants (dopamine and norepinephrine enhancers)

The term "psychostimulant" is used particularly to denote those drugs that elicit direct neurologic effects, typically defined as heightened alertness, enhanced energy, appetite suppression, and, occasionally, euphoria [221,222]. Psychostimulants represent a heterogeneous group of drug agents with initial clinical actions by modulation of the neurotransmission system of the CNS via modulation of synaptic levels of dopamine and norepinephrine [223]. These agents have clinical utility in the treatment of conditions such as narcolepsy, ADHD, and even certain treatment-resistant depressions. For example, amphetamine (Figure 1.7) derivatives like mixed amphetamine salts, dextroamphetamine, and the prodrug lisdexamfetamine primarily work through the stimulation of monoamine release from presynaptic neurons, reversing the transport operations of dopamine and norepinephrine, and the inhibition of vesicular storage of monoamines, causing prolonged elevation of extracellular catecholamines to enhance neural signaling in brain regions responsible for executive function and reward [224–226]. Conversely, methylphenidate-type stimulants, including those sold as Ritalin (Figure 1.7), Concerta, and Focalin, work primarily as reuptake inhibitors of norepinephrine and dopamine, thus making them more available in the synaptic cleft and providing both immediate-release and extended-release preparations to follow the circadian rhythms of symptom presentation more closely in ADHD [226]. Modafinil and armodafinil differ structurally from traditional amphetamines, promoting alertness by mildly inhibiting dopamine reuptake and influencing neurotransmitters like glutamate and orexin [227,228]. Both reduce excessive daytime sleepiness in conditions like narcolepsy and shift work sleep disorder, with less potential for abuse. Moreover, bupropion, extensively employed as an antidepressant and smoking cessation agent, exerts less powerful stimulant action by selectively blocking the reuptake of dopamine and norepinephrine [229]. Bupropion offers a clinical substitute where there are desired effects without the entire range of side effects of more potent stimulants [230]. Despite their therapeutic benefits, the potent action of psychostimulants on brain reward pathways necessitates close regulation; many of these drugs are scheduled as controlled substances to reduce abuse, dependence, and potential cardiovascular effects and therefore require careful patient selection, individualized dosing schedules, and intensive monitoring of therapeutic effects and adverse effects.

1.4 CONCLUSION

Small molecules have revolutionized pain and mental health clinical management and outcomes by providing precise therapies with improved efficacy and reduced side effects. Advances in pharmacology, combined with insights into neurotransmitter systems and molecular targets, allow clinicians to tailor interventions to individual patient needs. A deeper understanding of disease pathology and molecular targets enhances the ability to manage mental health disorders and chronic pain effectively. Modern pain management embraces a multimodal approach that integrates pharmacologic, interventional, and psychological strategies. Amid growing concerns over opioid misuse, small-molecule drugs, such as NSAIDs, antidepressants, and anticonvulsants, have become essential alternatives for pain relief. The rise of precision medicine, driven by advances in genetics, biomarkers, and artificial intelligence, is further enhancing the role of small molecules in pain and mental healthcare. These breakthroughs enable clinicians to tailor treatment plans based on individual biological profiles, optimizing therapeutic success. For mental health, small molecules continue to transform treatment paradigms. The development of novel antidepressants such as esketamine, dextromethorphan-bupropion, gepirone, and zuranolone has introduced new mechanisms of action, offering rapid and targeted relief for patients unresponsive to conventional therapies. Psychostimulants, including amphetamines, methylphenidate, and modafinil, remain the cornerstone for ADHD and narcolepsy treatment, with prodrug formulations like lisdexamfetamine reducing the potential for abuse while maintaining efficacy. Careful monitoring remains essential to mitigate risks associated with these powerful agents. The psychopharmacology of small molecules remains essential in the management of pain and mental health disorders. These compounds target specific neurotransmitter systems and receptor pathways to modulate neural activity, providing effective symptom relief. Ongoing research continues to refine their efficacy, reduce side effects, and enhance patient outcomes. As our understanding of neurochemical interactions deepens, small molecules will drive advancements in personalized and innovative treatments, ensuring more precise and effective therapeutic strategies for pain and mental health management.

AUTHOR CONTRIBUTIONS

AB, AY, and AK wrote the first draft. AB and AK created the illustration. All the authors participated in editing the chapter.

DECLARATION OF GENERATIVE AI AND AI-ASSISTED TECHNOLOGIES IN THE WRITING PROCESS

The author(s) utilized AI-assisted tools, ChatGPT, and Microsoft Word's Grammar suggestions to enhance the grammatical accuracy and stylistic coherence of this work. Following the use of these tools, the author(s) thoroughly reviewed and refined the content as necessary and assumed full responsibility for the final publication.

REFERENCES

1. Cohen SP, Vase L, Hooten WM. Chronic Pain: An Update on Burden, Best Practices, and New Advances. *Lancet* 2021;397:2082–97.
2. Starobova H, Alshammari A, Winkler IG, Vetter I. The Role of the Neuronal Microenvironment in Sensory Function and Pain Pathophysiology. *J Neurochem* 2024;168:3620–43.
3. Dorkenwald S, Matsliah A, Sterling AR, Schlegel P, Yu S-C, McKellar CE, et al. Neuronal Wiring Diagram of an Adult Brain. *Nature* 2024;634:124–38.
4. Woolf CJ. Central Sensitization: Implications for the Diagnosis and Treatment of Pain. *Pain* 2011;152:S2–15.

5. Apfelbaum JL, Chen C, Mehta SS, Gan TJ. Postoperative Pain Experience: Results from a National Survey Suggest Postoperative Pain Continues to Be Undermanaged. *Anesth Analg* 2003;97:534–40.
6. Collier R. A Short History of Pain Management. *J L'Association Medicale Can Can Med Assoc J* 2018;190:26–7.
7. Meldrum ML. A Capsule History of Pain Management. *JAMA J Am Med Assoc* 2003;290:2470–5.
8. Desborough MJR, Keeling DM. The Aspirin Story: From Willow to Wonder Drug. *Br J Haem Atol* 2017;177:674–83.
9. Vane JR, Botting RM. Anti-Inflammatory Drugs and Their Mechanism of Action. *Al Inflamm Res* 1998;47:78–87.
10. Rao P, Knaus EE. Evolution of Nonsteroidal Anti-Inflammatory Drugs (NSAIDs): Cyclooxygenase (COX) Inhibition and Beyond. *J Pharm Pharm Sci Publ Can Soc Pharm Sci Soc Can Sci Pharm* 2008;11:81–110.
11. Buttgereit F, Burmester GR, Simon LS. Gastrointestinal Toxic Side Effects of Nonsteroidal Anti-Inflammatory Drugs and Cyclooxygenase-2-Specific Inhibitors. *Am J Med* 2001;110(Suppl 3A):13–9.
12. Peterson WL, Cryer B. COX-1-Sparing NSAIDs–Is the Enthusiasm Justified? *JAMA J Am Med Assoc Am Med Assoc AMA* 1999;282:1961–3.
13. Hawkey CJ. COX-2 Inhibitors. *Lancet* 1999;353:307–14.
14. Shah AA, Thjodleifsson B, Murray FE, Kay E, Barry M, Sigthorsson G, et al. Selective Inhibition of COX-2 in Humans Is Associated with Less Gastrointestinal Injury: A Comparison of Nimesulide and Naproxen. *Gut* 2001;48:339–46.
15. Graham DJ. COX-2 Inhibitors, Other NSAIDs, and Cardiovascular Risk: The Seduction of Common Sense. *JAMA J Am Med Assoc Am Med Assoc AMA* 2006;296:1653–6.
16. FitzGerald GA, Patrono C. The Coxibs, Selective Inhibitors of Cyclooxygenase-2. *N Engl J Med* 2001;345:433–42.
17. Sinatra R. Role of COX-2 Inhibitors in the Evolution of Acute Pain Management. *J Pain Symptom Manage* 2002;24:18–27.
18. Camu F, Shi L, Vanlersberghe C. The Role of COX-2 Inhibitors in Pain Modulation. *Drugs* 2003;63(Suppl 1):1–7.
19. Donnelly MT, Hawkey CJ. Review Article: COX-II Inhibitors: A New Generation of Safer NSAIDs? *Aliment Pharmacol Ther* 1997;11:227–36.
20. Dogné J-M, Supuran CT, Pratico D. Adverse Cardiovascular Effects of the Coxibs. *J Med Chem* 2005;48:2251–7.
21. Stiller C-O, Hjemdahl P. Lessons from 20 Years with COX-2 Inhibitors: Importance of Dose-Response Considerations and Fair Play in Comparative Trials. *J Intern Med* 2022;292:557–74.
22. Costa BR da, Pereira TV, Saadat P, Rudnicki M, Iskander SM, Bodmer NS, et al. Effectiveness and Safety of Non-Steroidal Anti-Inflammatory Drugs and Opioid Treatment for Knee and Hip Osteoarthritis: Network Meta-Analysis. *BMJ* 2021;375:n2321.
23. Botting RM. Mechanism of Action of Acetaminophen: Is There a Cyclooxygenase 3?" *Clin Infect Dis* 2000;5:202–10.
24. Anderson BJ. Paracetamol (Acetaminophen): Mechanisms of Action. *Paediatr Anaesth* 2008;18:915–21.

25. Graham GG. Mechanism of Action of Paracetamol. *Am J Ther* 2005;12:46–55.
26. Graham GG, Scott KF. Mechanisms of Action of Paracetamol and Related Analgesics. *Inflammopharmacology* 2003;11:401–13.
27. Przybyła GW, Szychowski KA, Gmiński J. Paracetamol: An Old Drug with New Mechanisms of Action. *Clin Exp Pharmacol Physiol* 2021;48:3–19.
28. Stein C. Opioid Receptors. *Annu Rev Med* 2016;67:433–51.
29. Che T, Roth BL. Molecular Basis of Opioid Receptor Signaling.*Cell* 2023; 186:5203–19.
30. Corder G, Castro DC, Bruchas MR, Scherrer G. Endogenous and Exogenous Opioids in Pain. *Annu Rev Neurosci* 2018;41:453–73.
31. Zhang X-Y, Dou Y-N, Yuan L, Li Q, Zhu Y-J, Wang M, et al. Different Neuronal Populations Mediate Inflammatory Pain Analgesia by Exogenous and Endogenous Opioids. *ELife* 2020;9. https://doi.org/10.7554/eLife.55289.
32. Uniyal A, Tiwari V, Tsukamoto T, Dong X, Guan Y, Raja SN. Targeting Sensory Neuron GPCRs for Peripheral Neuropathic Pain. *Trends Pharmacol Sci* 2023;44:1009–27.
33. Rusin KI, Giovannucci DR, Stuenkel EL, Moises HC. Kappa-Opioid Receptor Activation Modulates Ca^{2+} Currents and Secretion in Isolated Neuroendocrine Nerve Terminals. *J Neurosci Off J Soc Neurosci* 1997;17:6565–74.
34. Dembla S, Behrendt M, Mohr F, Goecke C, Sondermann J, Schneider FM, et al. Anti-Nociceptive Action of Peripheral Mu-Opioid Receptors by G-Beta-Gamma Protein-Mediated Inhibition of TRPM3 Channels. *ELife* 2017;6. https://doi.org/10.7554/eLife.26280.
35. Clapham DE, Neer EJ. G Protein Beta Gamma Subunits. *Annu Rev Pharmacol Toxicol* 1997;37:167–203.
36. Krapivinsky G, Krapivinsky L, Wickman K, Clapham DE. G Beta Gamma Binds Directly to the G Protein-Gated K+ Channel, IKACh. *J Biol Chem* 1995;270:29059–62.
37. Rhim H, Miller RJ. Opioid Receptors Modulate Diverse Types of Calcium Channels in the Nucleus Tractus Solitarius of the Rat. *J Neurosci Off J Soc Neurosci* 1994;14:7608–15.
38. Ingram S, Wilding TJ, McCleskey EW, Williams JT. Efficacy and Kinetics of Opioid Action on Acutely Dissociated Neurons. *Mol Pharmacol* 1997;52:136–43.
39. Velayudhan A, Bellingham G, Morley-Forster P. Opioid-Induced Hyperalgesia. *Contin Educ Anaesth Crit Care Pain* 2014;3:125–9.
40. Maksymowicz R, Strączek CN, Matysek JJ, Lech DM, Nosek K. Case Report: Comprehensive Management of Painful Diabetic Neuropathy-Addressing Opioid-Induced Hyperalgesia through Multimodal Approaches. *Medicine (Baltimore)* 2024;103:39126.
41. Ingram SL. Toward Understanding the Opioid Paradox: Cellular Mechanisms of Opioid-Induced Hyperalgesia. *Neuropsychopharmacol* 2022;47:427–428.
42. Cox BM. Recent Developments in the Study of Opioid Receptors. *Mol Pharmacol* 2013;83:723–8.
43. Henriksen G, Willoch F. Imaging of Opioid Receptors in the Central Nervous System. *Brain J Neurol* 2008;131:1171–96.
44. Nummenmaa L, Tuominen L. Opioid System and Human Emotions. *Br J Pharmacol* 2018;175:2737–49.

45. White JM. Pleasure into Pain: The Consequences of Long-Term Opioid Use. *Addict Behav* 2004;29:1311–24.
46. Martínez V, Abalo R. Peripherally Acting Opioid Analgesics and Peripherally-Induced Analgesia. *Behav Pharmacol* 2020;31:136–58.
47. Sehgal N, Smith HS, Manchikanti L. Peripherally Acting Opioids and Clinical Implications for Pain Control. *Pain Physician* 2011;14:249–58.
48. Erstad BL, Puntillo K, Gilbert HC, Grap MJ, Li D, Medina J, et al. Pain Management Principles in the Critically Ill. *Chest* 2009;135:1075–86.
49. Ballantyne JC, Shin NS. Efficacy of Opioids for Chronic Pain: A Review of the Evidence. *Clin J Pain* 2008;24:469–78.
50. Schug SA, Goddard C. Recent Advances in the Pharmacological Management of Acute and Chronic Pain. *Ann Palliat Med* 2014;3:263–75.
51. Martell BA, O'Connor PG, Kerns RD, Becker WC, Morales KH, Kosten TR, et al. Systematic Review: Opioid Treatment for Chronic Back Pain: Prevalence, Efficacy, and Association with Addiction. *Ann Intern Med* 2007;146:116–27.
52. Macdonald RL, McLean MJ. Anticonvulsant Drugs: Mechanisms of Action. *Adv Neurol* 1986;44:713–36.
53. Stefan H, Feuerstein TJ. Novel Anticonvulsant Drugs. *Pharmacol Ther* 2007;113:165–83.
54. Hakami T. Neuropharmacology of Antiseizure Drugs. *Neuropsychopharmacol Rep* 2021;41:336–51.
55. McQuay H, Carroll D, Jadad AR, Wiffen P, Moore A. Anticonvulsant Drugs for Management of Pain: A Systematic Review. *BMJ* 1995;311:1047–52.
56. Wiffen P, Collins S, McQuay H, Carroll D, Jadad A, Moore A. Anticonvulsant Drugs for Acute and Chronic Pain. *Cochrane Database Syst Rev* 2005; 3:CD001133.
57. Mula M, Pini S, Cassano GB. The Role of Anticonvulsant Drugs in Anxiety Disorders: A Critical Review of the Evidence. *J Clin Psychopharmacol* 2007;27:263–72.
58. Kaufman KR. Antiepileptic Drugs in the Treatment of Psychiatric Disorders. *Epilepsy Behav EB* 2011;21:1–11.
59. Söderpalm B. Anticonvulsants: Aspects of Their Mechanisms of Action. *Eur J Pain* 2002;6:3–9.
60. Kwan P, Sills GJ, Brodie MJ. The Mechanisms of Action of Commonly Used Antiepileptic Drugs. *Pharmacol Ther* 2001;90:21–34.
61. Rogawski MA, Löscher W. The Neurobiology of Antiepileptic Drugs for the Treatment of Nonepileptic Conditions. *Nat Med* 2004;10:685–92.
62. Guy S, Mehta S, Leff L, Teasell R, Loh E. Anticonvulsant Medication Use for the Management of Pain Following Spinal Cord Injury: Systematic Review and Effectiveness Analysis. *Spinal Cord* 2014;52:89–96.
63. Dickenson AH, Matthews EA, Suzuki R. Neurobiology of Neuropathic Pain: Mode of Action of Anticonvulsants. *Eur J Pain* 2002;6:51–60.
64. Calabresi P, Galletti F, Rossi C, Sarchielli P, Cupini LM. Antiepileptic Drugs in Migraine: From Clinical Aspects to Cellular Mechanisms. *Trends Pharmacol Sci* 2007;28:188–95.
65. Moore A, Wiffen P, Kalso E. Antiepileptic Drugs for Neuropathic Pain and Fibromyalgia. *JAMA J Am Med Assoc* 2014;312:182–3.

66. Wiffen PJ, Sheena Derry RAM, Aldington D, Cole P, Rice ASC, Lunn MPT, et al. Antiepileptic Drugs for Neuropathic Pain and Fibromyalgia: An Overview of Cochrane Reviews. *Cochrane Database Syst Rev* 2013;2013:CD010567.
67. Stahl SM, Porreca F, Taylor CP, Cheung R, Thorpe AJ, Clair A. The Diverse Therapeutic Actions of Pregabalin: Is a Single Mechanism Responsible for Several Pharmacological Activities? *Trends Pharmacol Sci* 2013;34:332–9.
68. Robertson K, Marshman LAG, Plummer D, Downs E. Effect of Gabapentin vs Pregabalin on Pain Intensity in Adults with Chronic Sciatica: A Randomized Clinical Trial. *JAMA Neurol* 2019;76:28–34.
69. Markman JD, Jensen TS, Semel D, Li C, Parsons B, Behar R, et al. Effects of Pregabalin in Patients with Neuropathic Pain Previously Treated with Gabapentin: A Pooled Analysis of Parallel-Group, Randomized, Placebo-Controlled Clinical Trials. *Pain Pract Off J World Inst Pain* 2017;17:718–28.
70. Richter RW, Portenoy R, Sharma U, Lamoreaux L, Bockbrader H, Knapp LE. Relief of Painful Diabetic Peripheral Neuropathy with Pregabalin: A Randomized, Placebo-Controlled Trial. *J Pain Off J Am Pain Soc* 2005;6:253–60.
71. Quilici S, Chancellor J, Löthgren M, Simon D, Said G, Le TK, et al. Meta-Analysis of Duloxetine vs. Pregabalin and Gabapentin in the Treatment of Diabetic Peripheral Neuropathic Pain. *BMC Neurol* 2009;9:6.
72. Freeman R, Durso-Decruz E, Emir B. Efficacy, Safety, and Tolerability of Pregabalin Treatment for Painful Diabetic Peripheral Neuropathy: Findings from Seven Randomized, Controlled Trials across a Range of Doses. *Diabetes Care* 2008;31:1448–54.
73. Kruszewski SP, Shane JA. Pregabalin in Central Neuropathic Pain Associated with Spinal Cord Injury: A Placebo-Controlled Trial. *Neurology* 2007;68:2158–9.
74. Freynhagen R, Strojek K, Griesing T, Whalen E, Balkenohl M, editors. Efficacy of Pregabalin in Neuropathic Pain Evaluated in a 12-Week, Randomised, Double-Blind, Multicentre, Placebo-Controlled Trial of Flexible- and Fixed-Dose Regimens. *Pain* 2005;115:254–63.
75. Stefano, Giulia SLC, Truini A, Cruccu G. Natural History and Outcome of 200 Outpatients with Classical Trigeminal Neuralgia Treated with Carbamazepine or Oxcarbazepine in a Tertiary Centre for Neuropathic Pain. *J Headache Pain* 2014;15:34.
76. Ashina S, Robertson CE, Srikiatkhachorn A, Stefano G, Donnet A, Hodaie M, et al. Trigeminal Neuralgia. *Nat Rev Dis Primer* 2024;10:39.
77. Dharmshaktu P, Tayal V, Kalra BS. Efficacy of Antidepressants as Analgesics: A Review. *J Clin Pharmacol* 2012;52:6–17.
78. Urits I, Peck J, Orhurhu MS, Wolf J, Patel R, Orhurhu V, et al. Off-Label Antidepressant Use for Treatment and Management of Chronic Pain: Evolving Understanding and Comprehensive Review. *Curr Pain Headache Rep* 2019;23:66.
79. Colloca L, Ludman T, Bouhassira D, Baron R, Dickenson AH, Yarnitsky D, et al. Neuropathic Pain. *Nat Rev Dis Primer* 2017;3:17002.
80. Häuser W, Ablin J, Fitzcharles M-A, Littlejohn G, Luciano JV, Usui C, et al. Fibromyalgia. *Nat Rev Dis Primer* 2015;1:15022.

81. Micó JA, Ardid D, Berrocoso E, Eschalier A. Antidepressants and Pain. *Trends Pharmacol Sci* 2006;27:348–54.
82. McCleane G. Antidepressants as Analgesics. *CNS Drugs* 2008;22:139–56.
83. Jackson 2nd Kenneth C, Onge ELS. Antidepressant Pharmacotherapy: Considerations for the Pain Clinician. *Pain Pract Off J World Inst Pain* 2003;3:135–43.
84. Lee Y-C, Chen P-P. A Review of SSRIs and SNRIs in Neuropathic Pain. *Expert Opin Pharmacother* 2010;11:2813–25.
85. Bonilla-Jaime H, José Armando Sánchez-Salcedo MME-C, Molina-Jiménez T, Cortes-Altamirano JL, Alfaro-Rodríguez A. Depression and Pain: Use of Antidepressants. *Curr Neuropharmacol* 2022;20:384–402.
86. Arnold LM. Duloxetine and Other Antidepressants in the Treatment of Patients with Fibromyalgia. *Pain Med Malden Mass* 2007;2:63–74.
87. Riediger C, Schuster T, Barlinn K, Maier S, Weitz J, Siepmann T. Adverse Effects of Antidepressants for Chronic Pain: A Systematic Review and Meta-Analysis. *Front Neurol* 2017;8:307.
88. Goldstein DJ, Lu Y, Detke MJ, Lee TC, Iyengar S. Duloxetine vs. Placebo in Patients with Painful Diabetic Neuropathy. *Pain* 2005;116:109–18.
89. Anderson IM. Selective Serotonin Reuptake Inhibitors versus Tricyclic Antidepressants: A Meta-Analysis of Efficacy and Tolerability. *J Affect Disord* 2000;58:19–36.
90. Catterall WA. Common Modes of Drug Action on Na+ Channels: Local Anesthetics, Antiarrhythmics and Anticonvulsants. *Trends Pharmacol Sci* 1987;8:57–65.
91. Heavner JE. Local Anesthetics. *Curr Opin Anaesthesiol* 2007;20:336–42.
92. Fozzard HA, Lee PJ, Lipkind GM. Mechanism of Local Anesthetic Drug Action on Voltage-Gated Sodium Channels. *Curr Pharm Des* 2005;11:2671–86.
93. Nau C, Wang GK. Interactions of Local Anesthetics with Voltage-Gated Na+ Channels. *J Membr Biol* 2004;201:1–8.
94. Lirk P, Hollmann MW, Strichartz G. The Science of Local Anesthesia: Basic Research, Clinical Application, and Future Directions. *Anesth Analg* 2018;126:1381–92.
95. Binder A, Koroschetz J, Baron R. Disease Mechanisms in Neuropathic Itch. *Nat Clin Pract Neurol* 2008;4:329–37.
96. Kravitz ND. The Use of Compound Topical Anesthetics: A Review. *J Am Dent Assoc* 2007;10:138.
97. Bräu ME, Nau C, Hempelmann G, Vogel W. Local Anesthetics Potently Block a Potential Insensitive Potassium Channel in Myelinated Nerve. *J Gen Physiol* 1995;105:485–505.
98. Suzuki S, Gerner P, Lirk P. Local Anesthetics. *Pharmacol Physiol Anesth* 2019;20:390–411.
99. Tremont-Lukats IW, Challapalli V, McNicol ED, Lau J, Carr DB. Systemic Administration of Local Anesthetics to Relieve Neuropathic Pain: A Systematic Review and Meta-Analysis. *Anesth Analg* 2005;101:1738–49.
100. Wiles MD, Nathanson MH. Local Anaesthetics and Adjuvants–Future Developments. *Anaesthesia* 2010;1:22–37.
101. Jeong J-O, Kim M, Kim S, Lee KK, Choi H. Advanced Hydrogel Systems for Local Anesthetic Delivery: Toward Prolonged and Targeted Pain Relief. *Gels* 2025;11:131. https://doi.org/10.3390/gels11020131.

102. Ciechanowicz S, Patil V. Lipid Emulsion for Local Anesthetic Systemic Toxicity. *Anesthesiol Res Pract* 2012; 2012:131784.
103. Zhou Q, Sheng M. NMDA Receptors in Nervous System Diseases. *Neuropharmacology* 2013;74:69–75.
104. Aiyer R, Mehta N, Gungor S, Gulati A. A Systematic Review of NMDA Receptor Antagonists for Treatment of Neuropathic Pain in Clinical Practice. *Clin J Pain* 2018;34:450–67.
105. Parsons CG. NMDA Receptors as Targets for Drug Action in Neuropathic Pain. *Eur J Pharmacol* 2001;429:71–8.
106. Collins S, Sigtermans MJ, Dahan A, Zuurmond WWA, Perez RSGM. NMDA Receptor Antagonists for the Treatment of Neuropathic Pain. *Pain Med Malden Mass* 2010;11:1726–42.
107. Mason OJ, Morgan CJM, Stefanovic A, Curran HV. The Psychotomimetic States Inventory (PSI): Measuring Psychotic-Type Experiences from Ketamine and Cannabis. *Schizophr Res* 2008;103:138–42.
108. Pickering G, Morel V, Joly D, Villatte C, Roux D, Dubray C, et al. Prevention of Post-Mastectomy Neuropathic Pain with Memantine: Study Protocol for a Randomized Controlled Trial. *Trials* 2014;15:331.
109. Pickering G, Morel V. Memantine for the Treatment of General Neuropathic Pain: A Narrative Review. *Fundam Clin Pharmacol* 2018;32:4–13.
110. Sani G, Serra G, Kotzalidis GD, Romano S, Tamorri SM, Manfredi G, et al. The Role of Memantine in the Treatment of Psychiatric Disorders Other than the Dementias: A Review of Current Preclinical and Clinical Evidence. *CNS Drugs* 2012;26:663–90.
111. Sirinathsinghji DJS, Hill RG, editors. *NMDA Antagonists as Potential Analgesic Drugs. New York, NY*: Springer; 2014.
112. Edvinsson L, Edvinsson JCA, Haanes KA. Biological and Small Molecule Strategies in Migraine Therapy with Relation to the Calcitonin Gene-Related Peptide Family of Peptides. *Br J Pharmacol* 2022;179:371–80.
113. Li D, Abreu J, Tepper SJ. A Brief Review of Gepants. *Curr Pain Headache Rep* 2023;27:479–88.
114. Dhillon S. Correction to: Zavegepant: First Approval. *Drugs* 2023;83:1063.
115. Hutchings CJ, Koglin M, Olson WC, Marshall FH. Opportunities for Therapeutic Antibodies Directed at G-Protein-Coupled Receptors. *Nat Rev Drug Discov* 2017;16:661.
116. Robblee J, Harvey LK. Cardiovascular Disease and Migraine: Are the New Treatments Safe? *Curr Pain Headache Rep* 2022;26:647–55.
117. Nguyen PT, Yarov-Yarovoy V. Towards Structure-Guided Development of Pain Therapeutics Targeting Voltage-Gated Sodium Channels. *Front Pharmacol* 2022;13:842032.
118. Osteen JD, Immani S, Tapley TL, Indersmitten T, Hurst NW, Healey T, et al. Pharmacology and Mechanism of Action of Suzetrigine, a Potent and Selective NaV1.8 Pain Signal Inhibitor for the Treatment of Moderate to Severe Pain. *Pain Ther* 2025. https://doi.org/10.1007/s40122-024-00697-0.
119. Zimmermann K, Leffler A, Babes A, Cendan CM, Carr RW, Kobayashi J-I, et al. Sensory Neuron Sodium Channel Nav1.8 Is Essential for Pain at Low Temperatures. *Nature* 2007;447:855–8.

120. Xiao Y, Barbosa C, Pei Z, Xie W, Strong JA, Zhang J-M, et al. Increased Resurgent Sodium Currents in Nav1.8 Contribute to Nociceptive Sensory Neuron Hyperexcitability Associated with Peripheral Neuropathies. *J Neurosci Off J Soc Neurosci* 2019;39:1539–50.
121. Benn SC, Costigan M, Tate S, Fitzgerald M, Woolf CJ. Developmental Expression of the TTX-Resistant Voltage-Gated Sodium Channels Nav1.8 (SNS) and Nav1.9 (SNS2) in Primary Sensory Neurons. *J Neurosci Off J Soc Neurosci* 2001;21:6077–85.
122. Hameed S. Nav1.7 and Nav1.8: Role in the Pathophysiology of Pain. *Mol Pain* 2019;15:1744806919858801.
123. Alsaloum M, Higerd GP, Effraim PR, Waxman SG. Status of Peripheral Sodium Channel Blockers for Non-Addictive Pain Treatment. *Nat Rev Neurol* 2020;16:689–705.
124. Cheng EL, Kong AYH, Tan HS, Habib AS. Suzetrigine. *ASA Monit* 2025;89:16–8.
125. Freeman M. The World Mental Health Report: Transforming Mental Health for All. *World Psychiatry Off J World Psychiatr Assoc WPA* 2022;21:391–2.
126. McGrath JJ, Al-Hamzawi A, Alonso J, Altwaijri Y, Andrade LH, Bromet EJ, et al. Age of Onset and Cumulative Risk of Mental Disorders: A Cross-National Analysis of Population Surveys from 29 Countries. *Lancet Psychiatry* 2023;10:668–81.
127. Penninx BWJH, Benros ME, Klein RS, Vinkers CH. How COVID-19 Shaped Mental Health: From Infection to Pandemic Effects. *Nat Med* 2022;28:2027–37.
128. Georgieva I, Lepping P, Bozev V, Lickiewicz J, Pekara J, Wikman S, et al. Prevalence, New Incidence, Course, and Risk Factors of PTSD, Depression, Anxiety, and Panic Disorder during the Covid-19 Pandemic in 11 Countries. *Healthcare (Basel)* 2021;9(6):664.
129. Cohen S. The Mental Hygiene Movement, the Development of Personality and the School: The Medicalization of American Education. *Hist Educ Q* 1983;23:123–49.
130. Meyer A. The Mental Hygiene Movement. *Can Med Assoc J* 1918;8:632–4.
131. Wong M-L, Licinio J. From Monoamines to Genomic Targets: A Paradigm Shift for Drug Discovery in Depression. *Nat Rev Drug Discov* 2004;3:136–51.
132. Jilek WG. Emil Kraepelin and Comparative Sociocultural Psychiatry. *Eur Arch Psychiatry Clin Neurosci* 1995;245:231–8.
133. Blum GS. *Psychoanalytic Theories of Personality. New York, NY*: McGraw-Hill; 1953.
134. Sperling MB, Berman WH, Fagen G. Classification of Adult Attachment: An Integrative Taxonomy from Attachment and Psychoanalytic Theories. *J Pers Assess* 1992;59:239–47.
135. Braslow JT, Marder SR. History of Psychopharmacology. *Annu Rev Clin Psychol* 2019;15:25–50.
136. Walk A. Medico-Psychologists, Maudsley and the Maudsley. *Br J Psychiatry J Ment Sci* 1976;128:19–30.
137. López-Muñoz F, Ucha-Udabe R, Alamo C. The History of Barbiturates a Century after Their Clinical Introduction. *Neuropsychiatr Dis Treat* 2005;1:329–43.

138. Schneider-Thoma J, Hamza T, Chalkou K, Siafis S, Dong S, Bighelli I, et al. Efficacy of Clozapine versus Second-Generation Antipsychotics in People with Treatment-Resistant Schizophrenia: A Systematic Review and Individual Patient Data Meta-Analysis. *Lancet Psychiatry* 2025. https://doi.org/10.1016/S2215-0366(25)00001-X.
139. Seeman P. Atypical Antipsychotics: Mechanism of Action. *Can J Psychiatry Rev Can Psychiatr* 2002;47:27–38.
140. Deutch AY, Moghaddam B, Innis RB, Krystal JH, Aghajanian GK, Bunney BS, et al. Mechanisms of Action of Atypical Antipsychotic Drugs. Implications for Novel Therapeutic Strategies for Schizophrenia. *Schizophr Res* 1991;4:121–56.
141. Leucht S, Corves C, Arbter D, Engel RR, Li C, Davis JM. Second-Generation versus First-Generation Antipsychotic Drugs for Schizophrenia: A Meta-Analysis. *Lancet* 2009;373:31–41.
142. Leucht S, Priller J, Davis JM. Antipsychotic Drugs: A Concise Review of History, Classification, Indications, Mechanism, Efficacy, Side Effects, Dosing, and Clinical Application. *Am J Psychiatry* 2024;181:865–78.
143. McCutcheon RA, Harrison PJ, Howes OD, McGuire PK, Taylor DM, Pillinger T. Data-Driven Taxonomy for Antipsychotic Medication: A New Classification System. *Biol Psychiatry* 2023;94:561–8.
144. Meltzer HY. The Mechanism of Action of Novel Antipsychotic Drugs. *Schizophr Bull* 1991;17:263–87.
145. Kapur S, Mamo D. Half a Century of Antipsychotics and Still a Central Role for Dopamine D2 Receptors. *Prog Neuropsychopharmacol Biol Psychiatry* 2003;27:1081–90.
146. Meltzer HY. Update on Typical and Atypical Antipsychotic Drugs. *Annu Rev Med* 2013;64:393–406.
147. Keefe RSE, Seidman LJ, Christensen BK, Hamer RM, Sharma T, Sitskoorn MM, et al. Comparative Effect of Atypical and Conventional Antipsychotic Drugs on Neurocognition in First-Episode Psychosis: A Randomized, Double-Blind Trial of Olanzapine versus Low Doses of Haloperidol. *Am J Psychiatry* 2004;161:985–95.
148. Beach SR, Gross AF, Hartney KE, Taylor JB, Rundell JR. Intravenous Haloperidol: A Systematic Review of Side Effects and Recommendations for Clinical Use. *Gen Hosp Psychiatry* 2020;67:42–50.
149. Beresford R, Ward A. Haloperidol Decanoate. *Drugs* 1987;33:31–49.
150. Kurz M, Hummer M, Oberbauer H, Fleischhacker WW. Extrapyramidal Side Effects of Clozapine and Haloperidol. *Psychopharmacology (Berl)* 1995;118:52–6.
151. Pierre JM. Extrapyramidal Symptoms with Atypical Antipsychotics. *Drug Saf Int J Med Toxicol Drug Exp* 2005;28:191–208.
152. Stroup TS, Gray N. Management of Common Adverse Effects of Antipsychotic Medications. *World Psychiatry Off J World Psychiatr Assoc WPA* 2018;17:341–56.
153. Horacek J, Bubenikova-Valesova V, Kopecek M, Palenicek T, Dockery C, Mohr P, et al. Mechanism of Action of Atypical Antipsychotic Drugs and the Neurobiology of Schizophrenia. *CNS Drugs* 2006;20:389–409.

154. Meltzer HY. Treatment-Resistant Schizophrenia–the Role of Clozapine. *Curr Med Res Opin* 1997;14:1–20.
155. Lobos A, Claudia KK, Rummel-Kluge C, Hunger H, Schmid F, Schwarz S, et al. Clozapine versus Other Atypical Antipsychotics for Schizophrenia. *Cochrane Database Syst Rev* 2010;11:CD006633.
156. Frazier JA, Meyer MC, Biederman J, Wozniak J, Wilens TE, Spencer TJ, et al. Risperidone Treatment for Juvenile Bipolar Disorder: A Retrospective Chart Review. *J Am Acad Child Adolesc Psychiatry* 1999;38:960–5.
157. Fenton C, Scott LJ. Risperidone. *CNS Drugs* 2005;19:429–44.
158. Eder U, Mangweth B, Ebenbichler C, Weiss E, Hofer A, Hummer M, et al. Association of Olanzapine-Induced Weight Gain with an Increase in Body Fat. *Am J Psychiatry* 2001;158:1719–22.
159. He M, Deng C, Huang X-F. The Role of Hypothalamic H1 Receptor Antagonism in Antipsychotic-Induced Weight Gain. *CNS Drugs* 2013;27:423–34.
160. Lieberman JA. Dopamine Partial Agonists. *CNS Drugs* 2004;18:251–67.
161. Mailman RB, Murthy V. Third Generation Antipsychotic Drugs: Partial Agonism or Receptor Functional Selectivity? *Curr Pharm Des* 2010;16:488–501.
162. Kuroki T, Nagao N, Nakahara T. Neuropharmacology of Second-Generation Antipsychotic Drugs: A Validity of the Serotonin-Dopamine Hypothesis. *Prog Brain Res* 2008;172:199–212.
163. Cummings J, Isaacson S, Mills R, Williams H, Chi-Burris K, Corbett A, et al. Pimavanserin for Patients with Parkinson's Disease Psychosis: A Randomised, Placebo-Controlled Phase 3 Trial. *Lancet* 2014;383:533–40.
164. Feighner JP. Mechanism of Action of Antidepressant Medications. *J Clin Psychiatry* 1999;60(Suppl 4):4–11, 12–3.
165. Castrén E. Neurotrophic Effects of Antidepressant Drugs. *Curr Opin Pharmacol* 2004;4:58–64.
166. Duman RS, Heninger GR, Nestler EJ. A Molecular and Cellular Theory of Depression. *Arch Gen Psychiatry* 1997;54:597–606.
167. Duman RS, Deyama S, Fogaça MV. Role of BDNF in the Pathophysiology and Treatment of Depression: Activity-Dependent Effects Distinguish Rapid-Acting Antidepressants. *Eur J Neurosci* 2021;53:126–39.
168. Duman RS, Monteggia LM. A Neurotrophic Model for Stress-Related Mood Disorders. *Biol Psychiatry* 2006;59:1116–27.
169. Stahl SM. Mechanism of Action of Serotonin Selective Reuptake Inhibitors. *J Affect Disord* 1998;51:215–35.
170. Fava GA, Gatti A, Belaise C, Guidi J, Offidani E. Withdrawal Symptoms after Selective Serotonin Reuptake Inhibitor Discontinuation: A Systematic Review. *Psychother Psychosom* 2015;84:72–81.
171. Chouinard G, Chouinard V-A. New Classification of Selective Serotonin Reuptake Inhibitor Withdrawal. *Psychother Psychosom* 2015;84:63–71.
172. Fava M, Rosenbaum JF, Hoog SL, Tepner RG, Kopp JB, Nilsson ME. Fluoxetine versus Sertraline and Paroxetine in Major Depression: Tolerability and Efficacy in Anxious Depression. *J Affect Disord* 2000;59:119–26.

173. Ball SG, Kuhn A, Wall D, Shekhar A, Goddard AW. Selective Serotonin Reuptake Inhibitor Treatment for Generalized Anxiety Disorder: A Double-Blind, Prospective Comparison between Paroxetine and Sertraline. *J Clin Psychiatry* 2005;66:94–9.
174. Carvalho AF, Sharma MS, Brunoni AR, Vieta E, Fava GA. The Safety, Tolerability and Risks Associated with the Use of Newer Generation Antidepressant Drugs: A Critical Review of the Literature. *Psychother Psychosom* 2016;85:270–88.
175. Edinoff AN, Akuly HA, Hanna TA, Ochoa CO, Patti SJ, Ghaffar YA, et al. Selective Serotonin Reuptake Inhibitors and Adverse Effects: A Narrative Review. *Neurol Int* 2021;13:387–401.
176. Lambert O, Bourin M. SNRIs: Mechanism of Action and Clinical Features. *Expert Rev Neurother* 2002;2:849–58.
177. Trouvin A-P, Perrot S, Lloret-Linares C. Efficacy of Venlafaxine in Neuropathic Pain: A Narrative Review of Optimized Treatment. *Clin Ther* 2017;39:1104–22.
178. Calvi A, Fischetti I, Verzicco I, Murri MB, Zanetidou S, Volpi R, et al. Antidepressant Drugs Effects on Blood Pressure. *Front Cardiovasc Med* 2021;8:704281.
179. Wang S-M, Han C, Bahk W-M, Lee S-J, Patkar AA, Masand PS, et al. Addressing the Side Effects of Contemporary Antidepressant Drugs: A Comprehensive Review. *Chonnam Med J* 2018;54:101–12.
180. Lunn MPT, Hughes RAC, Wiffen PJ. Duloxetine for Treating Painful Neuropathy, Chronic Pain or Fibromyalgia. *Cochrane Database Syst Rev* 2014;2014:CD007115.
181. Staiger TO, Gaster B, Sullivan MD, Deyo RA. Systematic Review of Antidepressants in the Treatment of Chronic Low Back Pain. *Spine* 2003;28:2540–5.
182. Fasipe O. Neuropharmacological Classification of Antidepressant Agents Based on Their Mechanisms of Action. *Arch Med Health Sci* 2018;6:81.
183. Parekh PK, Johnson SB, Liston C. Synaptic Mechanisms Regulating Mood State Transitions in Depression. *Annu Rev Neurosci* 2022;45:581–601.
184. Ascher JA, Cole JO, Colin JN, Feighner JP, Ferris RM, Fibiger HC, et al. Bupropion: A Review of Its Mechanism of Antidepressant Activity. *J Clin Psychiatry* 1995;56:395–401.
185. Vanderkooy JD, Kennedy SH, Bagby RM. Antidepressant Side Effects in Depression Patients Treated in a Naturalistic Setting: A Study of Bupropion, Moclobemide, Paroxetine, Sertraline, and Venlafaxine. *Can J Psychiatry Rev Can Psychiatr* 2002;47:174–80.
186. Amsterdam JD, Hornig-Rohan M. Treatment Algorithms in Treatment-Resistant Depression. *Psychiatr Clin North Am* 1996;19:371–86.
187. Fava M. Diagnosis and Definition of Treatment-Resistant Depression. *Biol Psychiatry* 2003;53:649–59.
188. Remick RA. Anticholinergic Side Effects of Tricyclic Antidepressants and Their Management. *Prog Neuropsychopharmacol Biol Psychiatry* 1988;12:225–31.
189. Moraczewski J, Awosika AO, Aedma KK. *Tricyclic Antidepressants*. StatPearls, Treasure Island, *FL*: StatPearls Publishing; 2025.

190. Sindrup SH, Otto M, Finnerup NB, Jensen TS. Antidepressants in the Treatment of Neuropathic Pain. *Basic Clin Pharmacol Toxicol* 2005;96:399–409.
191. Mavissakalian M, Turner SM, Michelson L, Jacob R. Tricyclic Antidepressants in Obsessive-Compulsive Disorder: Antiobsessional or Antidepressant Agents? II. *Am J Psychiatry* 1985;142:572–6.
192. Monda JM, Husmann DA. Primary Nocturnal Enuresis: A Comparison among Observation, Imipramine, Desmopressin Acetate and Bed-Wetting Alarm Systems. *J Urol* 1995;154:745–8.
193. Youdim MBH, Bakhle YS. Monoamine Oxidase: Isoforms and Inhibitors in Parkinson's Disease and Depressive Illness. *Br J Pharmacol* 2006;1:287–96.
194. Clarke A, Brewer F, Johnson ES, Mallard N, Hartig F, Taylor S, et al. A New Formulation of Selegiline: Improved Bioavailability and Selectivity for MAO-B Inhibition. *J Neural Transm Vienna Austria* 2003;110:1241–55.
195. Cristea IA, Naudet F. US Food and Drug Administration Approval of Esketamine and Brexanolone. *Lancet Psychiatry* 2019;6:975–7.
196. Schatzberg AF. Mechanisms of Action of Ketamine and Esketamine. *Am J Psychiatry* 2021;178:1130.
197. Daly EJ, Singh JB, Fedgchin M, Cooper K, Lim P, Shelton RC, et al. Efficacy and Safety of Intranasal Esketamine Adjunctive to Oral Antidepressant Therapy in Treatment-Resistant Depression: A Randomized Clinical Trial. *JAMA Psychiatry Chic Ill* 2018;75:139–48.
198. McCarthy B, Bunn H, Santalucia M, Wilmouth C, Muzyk A, Smith CM. Dextromethorphan-Bupropion (Auvelity) for the Treatment of Major Depressive Disorder. *Clin Psychopharmacol Neurosci Off Sci J Korean Coll Neuropsychopharmacol* 2023;21:609–16.
199. Phillips B, O'Connor C, Onge E. Gepirone: A New Extended-Release Oral Selective Serotonin Receptor Agonist for Major Depressive Disorder. *J Pharm Technol JPT Off Publ Assoc Pharm Tech* 2024;40:230–5.
200. Barnes KN, Vogl CM, Nelson LA. Zuranolone: The First FDA-Approved Oral Treatment Option for Postpartum Depression. *Ann Pharmacother* 2024;58:728–34.
201. Maguire JL, Mennerick S. Neurosteroids: Mechanistic Considerations and Clinical Prospects. *Neuropsychopharmacol* 2023. https://doi.org/10.1038/s41386-023-01626-z.
202. Schou M, Juel-Nielsen N, Stromgren E, Voldby H. The Treatment of Manic Psychoses by the Administration of Lithium Salts. *J Neurol Neurosurg Psychiatry* 1954;17:250–60.
203. Phiel CJ, Klein PS. Molecular Targets of Lithium Action. *Annu Rev Pharmacol Toxicol* 2001;41:789–813.
204. Hakami T. Efficacy and Tolerability of Antiseizure Drugs. *Ther Adv Neurol Disord* 2021;14:17562864211037430.
205. Yatham LN, Kusumakar V, Calabrese JR, Rao R, Scarrow G, Kroeker G. Third Generation Anticonvulsants in Bipolar Disorder: A Review of Efficacy and Summary of Clinical Recommendations. *J Clin Psychiatry* 2002;63:275–83.
206. Strawbridge R, Kurana S, Kerr-Gaffney J, Jauhar S, Kaufman KR, Yalin N, et al. A Systematic Review and Meta-Analysis of Treatments for Rapid Cycling Bipolar Disorder. *Acta Psychiatr Scand* 2022;146:290–311.

207. Stoner SC, Nelson LA, Lea JW, Marken PA, Sommi RW, Dahmen MM. Historical Review of Carbamazepine for the Treatment of Bipolar Disorder. *Pharmacotherapy* 2007;27:68–88.
208. Moreno RA, Moreno DH, Macedo Soares MB, Ratzke R. Anticonvulsants and antipsychotics in the treatment of bipolar disorder. *Rev Bras Psiquiatr* 2004;26:37–43.
209. Sills GJ. The Mechanisms of Action of Gabapentin and Pregabalin. *Curr Opin Pharmacol* 2006;6:108–13.
210. Taylor CP. Mechanisms of Analgesia by Gabapentin and Pregabalin–Calcium Channel Alpha2-Delta [Cavalpha2-Delta] Ligands. *Pain* 2009;142:13–6.
211. Griebel G, Holmes A. 50 Years of Hurdles and Hope in Anxiolytic Drug Discovery. *Nat Rev Drug Discov* 2013;12:667–87.
212. Sandford JJ, Argyropoulos SV, Nutt DJ. The Psychobiology of Anxiolytic Drugs. Part 1: Basic Neurobiology. *Pharmacol Ther* 2000;88:197–212.
213. Sigel E, Steinmann ME. Structure, Function, and Modulation of GABA(A) Receptors. *J Biol Chem* 2012;287:40224–31.
214. Miller PS, Aricescu AR. Crystal Structure of a Human GABAA Receptor. *Nature* 2014;512:270–5.
215. Froestl W. Chemistry and Pharmacology of GABAB Receptor Ligands. *Adv Pharmacol San Diego Calif* 2010;58:19–62.
216. Möhler H. GABA(A) Receptor Diversity and Pharmacology. *Cell Tissue Res* 2006;326:505–16.
217. Chebib M, Johnston GA. GABA-Activated Ligand Gated Ion Channels: Medicinal Chemistry and Molecular Biology. *J Med Chem* 2000;43:1427–47.
218. Baldwin DS, Aitchison K, Alan Bateson HVC, Davies S, Leonard B, Nutt DJ, et al. Benzodiazepines: Risks and Benefits. A Reconsideration. *J Psychopharmacol Oxf Engl* 2013;27:967–71.
219. Reed CE, Driggs MF, Foote CC. Acute Barbiturate Intoxication: A Study of 300 Cases Based on a Physiologic System of Classification of the Severity of the Intoxication. *Ann Intern Med* 1952;37:290–303.
220. Brandt J, Leong C. Benzodiazepines and Z-Drugs: An Updated Review of Major Adverse Outcomes Reported on in Epidemiologic Research. *Drugs RD* 2017;17:493–507.
221. Rothman RB, Baumann MH. Monoamine Transporters and Psychostimulant Drugs. *Eur J Pharmacol* 2003;479:23–40.
222. Pérez-Mañá C, Castells X, Torrens M, Capellà D, Farre M. Efficacy of Psychostimulant Drugs for Amphetamine Abuse or Dependence. *Cochrane Database Syst Rev* 2013;2013(9):CD009695.
223. Cunha-Oliveira T, Rego AC, Oliveira CR. Cellular and Molecular Mechanisms Involved in the Neurotoxicity of Opioid and Psychostimulant Drugs. *Brain Res Rev* 2008;58:192–208.
224. Heal DJ, Smith SL, Gosden J, Nutt DJ. Amphetamine, Past and Present: A Pharmacological and Clinical Perspective. *J Psychopharmacol Oxf Engl* 2013;27:479–96.
225. Quintero J, Gutiérrez-Casares JR, Álamo C. Molecular Characterisation of the Mechanism of Action of Stimulant Drugs Lisdexamfetamine and Methylphenidate on ADHD Neurobiology: A Review. *Neurol Ther* 2022;11:1489–517.

226. Sulzer D, Sonders MS, Poulsen NW, Galli A. Mechanisms of Neurotransmitter Release by Amphetamines: A Review. *Prog Neurobiol* 2005;75:406–33.
227. Ballon JS, Feifel D. A Systematic Review of Modafinil: Potential Clinical Uses and Mechanisms of Action. *J Clin Psychiatry* 2006;67:554–66.
228. Loland CJ, Mereu M, Okunola OM, Cao J, Prisinzano TE, Mazier S, et al. R-Modafinil (Armodafinil): A Unique Dopamine Uptake Inhibitor and Potential Medication for Psychostimulant Abuse. *Biol Psychiatry* 2012;72:405–13.
229. Patel K, Allen S, Haque MN, Angelescu I, Baumeister D, Tracy DK. Bupropion: A Systematic Review and Meta-Analysis of Effectiveness as an Antidepressant. *Ther Adv Psychopharmacol* 2016;6:99–144.
230. Ross S, Williams D. Bupropion: Risks and Benefits. *Expert Opin Drug Saf* 2005;4:995–1003.

Chapter 2

The evolution of medicinal compounds

Historical perspective and modern innovation

Shivani Tomar[1] #, *Bhavya Kapoor*[1] #, *Rajiv Pathak*[2], *and Vivek Srivastava*[1] *

[1]Department of Chemistry & Biochemistry, Sharda School of Engineering & Science, Sharda University, Greater Noida, Uttar Pradesh, India

[2]Department of Genetics, Albert Einstein College of Medicine, New York, USA

# Author's contributed equally

* Corresponding author

2.1 INTRODUCTION

Modern medicine has significantly contributed to treating infectious conditions and crises. In most other disciplines, the primary goal is control, often known as palliation. Pharmacology, including psychopharmacology, is primarily concerned with controlling and alleviating such symptoms. Clinicians and researchers must now focus their efforts resolutely on prevention and cure. Health and wellness are also significant concerns in modern medicine. Advances in vaccinations for hypertension, diabetes, cancer, and other conditions merit attention, as does the importance of meditation, yoga, and spirituality in illness prevention at various levels [1]. A detailed examination of alternative and complementary therapies is required to identify any relevant models they may have, putting aside their loud rhetoric and/or antipathy toward mainstream medical care. Medicine is a sign of human Eros; it should not be used to achieve Thanatos. It must reach its fullest potential so that Eros and Thanatos triumph only eventually, not prematurely [1]. Research on life expectancy, lifestyle adjustments, and healthy centenarians demands special attention to determine what promotes lifespan and well-being [2]. Modern medicine has created four advances that have dramatically altered human life. First, preventive medicine had phenomenal success with the baby welfare movement, and in the early 19th century, it won a series of victories against infectious diseases. These

DOI: 10.1201/9781003733010-2

achievements significantly impacted the population growth and history of the United States before World War I. Second, surgery launched the first successful therapeutic attack on disease in the final part of the 19th century. Third, with a scientific understanding of the etiology of end development and the etiology of endocrine and metabolic disease, significant progress was made in preventing, managing, and curing these disorders throughout the beginning quarter of the 20th century. Considerable progress was made in preventing, managing, and curing endocrine and metabolic diseases throughout the beginning quarter of the 20th century. Fourth, the discovery of non-toxic, systemic chemotherapeutic medicines that effectively eliminate microbial infections over the last 25 years has significantly lowered infectious disease mortality. Compared to the 3,500 years of recorded medical history, these conquests mark the commencement of a new era. Medicine is on the verge of entering a vast, nearly entirely new field [1,3].

The ways of identifying natural medications employ current processes, such as high-throughput screening (HTS), in which several hundred molecules can be examined in many experiments quickly and with a small number of chemicals. The most effective outcomes will likely be acquired from a collection of entirely organically recognized compounds that allow scientists to easily extract the lead ingredient to rapidly develop innovative medication compositions, such as *in vivo* investigations, full or partial synthesis, and clinical testing [4,5]. Drug development is a highly complex procedure that incorporates numerous fields, such as computer science, proteomics, metabolomics, and structural biology, to mention a few. Discovering candidates, synthesizing, characterizing, screening, and testing for therapeutic efficacy are all stages in the drug discovery process. When a molecule has demonstrated its worth in the tests above, the medication development procedure starts before clinical trials begin. In general, the procedure is costly and time-consuming [6,7]. Artificial intelligence has revolutionized the pharmaceutical sector in multiple drug discovery, development, and commercialization phases. Artificial intelligence (AI)-driven technologies have greatly improved the drug development process and provided creative answers to the industry's problems [8]. The pharmaceutical industry benefits significantly from using AI techniques and instruments since they may employ automated algorithms to perform tasks that traditionally require human intellect. The pharmaceutical sector has dramatically transformed because scientists use AI to create new medications, discover ways to treat illnesses, and more [9]. One area that is expanding quickly is the application of nanotechnology to natural products. Nanotechnology offers several benefits when it comes to delivering natural substances for the therapeutic management of cancer and other chronic human diseases. Adding nanoparticles to natural goods can improve their targeting, controlled release, and bioavailability. Some historical applications of medicinal plants can be explained, and possibly new uses might be found using molecular docking in organic product-based drug development programs [10,11]. We will highlight contemporary advances in science and technology that can help conquer difficulties in the nanoparticle-based discovery of drugs, concentrating on three areas: cultivation systems, genome mining and engineering, and analytical techniques. In the final section, we identify possible future approaches for nanoparticle drug development. A detailed analysis of 586 phytochemicals was conducted to assess their drug-likeness, pharmacological characteristics, and structure-based action against viral proteins (spike glycoprotein, papain-like protease, and main protease) and host proteins (ACE2, importin-subunit α-5, and β-1). Several traditional Ayurvedic and Western natural treatments have already been reported to be effective against a variety of viral infections, including Ebola, dengue, human immunodeficiency virus, chikungunya, and others, and they may also be an alternate choice against SARS-CoV-2 [12,13]. Future perspectives on aromatic herbs, essential oils derived from medicinal plants, and phytochemicals extracted as potential coronavirus therapeutics. The multidisciplinary group will bring Ayurveda into the national clinical management plan for COVID-19 [14]. The Ministry of AYUSH of India has initiated a research study for COVID-19 using four ways of Ayurvedic medicinal medicines: guduchi (Tinospora cordifolia (Willd.) Miers), ashwagandha (Withania somnifera (L.) Dunal), pipli (Piper longum L.), and yasthimadhu (Glycyrrhiza glabra L.). The World Health Organization promotes the development of products derived from nature-based treatments for almost any medical condition or disease [15,16]. Natural products account for approximately 70%–80% of conventional medicines. Thus, it is worthwhile to look for bioactive natural compounds with known antiviral effectiveness against CoV [12–16].

This chapter will discuss the impacts and historical perspective of natural-based medicines, which also play an essential role in modern medicine. We will also discuss how AI technology helped us with drug development processes and discoveries using nanoparticle-based drug delivery systems in anticancer therapeutics.

2.2 HISTORICAL PERSPECTIVE OF MEDICINAL COMPOUNDS

Herbal medicine's historical significance is an excellent demonstration of this centuries-old relationship between humanity and the natural world in search of wellness and good health. Numerous cultures worldwide have recognized and developed the use of plants' therapeutic qualities throughout generations. An overview of this persistent relationship is provided by the study of significant turning points in the establishment of plant-based medicine:

2.2.1 Sushruta's surgical legacy: a historical review of his contributions to the development of surgical science and practice

Ancient India's rich customs and myths were deeply rooted in Sushruta's childhood. The history of Indian medicine refers to the 6th century BCE, when he lived, as the "Golden Age." His place of birth was the ancient village of Kashi, one of the oldest continuously inhabited cities in the world [17]. Sushruta is a titan of surgical history who has accomplished medical achievements beyond time and location. The discoveries he made revolutionized ancient Indian medicine and have impacted Greek, Persian, and European medical practices beyond generations and geographical boundaries [18]. European surgeons began implementing Sushruta's approaches into the expanding field of plastic surgery in the 18th century. His emphasis on preciseness, sanitation, and an integrative approach to patient care is still significant in today's medical practice. Furthermore, his scientific strategy for providing instruction in surgery, which was founded on integrating theoretical knowledge and hands-on experience, has served as a foundation for modern surgical education. The ageless essence of ancient Indian medical traditions is proven by Sushruta's remarkable findings, which significantly impacted the initial principles of systemic health and continue to motivate and challenge medical practitioners worldwide [19].

Sushruta's exceptional contributions to medicine and surgery are a lasting example, continually influencing current medical practices. His remarkable skill and precision in operations such as bone realignment, cataract extraction, and rhinoplasty transformed the surgical techniques of his time. Inspired by Ayurveda, his methods emphasized the importance of maintaining various bodily harmonies and balances. Even without modern anesthetics and with rudimentary tools, Sushruta applied his anatomical knowledge and delivered precise patient care through autopsy practices. His commitment to cleanliness and precision has been passed down through generations, impacting medical culture worldwide. Sushruta not only transformed ancient surgery but also established enduring standards that have stood the test of time for thousands of years.

Ayurveda, often called the "science of life," significantly influenced Sushruta's work by emphasizing the connection between body, mind, and spirit. His remarkable contributions were pivotal in establishing the foundations of surgical science, which have shaped traditional and contemporary medical practices.

2.2.2 Ancient Egyptian healing practices: the Ebers Papyrus and beyond

Plant-related books, like those found in the Ebers Papyrus, which date back to approximately 1500 B.C., were the first to be written about medicine [20]. An Egyptian manuscript called the "Ebers Papyrus" outlines a thousand various herbal remedies [21]. Eight hundred seventy-seven recipes and prescriptions for skin, eye, and internal medical issues and gynecological issues were discovered in the Ebers papyrus. Another document that discusses gynecological issues in detail comes from the Kahun papyrus, which dates to circa 1800 B.C. Herbal remedies like frankincense, fennel, myrrh, castor oil, sienna, aloe, thyme, and linseed were the primary sources of medicinal drugs [22]. Several studies have published and translated this papyrus; the most valuable is the German edition, 'Grundriss de Medizin de alten ägypter,' based on this Paul Ghalioungui edition. According to Grundriss, the disorganized format of the medical advice in the papyrus suggests that they were drawn from many originals [23]. A sequence of prescriptions that are categorized by various illnesses, injuries, and diseases makes up the text of the Ebers Papyrus. Most of those groups begin with the phrase "Here begins," which is used 36 times. However, they are frequently disorganized and diversified. Given that "physician secrets" are mentioned in the text, the papyrus's owner was probably a doctor. Herodotus claims Egyptian doctors were highly skilled, and the Ebers Papyrus supports this assertion [23].

2.2.3 Chinese herbal medicine (ancient times) [2000 BCE]

The Chinese Herbal Guide claims that herbal treatment has been practiced since 2000 B.C. The Chinese medicine encyclopedia "Pen-ts'ao kang mu" lists 8,160 prescriptions and approximately 1,898 herbal medications [21–23]. The application of natural products is documented in another document named Tang Herbal (~659 A.D., 850 medications), Shennong Herbal (~100 B.C., 365 drugs), and the Chinese Materia Medica (1100 B.C.) (Wu Shi Er Bing Fang contains 52 prescriptions) [24]. China has a long history of utilizing herbal medicine, which has been used for millennia. The foundation of this tradition is found in the principles of traditional Chinese medicine (TCM), a compassionate system that uses botanical remedies to restore balance and promote broad good health [25,26]. Chinese herbal medicine is based on a meticulous understanding of many plant, mineral, and animal substances, which are still extensively relevant in present-day TCM [27]. TCM specialists use botanicals and herbal formulas to encourage good health and improve organ function; understanding the nature of multiple herbal constituents permits the TCM to produce a medicinal effect beyond the herbs' physical characteristics and chemical composition. The ingredients of Chinese herbal formulations, some of which have been used for more than 2200 years, are picked for their complementary roles. Unlike Western medicine, which usually takes an individualized approach, TCM often combines botanicals to achieve a synergistic effect [28,29].

2.2.4 Quinine: a malaria solution via the Cinchona tree (17th century)

Malaria has long been one of the deadliest diseases caused by the Plasmodium parasite and has claimed countless lives over time. A significant advancement in the fight against malaria came from the knowledge of indigenous South Americans, who used the bark of the Cinchona officinalis tree to treat fevers [30]. This traditional practice led to the discovery of quinine, the bark's active ingredient, which revolutionized malaria treatment and remained the primary therapy until the development of modern medicines [31,32].

This narrative showcases the deep wisdom present in traditional healing practices [33]. The indigenous people's mastery of nature's medicinal properties was not by chance but was rooted in generations of careful observation and experience. Their knowledge provided the essential foundation for what would become modern pharmaceutical science. This historical example highlights two important ideas: the immense healing potential still waiting to be uncovered in nature and our duty to protect these natural resources. As medical research progresses, safeguarding and exploring the planet's biodiversity is vital for future therapeutic discoveries [34].

2.2.5 Aspirin: a present given by the willow tree (19th century)

The history of botanic-based medical treatments has significantly impacted the development of pharmaceutical sciences. The most well-known is potentially aspirin, a chemical widely used as a pain reliever in all parts of the human body. The bark of the Salix alba, or willow tree, was first used to extract aspirin [35]. This type of natural medicine progressed into synthetical synthesis, making it possible to manufacture synthetic forms. As a result, it is now widely acknowledged and utilized as a key element in pain treatment. As a result, natural herbal remedies have significantly impacted the pharmaceutical business today, with aspirin a good illustration of the transition from herbal remedies to synthetic pharmaceuticals [36]. The Greeks and Egyptians utilized willow bark as a pain reliever, as did other ancient societies. Acetylsalicylic acid, the actual active element that improved pain management and inflammatory control during the 19th century, was finally extracted by scientists [37].

2.2.6 Penicillin

Penicillium mold was discovered by Alexander Fleming in 1928 to be effective against staphylococcus bacteria. In collaboration with Howard Florey, Ernst Chain returned to this discovery around 10 years later. Because of Florey and Chain's work, penicillin was being manufactured substantially by 1944. Using microorganisms to produce pharmaceuticals was the first step toward biotechnology [24,38]. Florey, Chain, and Fleming were awarded the Nobel Prize in Medicine and Physiology in 1945 for their significant contribution to the development of penicillin. The word "antibiotics" was coined when penicillin was separated from other microbes. Many more antibiotics, such as streptomycin, chloramphenicol, erythromycin, vancomycin, and others, were discovered by researchers in the 1940s and 1950s using analogous methods of discovery and biosynthesis [36].

2.2.7 Evolution and integration of AYUSH in India's healthcare system

AYUSH stands for Ayurveda, Yoga, Unani, Naturopathy, Siddha, and Homeopathy. India benefits significantly from the collaboration of the six traditional medical systems [39]. The Department of Indian Systems of Medicine was established in March 1995, and its name was changed to AYUSH in November 2003 to better focus on advancing these systems. The most recent change was the division of the government into autonomous ministries in 2014; these ministries are overseen by the Union Government of India [39]. The five-year planning procedure established in 1951 included planning these medical systems [40]. The National Rural Health Mission (NRHM) currently implements the AYUSH system as part of the mainline health system. Although it started in 2005, the "Mainstreaming of AYUSH and Rejuvenation of Local Health Traditions" plan was implemented in 2006 to bolster public health services [41]. According to the 12th 5-year plan, this initiative has officially begun its second phase as of April 1, 2012. The scheme was primarily started to accomplish three critical goals: giving patients a wide range of treatment options, bolstering operational facilities, and bolstering the execution of national health initiatives [42]. However, in several locations, it seems to be forced medical pluralism because the union government took a top-down strategy without any community input. To consider and provide some insight into the future directions of the AYUSH system, the 5-year planning documents from the first to the twelfth will be evaluated [43]. Several problems, including species-specific variations in potency; the lack of an integrated coding system for all species commonly used in TSMs; different growth locations; inaccurate drug identification and adulteration; inconsistent quality control standards; and disparities in processing techniques, highlight the urgent need for a comparative analysis of medications used in these two medical systems [44].

2.2.8 Advanced antiviral approaches: phytochemical inhibitors and molecular docking in COVID-19 research

In reality, no one anticipated the coronavirus illness 2019 (COVID-19) would be more catastrophic than any past coronavirus-caused epidemics, including SARS and MERS. Originating in Wong's Hospital in Wuhan, the new coronavirus, SARS-CoV-2, has wrought unforeseen devastation on viral pandemic proportions, serving as a severe alert to the scientific community [45]. Studies have shown that monoclonal antibodies, vaccinations, and treatments are ineffective at preventing CoV infection. Here's where we'll assume that fresh approaches are needed to deal with this problem: As of July 23, 2020, there were over 619,150 recorded death cases worldwide, indicating that the scientific community was ill-prepared. However, there were no approved medications for the therapy against MERS-CoV and SARS-CoV until 2020, so broad-spectrum antiviral inhibitors received special attention [46]. Alpha, beta, gamma, and delta coronaviruses are the four variations. Only the beta and alpha varieties are contagious to humans. Coronaviruses are uncontrollable because of their higher nucleotide mutation rate than other single-stranded RNA viruses [45]. Potential targets for developing broad-spectrum anti-CoV drugs are anticipated to be found in the degree of conservation among the significant active domains of several human coronaviruses, such as RNA helicase and 3CLpro (3C-like protease) [46]. There are six coronavirus strains that infect people apart from the current SARS-CoV2: HCoVHKU1, HCoV-OC43, HCoV-NL63, HCoV-229E, MERS-CoV, and SARS-CoV. These enveloped viruses are members of the Coronaviridae family, which causes respiratory diseases [47]. A death rate of 35.4% as of December 2016 and 9% in 2002–2003 were the outcomes of epidemics brought on by the devastating SARS-CoV and MERS-CoV outbreaks in various countries [48]. One of the most crucial steps in the search for antiviral drugs is docking and structure-based virtual screening, which may be partially attributed to the genetic information gleaned from the new coronavirus COVID-19. Nine potential compounds (including myricitrin, licoleafol, and amaranthin) that may block the action of the SARS-CoV-2 3CLpro were identified by the screening of about 32,000 phytoconstituents using the molecular docking technique [49]. Using Lopinvatrium P and nelfinavir as standards, Flaven-3-olesicate demonstrated that compounds derived from medicinal plants, including epicatechin-gallate, catechin, curcumin, oleuropein, apigenin-7-glucoside, naringenin, demethoxycurcumin, quercetin, kaempferol, and luteolin-7-glucoside, may function as possible inhibitors against the primary COVID-19 protease [50,51] as shown in Table 2.1.

Table 2.1 Bioactive compounds from natural sources with their medicinal applications

Compounds	*Source*	*Therapeutic potential*	*References*
Paclitaxel	*Taxus brevifolia*	Widely used as a chemotherapeutic agent against various types of cancer	[52,53]
Quinine	Bark of the Cinchona tree	Reversing agent of MDR, which is used to treat resistant acute leukemias.	[53,54]
Silymarin	*Silybum marianum*	Anti-inflammatory, antioxidant qualities, and encouragement of liver cell regeneration	[53,55]
Artemisinin	*Artemisia annua*	Type I diabetes and cancer	[53,56]

2.3 UNLOCKING THE THERAPEUTIC POTENTIAL OF NATURAL PRODUCTS

In addition to their diverse array of multi-dimensional chemical structures, natural products have garnered significant interest for their potential as biological function modifiers. They have since significantly influenced chemobiology and have been effectively used in developing novel medications [27,57,58]. From the standpoint of physical chemistry, the tremendous structural variety of natural products has been recognized over the previous century. The intricacy of their well-structured, three-dimensional chemical and steric characteristics, which provide several benefits in molecular target efficiency and selectivity, is linked to their effectiveness.

The plant Digitalis lanata has a substance called digitoxin, which has been used to treat cardiac conditions, including congestive heart failure [57]. Cocaine, recognized for its anesthetic properties, has been traditionally employed as a local anesthetic and for its ability to constrict blood vessels. Pilocarpine, derived from the leaves of the jaborandi plant, has gained prominence in medicine for its capacity to stimulate salivation and perspiration [58]. On the other hand, pinoline, initially utilized as a narcotic, was overshadowed by cocaine due to its superior efficacy in pain relief. Quinine, extracted from the bark of the Cinchona tree, played a pivotal role in combating malaria [27,59,60].

Artemisinin and its analogs, a successful example of medication creation from natural chemicals, are currently widely used to treat malaria. This demonstrates how natural product-based research has significantly contributed to discovering new drugs [59]. The plant Taxus brevifolia yields paclitaxel, which treats breast, ovarian, and lung cancer [61]. The traditional Chinese herb Artemisia annua contains artemisinin, which is used to treat malaria and is resistant to several other medications. Another rapidly developing substance that is being researched for its potential to cure malaria is artemisinin. Sweet wormwood is the source of artemisinin, an herb utilized for thousands of years in TCM, and it is now being researched for its potential to cure drug-resistant variants of malaria [62]. Hepatic problems are treated with silymarin extracted from Silybum marianum seeds. Notably, many early plant-derived drugs, such as digitoxin and quinine, continue to be used in contemporary medicine, underscoring their enduring significance [63].

2.3.1 Medicinal plants and their secondary metabolites in traditional medicine and modern drug discovery

Drug discoveries must be made using natural sources to develop a more varied medicinal inventory for human consumption. Furthermore, active transporters recognize produced compounds as base materials, whereas natural

Table 2.2 Therapeutic potential of plant secondary metabolites

Compounds	*Secondary metabolite*	*Source*	*Therapeutic potential*	*References*
Reserpine	Alkaloid	*R. serpentine a*	Hypertension, psychoses	[65]
Vincamine	Alkaloid	*Vinca minor*	Vasodilator	[65]
Athel tamarisk	Flavonoid	*Tamarix aphylla L*	Antidiabetic, wound healing antibacterial, anti-inflammatory, hypolipidemic, antifungal, antioxidant	[69]
Black mulberry	Flavonoid	*Morus nigra L*	Antinociceptive, anti-inflammatory	[70]
Paeoniflorin	Terpenoid	*Paeonia lactiflora*	Anti-inflammatory	[71]
Betulinic acid	Terpenoid	*Syzigium claviforum*	Anti-HIV activity, antiviral	[71]

materials are more likely to resemble naturally generated metabolites and metabolic intermediates [64]. Despite improvements in discovery tactics, most notably the emergence of medications that originate from molecular biology, an obligation remains to establish natural product-based medicines, which have proven to be a very successful strategy [64,65].

Plants utilized for medical energy have secondary metabolites and therapeutically beneficial essential oils. The main benefits stated for the therapeutic use of plants for medicinal purposes in various diseases are their safety, cost-effectiveness, and ease of access [66]. Because of these benefits, medicinal plants are increasingly used by certain traditional medicine practitioners daily. Secondary metabolites constitute naturally occurring substances that do not immediately impact the core metabolic pathways of growth, development, or reproduction. Various secondary metabolites have been employed for medicinal purposes and are key components of traditional pharmacopoeia [65]. Alkaloids are nitrogen-based compounds in different plant species that exhibit various pharmacological actions. Morphine produced from the opium poppy, quinine obtained from cinchona plants, and caffeine taken from coffee beans are all examples of alkaloids [67]. For ages, traditional medicine has used these chemicals as analgesic, antimalarials, and stimulants. For millennia, conventional medicine has made use of these compounds because of their analgesics, antimalarial, and stimulant properties [68], as shown in Table 2.2.

2.3.1.1 *Alkaloids: a diverse group of compounds with antimicrobial and medicinal properties*

Alkaloids are essential to human medicine and an organism's inherent defense. Alkaloids account for about 20% of all recognized secondary metabolites in plants [72]. They provide plants with protection from predators and act as a growth regulator. As a therapeutic substance, they act as an anesthetic, have cardioprotective functions, and possess anti-inflammatory properties [73]. Alkaloids are organic molecules that contain nitrogen. These compounds are defined by the presence of a basic nitrogen atom, which may exist as a tertiary amine (R_3N), secondary amine (R_2NH), or primary amine (RMN_2) [74]. Alkaloids can be categorized into two primary groups based on their chemical structure and/or natural origins. The first category, referred to as typical alkaloids or non-heterocyclic (also referred to as proto-alkaloids or naturally occurring amines), has nitrogen atoms situated in the side chain. The second category comprises the true or typical alkaloids, which are characterized by their heterocyclic structures. Due to the complexity of this latter group, they are further classified into 14 subcategories based on the configurations of their ring structures. As a result, alkaloids exhibit various pharmacological effects [75–77].

Recent research has thoroughly documented the antimicrobial effects of alkaloids obtained from plants. These substances can inhibit the proliferation of different microorganisms, including fungi, bacteria, viruses, and protozoa, through multiple mechanisms. This property is notably essential in combating drug-resistant microbial strains [78]. For example, the alkaloid isoquinoline functions as an efflux pump inhibitor (EPI), thus demonstrating antimicrobial properties. Numerous other alkaloids operate through similar mechanisms, enhancing their broad-spectrum antimicrobial capabilities [79].

2.3.1.2 Flavonoids

Another secondary metabolite is flavonoids, which have a benzopyrene ring bearing a phenolic or polyphenolic group at separate positions [80]. Fruits, herbs, stems, cereals, nuts, vegetables, flowers, and seeds have an abundance of flavonoids [81]. These different plant parts' medicinal value and biological activities are attributed to bioactive phytochemical constituents [82]. Over 10,000 flavonoid compounds have been isolated and identified [83]. Many of them are recognized as therapeutic agents [84]. Flavonoids are produced in plants through the phenylpropanoid metabolic process, and their effectiveness depends on how well the body can absorb and utilize them [85]. These compounds have found their way into diverse commercial applications, from creating natural colorants [86,87] to formulating beauty and skincare solutions [87,88], including products that target aging skin [89]. While these uses are significant, flavonoids have significantly impacted medicine. Scientists have discovered their remarkable versatility as therapeutic agents, with research supporting their use in fighting cancer, infections (both bacterial and viral), and malaria [90–92]. They've also proven valuable in preventing abnormal blood vessel growth, protecting nerve cells, and stopping harmful cells from multiplying. Their powerful antioxidant properties are particularly valuable in medical treatments [93–95]. Another rich source of flavonoids is apple peel extracts, which inhibit acetylcholinesterase (ACE) in vitro and are an effective antihypertensive agent [96–100]. It is cardioprotective, prevents metabolic disorders [101,102], and has better preserved cognitive performance with senescence [87,102].

2.3.1.3 *Terpenoids*

Terpenoids are a broad group of secondary metabolites found in plants, making them the most prevalent substances in nature. Plants rely on terpenoid compounds to support their life cycle, playing crucial roles in everything from their maturation process to their responses to environmental stressors. Beyond their importance in plant biology, these versatile molecules have become key components across multiple industries, with manufacturers incorporating them into medicines, food products, and beauty formulations [71]. Terpenoids have anticancer, anti-inflammatory, antibacterial, antiviral, and antimalarial properties, enhance transdermal absorption, prevent and manage cardiovascular disease, and are hypoglycemic. Furthermore, earlier investigations have indicated that terpenoids have several potentials. Applications include insect resistance, immunoregulation, antioxidation, anti-aging, and neuroprotection [103]. Terpenoids have a complicated structure with multiple effects and modes of action. This study reviews the activities and mechanisms of terpenoids. Terpenoid compounds' potential for research and use was also [104]. This study explores terpenoid molecules' potential development and applications, serving as a valuable resource for drug discovery and design [71].

2.3.2 The emerging role of marine-derived compounds in modern medicine

Marine natural products are an extremely bright and effective source of pharmaceuticals of any plant, animal, or microbial origin. So far, medications for clinical use have been issued from 20 marine sources [105]. Most approved marine chemicals are antineoplastic, while others are utilized for heparin overdosage, haptens, and chronic neuropathic pain vaccines as carriers, for example, omega-3 fatty acid dietary supplements. They include an extensive spectrum of structural properties found in marine medications and various methods for their action in response to target cells [106]. Compared to the past, clinical approval for a growing number of marine medications has only recently occurred, owing to growing concern and study of marine substances in laboratories worldwide. In this publication, we give a complete overview of all marine medications that have ever been used in therapeutic settings [107]. Given the enormous number of marine natural chemicals being studied in clinical and preclinical research, researchers and physicians are expected to discover many more such medicines in the future [105].

2.3.2.1 Natural products from animals

Animal therapy, also known as zootherapy, comes in various forms and can be encountered in cultures worldwide. Any animal component or metabolites generated within a living organism, comprising excrements, saliva, secretions, and other substances, can be employed in animal-based compounds for therapy. Animal exploitation as a source of therapeutics for medicinal use is based on indigenous knowledge passed down from our forefathers, such as squeezing Pseudomyrmex genus ants for toothache and then allowing them to bite joints for joint pain or using rattlesnake body parts to treat boils and bronchitis [108]. Snake venoms have been used since the 7th century for treating arthritis and gastrointestinal ailments. Since the 1930s, cobra snake venom has been used to treat various conditions, including polio, pain, multiple sclerosis, and asthma. Captopril, an antihypertensive medication, was created in the 1970s from the venom of the Brazilian viper *Bothrops jararaca* [109]. According to reports, it reduces serum cholesterol levels and speeds up tissue recovery. They employed leeches—Hirudo medicinalis—to treat abnormal swellings, abscesses, piles, poisonous bites, rheumatoid arthritis, and other conditions [110]. Leech salivary glands include vasodilators, anticoagulants, local anesthetics, and antibiotics. Studies have shown that the anticoagulants secreted by leeches are more effective than heparin, another anticoagulant. They're smaller in measurement, go through the clot fast, and hinder fibrin synthesis [108].

2.3.2.2 Natural products from creatures that inhabit the sea

Oceans cover more than 70% of the Earth's surface, home to an extraordinary diversity of marine life, surpassing even that of tropical rainforests. This vast biodiversity results in numerous unique structures produced by marine microfauna and microflora, though only a tiny fraction of these have been studied for their bioactive properties [111]. Plitidepsin, a depsipeptide obtained from *Aplidium albicans*, has demonstrated promising anticancer activity against various types of cancers and is currently undergoing phase 2 of research studies. Another notable compound, ecteinascidin, referred to as trabectedin, was extracted from *Ecteinascidia turbinata*. It was the first marine-derived anticancer agent authorized in the European Union [24]. Prialt is an inhibitor of calcium channels peptide that was launched in 2004 and is currently being utilized to treat chronic pain. Furthermore, the ShK peptide from sea anemones is effective against multiple sclerosis and other autoimmune disorders [108].

2.3.2.2.1 Marine sponges

Marine sponges are mostly sessile creatures that can be assumed to probably be the earliest multicellular life forms, with very little diversification having taken place over some 500 million years [24]. These nucleosides showed significant antiviral and anticancer activity [111]. The development of the "Self-Contained Underwater Breathing Apparatus (SCUBA)," which took place in the 1970s, and the advancements of remotely operated vehicles (ROVs) in the 1990s enabled further exploration of marine ecosystems [24]. Another vital compound was "discodermolide" from *Discodermia dissoluta*, which has a taxol-like stabilizing effect and cytotoxic activity against tumors resistant to both drugs. At the same time, the two-treating strategy with discodermolide and taxol enhanced tumor reduction *in vivo* [112].

2.3.2.2.2 Algae in marine waters

There are more than 30,000 kinds of algae in marine habitats worldwide. Algae were used for millennia as a supply of medicine, fertilizers, and food by numerous societies. Contemporary research in algae is based on ancient knowledge, as there is a considerable supply of varied algal species. Some of the most important bioactive compounds isolated from algae comprise terpenoid sterols and similar terpenoid-like compounds, such as brominated oxygen, phenazine derivatives, amino acids, nitrogen heterocycles, and amines [24,113]. Most of these bioactive compounds isolated from algae show antioxidant, anti-inflammatory, and anticancer effects. For instance, *C. japanonica* activated antioxidant enzymes and thus prevented apoptosis in response to H_2O_2. Moreover, the extracts of G. *tenuistipitata* also triggered the repair of DNA damage in human lung carcinoma cells, H1299 [114]. Furthermore, *Ulva lactuca* and *Sargassum wightii* had flavonoids, which augment the body's antioxidant defense system [115]. Extracts from *Ulva reticulata* have been reported to show antihepatotoxic activity *in vivo* [116]. The brown alga, Gracilaria verrucosa, and *Gracilaria textorii* are also known to reduce inflammation. Porphyridium spp. is another example. It produces a polysaccharide that can stop retroviruses. Scientists have found that extracts from Neorhodomela aculeata using methanol can lower the production of reactive oxygen species (ROS), slow down lipid peroxidation, and reduce nitric oxide synthase activity.

Certain sea algae also possess cancer-fighting properties. For instance, Gracilaria tenuistipitata extracts impede the growth of Ca9–22 oral cancer cells by inducing oxidative stress [87]. Studies show that Plocamium telfairiae, another algal species, can induce cell death in HT-29 colon cancer cells. Green algae produce a unique sulfonium molecular structure called dimethylsulfoniopropionate, which has proven effective against Ehrlich ascites carcinoma in live organisms. Additionally, researchers have extracted heterofucans from Sargassum filipendula, which are effective against cervical, prostate, and liver cancer cells [100].

2.3.2.2.3 Marine bacteria

Marine microbes are an efficient source of antifungal, antiviral, and antibacterial compounds. Information regarding marine compounds' ethnomedical history is low, leading to the slow incorporation of these compounds in therapeutics. Many drug companies have launched dedicated research initiatives to discover and analyze beneficial compounds in ocean-dwelling organisms [105]. Accounting for 45% of total microbial metabolites, actinomycetes are recognized as the most abundant source among microorganisms that synthesize beneficial biological compounds. Within this category, Streptomyces species produce approximately 75% of these critical molecules, while the less prevalent types of actinomycetes contribute the remaining 25%. From actinomycetes, a plethora of antibiotics have been isolated, such as glycopeptides, aminoglycosides, anthracyclines, beta-lactams, polyketides, macrolides, tetracyclines, and actinomycins [106]. A novel maytansinoid antibiotic called "ansamitocins" has been produced from Nocardia spp. This antibiotic has shown noteworthy antitumor and antifungal properties. Antiviral compounds effective against enteroviruses have been reported to be produced by Vibrio marinus, an isolate of the Baltic Sea. From an isolate of ribbonfish of the Baluchistan coast, Pakistan, called Pseudomonas stutzeri, a novel antibiotic named "Bushrin" has been extracted. This chemical is particularly effective against gram-negative and positive bacteria [108]. A novel pyrrole antibiotic has been reported to be produced by Pseudomonas bromoutilis. This demonstrated significant suppression of gram-positive bacteria. *Pseudoalteromonas phenolica* produced the antibiotic MC21-A, which showed a bactericidal effect. An antibiotic called Altemicidin, isolated from *Streptomyces sioyaensis*, showed acaricidal properties while inhibiting Xanthomonas strains [105]. An apoptosis-inducing antibiotic in multiple myeloma cells, Salinosporamide A, was obtained from Salinispora tropica [106]. A genus, *Saccharomonospora*, provided another novel entity called lodopyridone, which showed cytotoxicity in human colon adenocarcinoma (HCT-116) cells [111]. In human adenocarcinoma cells, it induced cytotoxicity. Specific proteins called "bacteriocins" are produced by some bacteria such as Vibrio vulnificus. This protein is bactericidal toward closely related species and is used for preserving seafood [24]. Another significant discovery was the collection of the microorganism SCRC-2738 from Pacific mackerel intestines. This bacterium produces eicosapentaenoic acid, an inhibitor and therapeutic agent against thrombosis and atherosclerosis [105].

2.3.2.3 Natural products from microbes

Around the world, fungi and bacteria live in various environments. As a result, they have adapted to deal with stressful situations. They develop multiple secondary metabolites with biological characteristics, including antidiabetic, anti-inflammatory, anticancer, and antibacterial properties [112]. An estimated 0.1% of the bacterial and 5% of the fungal species in the world that man has recognized, and just a minority of these species had to undergo bioactivity screening [113]. More than one million natural chemicals have been identified, of which 50%–60% originated from plants and 5% are derived from microorganisms. Twenty to 25% of natural products have been identified as biologically active.

Ten percent have been separated by microorganisms, 48% from fungus, 17% from unicellular bacteria, and 45% from actinomycetes [114,117,118]. The distinguished chemical structures, environmental interactions, and microbial origin are some of the most significant features of bioactive microbial substances. Commercial applications for products from microbes include the following:

- Direct use of fermentation products in agriculture, medicine, or any other industry.
- The resulting compounds may be employed as leads in synthesizing analogs and rational drug design; the products can be utilized as the foundation for derivatization [115,116]. One advantage of employing microorganisms to produce distinct bioactive compounds is the science of modifying and enhancing microbial strains to reduce costs.

2.4 MODERN INNOVATIONS IN MEDICINAL COMPOUNDS

Developing innovative therapeutics requires a paradigm shift toward an integrated approach that combines natural products with advanced omics technologies. This multidisciplinary strategy will enable the rapid identification and optimization of bioactive compounds, leading to the adoption of combinatorial therapies that generate more effective treatments for complex diseases.

2.4.1 Advances in drug discovery and development

Natural products are a vital source of novel chemicals for the pharmaceutical industry, as the search for new therapeutic agents has resulted in notable advancements in natural product research. The discovery of novel therapeutics for multifactorial diseases, including infectious diseases, neurological disorders, hormonal imbalance, and oncological disease, necessitates strategically implementing cutting-edge pharmacological and biomedical research methodologies [119]. Using sophisticated discharge and cleaning techniques, scientists can now extract and describe the lead compositions from their natural sources while accounting for factors like abundance, stability, and structural variety. The next step is high-performance screening for pre-established targets, which helps find possible therapeutic compounds. Moreover, due to the remarkable developments in medicinal chemistry and drug development, the biotechnology and pharmaceutical industries have renewed interest in natural products [119].

2.4.1.1 Role of genomics

Genomics plays a critical role in discovering plant-derived drugs. Accurate species identification is an essential stage in plant-based medicine development. Recent advancements in genomes have enabled the creation of precise identification criteria for plants and other natural product sources [120]. For example, DNA barcoding can identify infertile, immature, or low-quality plant samples. A significant knowledge gap persists regarding the molecular characterization and taxonomic classification of medicinal plant species, as most have yet to be incorporated into large-scale DNA barcoding initiatives or integrated into user-friendly, open-access databases [121–123]. In addition, the short barcodes employed for identification are not universal and sometimes fail to distinguish between closely related species, emphasizing the need for more thorough and reliable identification procedures [123–125]. The accurate identification of botanical species, such as Artemisia spp., is crucial for ensuring the authenticity and safety of natural products, particularly in cases where closely related species, like Actaea racemosa (black cohosh), may be mistakenly substituted, posing potential health risks [126]. Recently, the importance of consistency in the chemical composition of natural products has become increasingly evident. Biofarming practices have been adopted to meet such demands that involve growing plants under standardized conditions up to maturity and extracting the bioactive compounds. Later on, these compounds are authenticated through DNA barcoding, thus helping to correctly identify and justify the correctness of the source of the natural product [127]. Biofarming is based on raising plants in controlled conditions before harvesting and extracting. Some strategies have been developed to ensure uniformity in the procurement of natural products. The isolated chemical is authenticated through DNA barcoding [128]. Genomic technologies allow for the production of plant-specific markers from different botanical sources. These markers can be used in genome chips for natural product identification and high-speed genotyping [120,128]. This allows for fast and accurate identification of the plant species, ensuring the quality and safety of the natural goods.

2.4.1.2 Proteomics in drug discovery and development

Proteomics analysis is an essential complement to genomes and transcriptomics, providing critical insight into the mechanism of action of natural compounds. Proteomic research can guide the quality assessment of natural products by giving detailed information on the biosynthesis of proteins, functional interactions, and expression patterns [129–131]. The first primary goal of preclinical analysis of natural products is the identification of their targets on proteins, by which their safety profile and tendency for off-target toxicity can be determined. For example, affinity chromatography can be exploited to study interactions between natural products and the target protein so that the natural product remains unaltered. Native configurations of natural products are employed in most analytical determinations without any prior modification of the natural products, requiring chemical modification. For this reason, these compounds are typically studied in their unaltered, native form to identify the therapeutic effects and mechanisms of action with utmost accuracy. The latest techniques, such as intracellular temperature shift assays, can stabilize these interactions, while compound-target protein binding is facilitated through bioinformatics and AI. Natural products have a vast repertoire of biological features because of their diverse compositions and ability to bind to various receptors. Therefore, identifying specific binding sites for proteins is essential in understanding their mechanisms of action and the associated side effects. In the drug development process, natural products that have the potential to be therapeutically beneficial must undergo rigorous safety studies, such as toxicity testing, monitoring for adverse events, and off-target investigations of pharmacodynamic effects. The other method to confirm target proteins includes the use of RNA interference (RNAi) techniques, such as small interfering RNA (siRNA) and short hairpin RNA (shRNA), to precisely manipulate and functionally characterize proteins of interest [132,133]. Target protein validation is possible by the use of RNA interference (RNAi) technologies, particularly small interfering RNA (siRNA) and short hairpin RNA (shRNA) [134,135]. Applying RNAi-mediated gene silencing to knockdown target protein enables the assessment of off-target effects elicited by natural products. Furthermore, the CRISPR-Cas9 system provides a new avenue for exploring and potentially mitigating off-target effects. Combining CRISPR-Cas9 genome editing with sequencing and bioinformatic analysis can determine drug resistance phenotypes and validate anticancer therapeutic agents [136].

2.4.1.3 Metabolomics in the drug discovery of natural products

The metabolomic analysis is a complementary approach to DNA barcoding that offers a finer resolution of biochemical variation within and between species. While it is less reliable for species identification due to the impact of ambient factors and processing on the chemical structure of plants, metabolomics has emerged as an innovative technique for identifying novel medicinal interactions. Metabolomics profiling of the compound can lead to the detection and quantitation of metabolites associated with that chemical [137–139]. The use of state-of-the-art analytical methods such as high-resolution mass spectrometry, ultra-high-performance liquid chromatography coupled with quadrupole time-of-flight QTOF detection, and NMR spectroscopy has been utilized for detailed metabolomic analysis of natural products. This has helped identify new bioactive molecules with therapeutic applications [140,141]. It has been reported that phytochemical research has led to isolating drugs of pharmaceutical importance from a wide range of plant species, such as *Wild senna, American ginseng*, and *Newbouldia laevis* [142].

2.4.1.4 The role of automation in natural product drug discovery

Automation has transformed drug development, allowing for the quick screening of thousands of natural products and chemicals. Most drug development businesses use HTS processes and assays, dramatically expediting compound analysis and evaluation [143]. Computational software has made designing and synthesizing synthetic chemicals easier, and tools like ADAM and EVE help identify therapeutic targets and possible candidates [144,145]. Despite the benefits of bioinformatics tools, false positives remain an issue. Advanced software tools are being developed to help with compound creation and biological testing [146,147]. Integrating advanced microfluidics systems has enabled various laboratories and companies to exploit the benefits of fluid-based manipulations and high-temperature processing in synthesizing and purifying synthetic compounds [148]. AI has transformed the drug development landscape by reducing human errors and shortening the time required for candidate connections [149,150]. A compound's pharmacokinetic and pharmacodynamic properties, encompassing absorption, distribution, metabolism, excretion, and toxicity (ADMET) and its efficacy, profoundly influence the design of composite therapeutic agents. The drug development process is inherently complex, involving multiple factors and dimensions. Notably, optimal structural characteristics do not necessarily guarantee the desired biological response. Therefore, a delicate balance must be struck between the biological activity of a natural product or compound and its physicochemical properties [147]. Automation enables drug makers to compromise composite design, biological activity, and ADMET characteristics, simplifying the pharmaceutical development process.

2.4.1.4.1 Artificial intelligence in drug design and optimizing medicinal compound formulation

AI has altered natural product research, allowing for the quick discovery of possible therapeutic candidates and predicting their pharmacological effects. Deep neural networks (DNNs) and AI-based generative models are critical in this effort, as they streamline the drug development pipeline by reducing the number of molecules that need to be validated experimentally. AI-based computational models have emerged as an essential tool for forecasting drug toxicity in preclinical studies, lowering failure rates, and increasing drug development efficiency [151]. Advanced machine learning (ML) and deep learning (DL) methodologies, including artificial neural networks, have been employed to analyze large-scale datasets and predict potential adverse outcomes. Fungi, such as mushrooms, have been investigated for their anticancer, immunomodulatory, and antioxidant effects. AI and ML techniques can classify new fungus species, identify natural compounds, optimize extraction procedures, and discover novel uses and qualities [152,153]. Antitumor drug discovery is the most popular application of AI in natural product research, followed by antiviral and antibacterial compounds. AI-based anticancer lead identification in plants and their associated microbes has shown potential [154]. Despite these gains, difficulties persist, such as repeatedly discovering the same molecules, such as dereplication. This challenge can be solved using comprehensive AI datasets and technologies. Traditional approaches to nanomedicine design and optimization are frequently time-consuming. The composition of nanoparticles for targeted cancer therapy can be optimized by applying AI and ML approaches, which rely on empirical data to support predictive design modeling [155]. AI systems can find ideal nanoparticle projects and anticipate the most successful ligand combinations based on experimental data, improving precision and efficiency.

2.4.1.4.2 Molecular docking

The application of Molecular docking techniques has revolutionized the development of phytochemical drugs, allowing researchers to use computational modeling to forecast the binding affinity and orientation of phytochemical compounds to their target proteins [156]. This methodology has proven to be vital in identifying phytochemicals with significant potential for future experimental exploration. Several computational tools and techniques, such as AutoDock Vina, AutoDock GOLD, Discovery Studio, and Glide, are used to carry out molecular docking simulations [157]. The technologies make it easier to identify molecular targets of nutraceuticals used to treat various ailments. The application of molecular docking has yielded valuable insights into discovering potential treatment targets for COVID-19. For example, molecular docking experiments have been used to determine the binding profiles of plant-based chemical ligands with primary pharmacological targets involved in the life cycle and disease pathology of SARS-CoV-2 [158]. In-depth bioinformatics research has identified 3CLpro, or the main protease, as the prime target in the SARS-CoV-2 virus. As such, 3CLpro represents a desirable target for creating new drugs because this protein has to be highly modified in such a way that the very survival of the virus would be hugely affected [159]. Molecular docking has recently established its success by discovering potential phytochemical drugs. Three associations of potentially promising binding scores were obtained on screening 408 phytochemicals against protein burin. Microbially produced Chouhan et al. [160], screened via computational analysis, potential therapeutic applications. Natural products as inhibitors of the RPFB protein in Mycobacterium tuberculosis. Molecular docking has revolutionized the field of phytochemical drug discovery since it is possible to quickly identify potential therapeutic sites and phytochemical ligands [160].

2.4.2 Function of nanoparticles to improve drug delivery

The conventional methods of delivering drugs are limited by low solubility, instability, and short residence time within the body. Nanotechnology has opened the door to new delivery concepts by encapsulating medicinal chemicals inside nanoparticles, liposomes, or polymeric micelles [161–163]. Nanoparticles have gained massive attention because of their unusual properties, such as their small size, high surface area-to-volume ratio, and potential to be functionalized with targeted ligands. These properties allow nanoparticles to accumulate preferentially at tumor locations through passive targeting (EPR effect) or active targeting (ligand-receptor interactions) [164]. These materials have stimuli-responsive properties that make them respond to specific triggers, for example, changes in pH, temperature, or enzyme activity that enable the therapeutic agents to be delivered at specific desired locations [165,166]. Nanotechnology-based drug delivery systems can be a valuable approach to anticancer drugs, not only those made from natural products but also because they can deliver medicines in targeted and controlled manners, and the possibility of delivering multiple therapies.

2.4.2.1 Integrating natural compounds and nanotechnology

Natural compounds like anthocyanins, berberine, curcumin, and quercetin have been gaining much attention due to their potential therapeutic application in cancer therapy. These compounds' poor solubility, limited stability, and poor transport to the tumor location remain a significant concern and hinder their efficacy [167]. To overcome such challenges, scientists have used nanoscale carriers, including nanoparticles, liposomes, and polymeric micelles, to encapsulate and protect natural compounds. This also improves the natural substance's solubility, stability, and delivery using nanotechnology [168,169]. Researchers formulated iron oxide nanoparticles that will specifically target the cancer cells, thereby reducing or preventing damage caused by it in healthy cells [170]. These nanoparticles encapsulated with the derivatives of eugenol of the Pimenta dioica medicinal plant are non-toxic. They can provide heat to die when exposed to particular laser light. This discovery may lead to more effective cancer treatments [171]. For example, anthocyanins, a group of naturally occurring pigments, are the chromatic compounds responsible for the deep colors of several fruits, flowers, and berries. Moreover, their potential as a wholesome, plant-based alternative to synthetic food dyes presents an appealing opportunity for food product enrichment [172]. The intake of anthocyanin-rich fruits and vegetables has been linked to a reduced risk of various cancers, such as hepatic [173], breast [174], prostate [175], pulmonary [176], and cutaneous malignancies [177]. Because of their sensitivity to environmental factors, anthocyanins undergo extensive degradation during intestinal digestion, compromising their biological effectiveness. Despite its potential, the emergence of anthocyanin as a pharmaceutical molecule has been hindered by several drawbacks; these include very poor absorption, low membrane permeability, as well as limited solubility, leading to its restricted bioavailability when it is orally used and, thereby, its diminished therapeutic practicality. Developing an effective delivery system for anthocyanin is essential for fully harnessing the intense anticancer properties of these antioxidants. By exploiting nanotechnology, one can shield them from breakdown, increase their absorption, and deliver them directly to cancer cells, maximizing their healing impact [178]. Encapsulation of anthocyanins enables targeted and efficient delivery to malignant cells, enhancing their anticancer efficacy and reducing systemic toxicity. Furthermore, nanoencapsulation has significantly enhanced drug uptake by cancer cells, potentiating their therapeutic impact [179]. For instance, researchers explored the chemopreventive potential of bilberry anthocyanidins encapsulated within exosomes originating from bovine milk, specific to colon cancers [180]. The results of the research indicate that the encapsulation of anthocyanin results in a substantially enhanced growth activity for the inhibition of colon cancer cells, particularly HCT-116 and HT-29, which is even better compared to its unencapsulated counterpart. Significantly, the IC_{50} value for encapsulated anthocyanin has been reduced, indicating a significant improvement in its therapeutic action. The IC_{50} value of encapsulated anthocyanins was considerably lower than that of non-encapsulated anthocyanins, indicating a marked increase in their inhibitory effectiveness [180]. The encapsulation of anthocyanin within exosomes profoundly enhances its antiproliferative properties, surpassing those of the free compound, attributed to improved stability and enhanced cellular internalization. The encapsulation of anthocyanin within exosomes profoundly enhances its antiproliferative properties, surpassing those of the free compound, attributed to improved stability and enhanced cellular internalization. Encapsulated anthocyanin showed precise targeting, destroying colon cancer cells while preserving the integrity of healthy colon cells [180].

PEGylated polymeric micelles conjugated with cholesterol and vitamin E were also used to transport curcumin efficiently and boost its anticancer potential [181]. Researchers explored the possibility of albumin nanoparticles guided by carbonic anhydrase IX to selectively target and destroy triple-negative breast cancer cells, particularly in low-oxygen environments, and to visualize tumors derived from patient samples [182]. Graphene oxide and graphene quantum dots have emerged as promising nanocarriers for delivering anticancer agents, such as curcumin (CUR), which is poorly soluble in water and challenging for the body to absorb [182]. A novel nanocomposite consisting of chitosan, carbon quantum dots, iron oxide, and curcumin was developed and characterized for targeted delivery of curcumin in breast cancer treatment, offering a promising approach for precision medicine [183]. Researchers designed zinc oxide nanoparticles conjugated with phenylboronic acid to deliver curcumin directly to tumor tissues, demonstrating their potential to induce programmed cell death in human breast cancer cells by generating toxic ROS and damaging mitochondrial function [184]. Plant-derived nanoparticles (PDNP) have been shown to limit the development of fungi, particularly harmful fungi. The antifungal mechanism of PDNP activity is the fungus' extracellular membrane breakdown via electrostatic interaction of nanoions with ROS. Other strategies include apoptosis induction via mitochondrial oxidative stress, suppression of ATP generation, and different signaling pathways. Previous and contemporary investigations have found significant antifungal properties in several PDNPs. Recent research demonstrated that silver nanoparticles from Ligustrum lucidum were efficient against Setosphaeria turcica, with an IC_{50} of 170.20 µg/mL [185]. Other research has tested nanoparticles other than the well-known AGNPs. Furthermore, Srihasam et al. [186] found that Nanoparticles of nickel from Stevia were efficient against several pathogenic Aspergillus strains [186]. Other plants extracted to synthesize PDNP with antifungal activity, such as the fruit of Passiflora Foetida [187] and the extract of Dimocarpus Longan [188] against Fusarium Spin., C. albicans, C. neoformans, and A. niger.

2.5 ROLE OF PLANT-DERIVED COMPOUNDS IN COMBATING CANCER

Phytochemicals with anticancer properties can be categorized into two main groups: blockers, which intercept the initiation of cancer, and tumor suppressors, which inhibit the progression and growth of tumors [189]. Various plant-derived compounds, such as tannins, flavonoids, glucosinolates, and curcumin, have been shown to exert anticancer effects by boosting the body's natural defense mechanisms, specifically by enhancing the production of enzymes that detoxify harmful substances and prevent tumor formation. Genetic mutations can initiate oncogenic transformation, leading to the development of cancer. However, certain phytochemicals derived from plants have been identified as possessing anticarcinogenic properties [189]. These bioactive compounds can be categorized into two distinct classes: blockers, which inhibit the initiation and progression of carcinogenesis, and tumor suppressors, which impede the proliferation and metastasis of malignant cells. Specific plant compounds, such as curcumin, flavonoids, and glucosinolates, have been shown to effectively prevent tumor growth by activating the body's natural detoxification processes, shielding DNA from damage, and maintaining genetic stability [189,190]. Mutations that cause cells to undergo neoplastic transformation. The stimulation of the Nrf2-ARE pathway has been demonstrated to induce phase 2 detoxification enzymes [189,190]. Tumor suppressors are critical in inhibiting the proliferation and dissemination of cancer cells, thereby preventing tumor progression and promoting apoptosis. Notably, specific phytochemicals, including isothiocyanates present in Brassica vegetables, flavonoids found in citrus fruits and berries, and coumarins identified in spices such as cinnamon, have been recognized as potent tumor suppressors, offering a promising therapeutic strategy for cancer prevention and treatment [189,191]. Tumor suppressors employ a multifaceted approach to inhibit cancerous growth, including induction of cell cycle arrest, inhibition of angiogenesis, and disruption of key signaling pathways that facilitate oncogenic progression, such as the NFkB pathway [189,191]. Phytochemicals trigger programmed cell death by disrupting the balance of cellular proteins, specifically by suppressing the activity of proteins that prevent cell death, such as Bcl-2 and Bcl-xL, while simultaneously increasing the production of proteins that promote cell death, including Bak and Bax. This process triggers a cascade of events within the cell, releasing a key protein called cytochrome c from the mitochondria. This, in turn, activates a molecular complex known as the apoptosome, which sets off a chain reaction that ultimately leads to the activation of powerful enzymes called caspases, specifically caspases 3, 6, and 7, which execute the apoptosis of cells [189–192]. This process leads to the breakdown of essential proteins within the cell, causing it to shrink, fragment, and ultimately disintegrate. Additionally, specific suppressors have been shown to disrupt hormone production and reception in breast cancer cells, effectively starving the tumor of the hormones it needs to grow and spread. Phytochemicals hold great potential in the fight against cancer, offering a safe, accessible, and multifaceted approach to prevention and treatment. By harnessing their powerful biological properties, phytochemicals can help slow cancer growth, trigger the death of cancer cells, protect against oxidative stress, halt abnormal cell division, choke off tumor blood supply, and calm inflammation [193–195]. Phytochemicals exert their anticancer effects by inhibiting tumor progression, inducing apoptosis in malignant cells, safeguarding normal cellular integrity, modulating aberrant cell signaling, disrupting tumor angiogenesis, and attenuating inflammatory responses [193–195], as shown in Table 2.3.

Table 2.3 Mechanisms of action of phytochemicals with anticancer properties

Category	*Mechanism of action*	*Representative phytochemicals*	*Effects on cancer cells*	*References*
Blockers	Inhibit the initiation of carcinogenesis by enhancing cellular detoxification pathways and preserving genomic integrity.	Tannins, flavonoids, glucosinolates, curcumin	Upregulate phase 2 detoxification enzymes, mitigate DNA damage, and maintain genetic stability.	[189,190]
Tumor suppressors	Suppress tumor growth and metastatic progression through multiple regulatory mechanisms.	Isothiocyanates (Brassica vegetables), flavonoids (citrus fruits, berries), coumarins (cinnamon)	Induce apoptosis, inhibit angiogenesis, and disrupt key oncogenic signaling cascades such as NF-κB.	[189,191]
Cell cycle arrest	Halt uncontrolled cell division	Various flavonoids and polyphenols	Prevent the unchecked proliferation of malignant cells	[189,191]
Apoptosis induction	Trigger programmed cell death via mitochondrial pathway	Curcumin, isothiocyanates, flavonoids	Suppress anti-apoptotic proteins (Bcl-2, Bcl-xL), increase pro-apoptotic proteins (Bak, Bax), activate caspases (3,6,7)	[189,191,192]
Oxidative stress protection	Reduce oxidative damage to cells	Flavonoids, polyphenols, glucosinolates	Neutralize free radicals, maintain cellular integrity	[193–195]
Anti-inflammatory effects	Suppress chronic inflammation linked to cancer progression	Curcumin, flavonoids, tannins	Reduce inflammatory signaling and cytokine production	[193–195]

2.5.1 Phytochemicals for the treatment of SARS-CoV-2

Alkaloids are among the most structurally varied classes of plant secondary metabolites. Research conducted in laboratory settings has revealed that certain plant-derived compounds, known as alkaloids, exhibit potential anti-COVID-19 properties by inhibiting specific components of the SARS-CoV-2 virus or reducing the severity of infection. Studies have demonstrated that berberine effectively blocks the entry of SARS-CoV-2 into host cells, inhibits viral replication, and reduces the production of infectious viruses, as evidenced by two independent experiments using Vero E6 cells [196]. Further analysis revealed that berbamine significantly reduced the functional activity of the SARS-CoV-2 E protein's ion channel at elevated concentrations. A modified version of berberine, known as BE-33, showed enhanced potency against SARS-CoV-2 infection, with a significantly lower concentration required to inhibit viral replication, and demonstrated a stronger binding affinity to the E protein's ion channel compared to its parent compound [197]. This derivative substantially reduced lung viral levels and decreased inflammatory responses in genetically engineered mice. This suggests that it may possess anti-inflammatory properties that could provide therapeutic benefits for patients with severe COVID-19. Berberine, a naturally occurring alkaloid, was found to inhibit the production of infectious SARS-CoV-2 particles but not the replication of viral genetic material, indicating that its antiviral effects occur at a later stage in the viral life cycle [198,199]. A comprehensive analysis of existing FDA-approved drugs identified several promising candidates, including bromhexine, reserpine, hydroquinidine, and quinidine, that demonstrate the ability to suppress SARS-CoV-2 replication, positioning them as potential treatments for COVID-19 [200]. Studies have further revealed that the antimalarial agents quinacrine and quinine possess robust antiviral properties, successfully blocking SARS-CoV-2 infection in cell culture experiments [201,202], as shown in Table 2.4.

Table 2.4 Anti-COVID-19 properties of alkaloids

Alkaloid/ compound	*Mechanism of action*	*Key findings*	*References*
Berberine	Inhibits viral entry, replication, and infectious virus production	Blocks SARS-CoV-2 entry into host cells inhibits viral replication, and reduces contagious virus production in Vero E6 cells	[196]
BE-33 (modified berbamine)	Inhibits viral replication and E protein function	More potent than berberine, with a stronger binding affinity to the SARS-CoV-2 E protein's ion channel, significantly lowers viral levels and inflammation in genetically engineered mice	[197]
Berberine	Inhibits production of infectious SARS-CoV-2 particles	It does not affect viral RNA replication but prevents the formation of infectious viral particles, indicating an effect at a later stage in the viral life cycle	[198,199]
Bromhexine, reserpine, hydroquinidine, quinidine	Suppress SARS-CoV-2 replication	Identified as potential antiviral treatments through FDA-approved drug analysis	[200]
Quinacrine, quinine	Block SARS-CoV-2 infection	Demonstrate strong antiviral activity in cell culture experiments	[201,202]

2.5.2 Recent development involving the expression of plant-based monoclonal antibody against cancer

Genetically engineered plants offer the most promising platform for large-scale production of monoclonal antibodies (MABs), providing a viable and efficient system for commercial-scale manufacturing. Alternatively, monoclonal antibodies (MABs) can be produced through temporary genetic modification using recombinant plant viruses and agroinfiltration techniques. This offers a faster and more flexible approach than traditional transgenic plant methods. Moreover, researchers utilized Tobacco Mosaic Virus (TMV) vectors, engineered to carry both light and heavy chain components, to successfully produce fully assembled, functional monoclonal antibodies (MABs) in Nicotiana benthamiana plants [203–205]. The MagnICON expression system, a commercially available platform, has demonstrated exceptional efficiency in producing high yields of monoclonal antibodies (MABs) in Nicotiana benthamiana plants, making it a highly effective tool for large-scale MAB production. A pioneering phase 1 clinical trial showcased the power of this technology, where patient-specific, chimeric antibody-based vaccines were successfully created to combat B cell follicular lymphoma. This approach pioneers precision oncology by using a patient's malignant cells to develop tailored treatments [206].

2.6 CHALLENGES IN THE DEVELOPMENT OF MEDICINAL COMPOUNDS

Herbal medications and their preparations have long been used in underdeveloped and developed nations due to their natural origins and lesser side effects or dissatisfaction with the outcomes of synthetic pharmaceuticals. Oriental herbal medicine has various distinguishing characteristics. Medicine preparations relate to all herbal treatments, whether they are provided as single herbs or plant combinations [207]. Traditional medicines contain medicinal substances; organic matter includes plants, minerals, and other substances. Herbal medicines are most commonly utilized in conventional medicine, which heals patients using medicinal plant mixes [208]. These drugs are created using environmentally friendly techniques, and renewable raw material supplies will be accessible, resulting in economic advantages. People who cultivate these fundamental resources should profit [209]. Because of the abundance of medicinal plants in diverse bioclimatic zones, India is known as the "Emporium of Medicinal Plants" [210]. Both contemporary and traditional medicine continue to rely on medicinal plants to provide crucial therapeutic ingredients. The evaluation of the efficacy of plant-based pharmaceuticals used in conventional medicine is being focused on since they are affordable, have minimal side effects, and, according to the WHO, nearly 80% of the world's population depends on herbal medicines [210]. The development and use of phytopharmaceuticals face numerous challenges. One primary concern is the inconsistent and often low yield of plant material, which hampers the production of phytomedicines. Additionally, the formulation of plant extracts is complicated by their poor solubility in water and other solvents. The presence of toxic compounds in plant extracts also poses serious health risks, while the low bioavailability of plant-derived samples negates their therapeutic value [211]. In addition, the abuse of phytopharmaceuticals has resulted in adverse health incidents, which necessitates proper education and training. The wrong identification of plant species and the unregulated use of traditional remedies also worsen these problems—household accidents involving ingesting ornamental plants containing toxic compounds. The existence of hormonally active compounds and allergens in some plants is a health hazard. To mitigate these challenges, it is essential to invest in rigorous research, quality control measures, and regulatory oversight to ensure the safe and effective use of phytopharmaceuticals [211].

2.7 CONCLUSION & FUTURE DIRECTIONS

The advancement of green nanoparticle synthesis represents a major step toward achieving sustainable materials and technologies. This innovative approach provides an eco-friendlier option compared to traditional chemical methods, facilitating the development of targeted and effective treatments for diseases such as cancer. The introduction of AI in drug discovery is set to transform the field, enabling scientists to accurately identify potential drug targets while deepening their understanding of disease biology. AI models trained on vast biomedical data will accelerate the identification of new therapeutic candidates, improve drug design, and lower the risk of clinical trial failures. Furthermore, plant-based biomanufacturing is revolutionizing the pharmaceutical industry by offering a more sustainable and efficient method for drug development. The biopharming platform has become a key resource for producing plant-derived cancer therapies, paving the way for groundbreaking treatments. With the rising global demand for sustainable technologies, green synthesis methods are emerging as essential solutions. Given its tremendous potential to create eco-friendly nanomaterials, green synthesis is set to leave a notable mark on various sectors, fostering innovation and sustainability in the future.

AUTHOR CONTRIBUTIONS

ST, BK, RP, and VS wrote the first draft. All the authors participated in revising and editing the chapter.

ACKNOWLEDGEMENT

The authors sincerely acknowledge Dr. Vaibhav Gautam and Dr. Rajiv Pathak for providing the opportunity to contribute to this book chapter. Their valuable support, encouragement, and trust throughout this endeavour are deeply appreciated. The authors also thank Sharda University, Greater Noida, for offering a wonderful academic environment, research facilities, and institutional support that facilitated in a great way to prepare this chapter. The Department of Chemistry & Biochemistry, Sharda School of Engineering & Science (SSES) has provided a conducive environment for collaborative research and to foster academic excellence.

LIST OF ABBREVIATIONS

ACE2:	angiotensin-converting enzyme
ADMET:	absorption, distribution, metabolism, excretion, and toxicity
AI:	artificial intelligence
AYUSH:	ayurveda, yoga, and naturopathy, unani, siddha, and homeopathy
CRISPR-Cas9:	clustered regularly interspaced short palindromic repeats
CUR:	curcumin
DL:	deep learning
DNNs:	deep neural networks
EPI:	efflux pump inhibitor
EPR:	enhanced permeability and retention effect
HCT:	human colon adenocarcinoma
MABs:	monoclonal antibody
ML:	machine learning
NMR:	nuclear magnetic resonance
PDN:	plant-derived nanoparticles
QTOF:	quadrupole time-of-flight
RNAi:	RNA interference
RPFB:	ribosomal protein family B
shRNA:	short hairpin RNA
siRNA	small interfering RNA
TCM:	traditional Chinese medicine
TMV:	tobacco mosaic virus

REFERENCES

1. Singh, A.R., Modern medicine: towards prevention, cure, well-being and longevity. *Mens Sana Monographs*, 2010. 8(1): p. 17.

2. Li, Y., et al., Healthy lifestyle and the likelihood of becoming a centenarian. *JAMA Network Open*, 2024. 7(6): p. e2417931–e2417931.
3. Kilgour, F.G., Modern medicine in historical perspective. *Bulletin of the Medical Library Association*, 1962. **50**(1): p. 42.
4. Atanasov, A.G., et al., Natural products in drug discovery: advances and opportunities. *Nature Reviews Drug Discovery*, 2021. **20**(3): p. 200–216.
5. Tiwari, H., et al., Deciphering the landscape of triple negative breast cancer from microenvironment dynamics and molecular insights to biomarker analysis and therapeutic modalities. *Medicinal Research Reviews*, 2025. **45**(3): p. 817–841.
6. Sahu, A., J. Mishra, and N. Kushwaha, Artificial intelligence (AI) in drugs and pharmaceuticals. *Combinatorial Chemistry & High Throughput Screening*, 2022. **25**(11): p. 1818–1837.
7. Bhaskar, P., et al., Navigating the Nanoscale Frontier: An In-Depth Introduction to the World of Nanomedicine, in *Nanomedicine: Innovations, Applications, and Breakthroughs in the Quest for Health and Medicine's Future*, V. Gautam, et al., Editors. 2024, Springer *Nature*: Cham. p. 531–559.
8. Schneider, G. and U. Fechner, Computer-based de novo design of drug-like molecules. *Nature Reviews Drug Discovery*, 2005. **4**(8): p. 649–663.
9. Vora, L.K., et al., Artificial intelligence in pharmaceutical technology and drug delivery design. *Pharmaceutics*, 2023. **15**(7): p. 1916.
10. Guerrero-Perilla, C., F.A. Bernal, and E.D. Coy-Barrera, Molecular docking study of naturally occurring compounds as inhibitors of N-myristoyl transferase towards antifungal agent discovery. *Revista Colombiana de Ciencias Químico-Farmacéuticas*, 2015. **44**(2): p. 162–178.
11. Rai, N., et al., Isolation and characterization of N-(2-hydroxyethyl) hexadecanamide from Colletotrichum gloeosporioides with apoptosis-inducing potential in breast cancer cells. *BioFactors*, 2023. **49**(3): p. 663–683.
12. Bhuiyan, F.R., et al., Plants metabolites: possibility of natural therapeutics against the COVID-19 pandemic. *Frontiers in Medicine*, 2020. 7: p. 444.
13. Boukhatem, M.N. and W.N. Setzer, Aromatic herbs, medicinal plant-derived essential oils, and phytochemical extracts as potential therapies for coronaviruses: future perspectives. *Plants*, 2020. **9**(6): p. 800.
14. Kotecha, R., The journey with COVID-19: initiatives by Ministry of AYUSH. *Journal of Ayurveda and Integrative Medicine*, 2021. **12**(1): p. 1.
15. Yuan, H., et al., The traditional medicine and modern medicine from natural products. *Molecules*, 2016. **21**(5): p. 559.
16. World Health Organization, *WHO Global Report on Traditional and Complementary Medicine 2019*. 2019: World Health Organization.
17. Dave, T., et al., Sushruta: the father of Indian surgical history. *Plastic and Reconstructive Surgery–Global Open*, 2024. **12**(4): p. e5715.
18. Shah, S., Ophthalmology in ancient time–the Sushruta Samhita. *Journal of Clinical Ophthalmology and Research*, 2018. **6**(3): p. 117–120.
19. Saniotis, A. and A. Saniotis, Sushruta on human dissection as pre-requisite to surgery. *International Journal of Morphology*, 2021. **39**(6): p. 1787–1790.
20. Jamshidi-Kia, F., Z. Lorigooini, and H. Amini-Khoei, Medicinal plants: past history and future perspective. *Journal of Herbmed Pharmacology*, 2017. 7(1): p. 1–7.

21. Mushtaq, S., et al., Natural products as reservoirs of novel therapeutic agents. *EXCLI Journal*, 2018. **17**: p. 420.
22. Li, J.J., History of drug discovery. in *Drug Discovery: Practices, Processes, and Perspectives*, J.J. Li and E.J. Corey, Editors. 2013, p. 1–42.
23. Hallmann-Mikołajczak, A., Ebers Papyrus. The book of medical knowledge of the 16th century BC Egyptians. *Archiwum historii i filozofii medycyny*, 2004. **67**(1): p. 5–14.
24. Dias, D.A., S. Urban, and U. Roessner, A historical overview of natural products in drug discovery. *Metabolites*, 2012. **2**(2): p. 303–336.
25. Xiao, L.-J. and R. Tao, Traditional Chinese medicine (TCM) therapy. In *Substance and Non-Substance Addiction, Advances in Experimental Medicine and Biology*, X. Zhang, J. Shi and R. Tao, Editors. 2017, Springer: Singapore. vol. 1010, p. 261–280.
26. Tang, J.-L., B.-Y. Liu, and K.-W. Ma, Traditional Chinese medicine. *The Lancet*, 2008. **372**(9654): p. 1938–1940.
27. Jia, W., et al., The rediscovery of ancient Chinese herbal formulas. *Phytotherapy Research: An International Journal Devoted to Pharmacological and Toxicological Evaluation of Natural Product Derivatives*, 2004. **18**(8): p. 681–686.
28. Fan, T.-P., et al., Future development of global regulations of Chinese herbal products. *Journal of Ethnopharmacology*, 2012. **140**(3): p. 568–586.
29. He, S., Chinese herbal dose in ancient and modern times: a comparative study. *Journal of Traditional Chinese Medicine*, 2013. **33**(2): p. 268–271.
30. Packard, R.M., *The Making of a Tropical Disease: A Short History of Malaria. 2021*: JHU Press.
31. Capasso, F., Phytotherapy: A Quick Reference to Herbal Medicine. 2003: Springer Science & Business Media.
32. Dagen, M., History of malaria and its treatment, in *Antimalarial Agents*, Graham L. Patrick, Editor. 2020, Elsevier. p. 1–48.
33. Rogers, K., *Out of Nature: Why Drugs from Plants Matter to the Future of Humanity*. 2012: University of Arizona Press.
34. David, B., J.-L. Wolfender, and D.A. Dias, The pharmaceutical industry and natural products: historical status and new trends. *Phytochemistry Reviews*, 2015. **14**: p. 299–315.
35. Montinari, M.R., S. Minelli, and R. De Caterina, The first 3500 years of aspirin history from its roots: a concise summary. *Vascular Pharmacology*, 2019. **113**: p. 1–8.
36. Ugurlucan, M., et al., Aspirin: from a historical perspective. *Recent Patents on Cardiovascular Drug Discovery*, 2012. **7**(1): p. 71–76.
37. Miner, J. and A. Hoffhines, The discovery of aspirin's antithrombotic effects. *Texas Heart Institute Journal*, 2007. **34**(2): p. 179.
38. Gaynes, R., The discovery of penicillin: new insights after more than 75 years of clinical use. *Emerging Infectious Diseases*, 2017. **23**(5): p. 849.
39. Samal, J., Role of AYUSH workforce, therapeutics, and principles in health care delivery with special reference to National Rural Health Mission. *AYU (An International Quarterly Journal of Research in Ayurveda)*, 2015. **36**(1): p. 5–8.
40. Delhi, N., Ministry of Health and family welfare. Government of India, 2001: p. 7–30.

41. Gupta, P.K., et al., Public health initiatives and Ayush: projects to policy. *International Journal of Ayurveda Research*, 2023. **4**(3): p. 146–154.
42. Faster, M.I. and S. Growth, Twelfth Five Year Plan (2012–2017). 2013.
43. Samal, J., Situational analysis and future directions of AYUSH: an assessment through 5-year plans of India. *Journal of Intercultural Ethnopharmacology*, 2015. **4**(4): p. 348.
44. Mohanty, P., et al., Utilization of ayurveda, yoga, naturopathy, unani, siddha, and homoeopathy (AYUSH) practitioners' services among older adults: results from the longitudinal aging study in India. *Cureus*, 2024. **16**(6): p. e62192.
45. Sharma, A., I. Ahmad Farouk, and S.K. Lal, COVID-19: a review on the novel coronavirus disease evolution, transmission, detection, control and prevention. *Viruses*, 2021. **13**(2): p. 202.
46. Friedman, N., et al., Human coronavirus infections in Israel: epidemiology, clinical symptoms and summer seasonality of HCoV-HKU1. *Viruses*, 2018. **10**(10): p. 515.
47. Liu, D.X., J.Q. Liang, and T.S. Fung, Human coronavirus-229E,-OC43,-NL63, and-HKU1 (Coronaviridae). *Encyclopedia of Virology*, 2021: p. 428.
48. De Wit, E., et al., SARS and MERS: recent insights into emerging coronaviruses. *Nature Reviews Microbiology*, 2016. **14**(8): p. 523–534.
49. ul Qamar, M.T., et al., Structural basis of SARS-CoV-2 3CLpro and anti-COVID-19 drug discovery from medicinal plants. *Journal of Pharmaceutical Analysis*, 2020. **10**(4): p. 313–319.
50. Pandey, A., et al., Natural plant products: a less focused aspect for the COVID-19 viral outbreak. *Frontiers in Plant Science*, 2020. **11**: p. 568890.
51. Khaerunnisa, S., Potential inhibitor of COVID-19 main protease (Mpro) from several medicinal plant compounds by molecular docking study-[v1]. 2020.
52. Sharifi-Rad, J., et al., Paclitaxel: application in modern oncology and nanomedicine-based cancer therapy. *Oxidative Medicine and Cellular Longevity*, 2021. **2021**(1): p. 3687700.
53. National Center for Biotechnology Information, *PubChem Compound Summary*. 2021, National Center for Biotechnology Information *Bethesda, MD*.
54. Solary, E., et al., Feasibility of using quinine, a potential multidrug resistance-reversing agent, in combination with mitoxantrone and cytarabine for the treatment of acute leukemia. *Journal of Clinical Oncology*, 1992. **10**(11): p. 1730–1736.
55. Jaffar, H.M., et al., Silymarin: unveiling its pharmacological spectrum and therapeutic potential in liver diseases: a comprehensive narrative review. *Food Science & Nutrition*, 2024. **12**(5): p. 3097–3111.
56. Nasim, N., I.S. Sandeep, and S. Mohanty, Plant-derived natural products for drug discovery: current approaches and prospects. *The Nucleus*, 2022. **65**(3): p. 399–411.
57. Shakya, A.K., Medicinal plants: future source of new drugs. *International Journal of Herbal Medicine*, 2016. **4**(4): p. 59–64.
58. Rates, S.M.K., Plants as source of drugs. *Toxicon*, 2001. **39**(5): p. 603–613.
59. Salim, A.A., Y.-W. Chin, and A.D. Kinghorn, Drug discovery from plants. *Bioactive Molecules and Medicinal Plants*, 2008. 1(1): p. 1–24.

60. Katiyar, C., et al., Drug discovery from plant sources: an integrated approach. *AYU (An International Quarterly Journal of Research in Ayurveda)*, 2012. **33**(1): p. 10–19.
61. Zhang, S., et al., Research advances in clinical applications, anticancer mechanism, total chemical synthesis, semi-synthesis and biosynthesis of paclitaxel. *Molecules*, 2023. **28**(22): p. 7517.
62. Balunas, M.J. and A.D. Kinghorn, Drug discovery from medicinal plants. *Life Sciences*, 2005. **78**(5): p. 431–441.
63. Gillessen, A. and H.H.J. Schmidt, Silymarin as supportive treatment in liver diseases: a narrative review. *Advances in Therapy*, 2020. **37**(4): p. 1279–1301.
64. Ganesan, A., The impact of natural products upon modern drug discovery. *Current Opinion in Chemical Biology*, 2008. **12**(3): p. 306–317.
65. Heinrich, M., J. Mah, and V. Amirkia, Alkaloids used as medicines: structural phytochemistry meets biodiversity: an update and forward look. *Molecules*, 2021. **26**(7): p. 1836.
66. Sofowora, A., E. Ogunbodede, and A. Onayade, The role and place of medicinal plants in the strategies for disease prevention. *African Journal of Traditional, Complementary and Alternative Medicines*, 2013. **10**(5): p. 210–229.
67. Alamgir, A.N.M., *Therapeutic Use of Medicinal Plants and Their Extracts: Volume* 1. Vol. 73. *2017*, Springer.
68. Ravina, E., *The Evolution of Drug Discovery: From Traditional Medicines to Modern Drugs*. 2011, John Wiley & Sons.
69. Ali, M., Hassan Ahmad Alhazmi, S. H. Ansari, Arshad Hussain, Sarfaraz Ahmad, Md Sarfaraz Alam, Md Sajid Ali, Karam A. El-Sharkawy, and Khalid Rehman Hakeem, Tamarix aphylla (L.) Karst. Phytochemical and bioactive profile compilations of less discussed but effective naturally growing Saudi plant. *Plant and Human Health*, Volume 3: Pharmacology and Therapeutic Uses, 2019, pp. 343–352. Springer International Publishing, Cham.
70. Chen, H., et al., Anti-inflammatory and antinociceptive properties of flavonoids from the fruits of black mulberry (Morus nigra L.). *PloS One*, 2016. **11**(4): p. e0153080.
71. Yang, W., et al., Advances in pharmacological activities of terpenoids. *Natural Product Communications*, 2020. **15**(3): p. 1934578X20903555.
72. Kaur, R. and S. Arora, Alkaloids-important therapeutic secondary metabolites of plant origin. *Journal of Critical Reviews*, 2015. **2**(3): p. 1–8.
73. Kurek, J., *Alkaloids: Their Importance in Nature and Human Life*. 2019, BoD–Books on Demand.
74. Parthasarathy, A., et al., Amino acid–derived defense metabolites from plants: a potential source to facilitate novel antimicrobial development. *Journal of Biological Chemistry*, 2021. **296**: p. 100438.
75. Cushnie, T.P.T., B. Cushnie, and A.J. Lamb, Alkaloids: an overview of their antibacterial, antibiotic-enhancing and antivirulence activities. *International Journal of Antimicrobial Agents*, 2014. **44**(5): p. 377–386.
76. Thawabteh, A., et al., The biological activity of natural alkaloids against herbivores, cancerous cells and pathogens. *Toxins*, 2019. **11**(11): p. 656.
77. Khameneh, B., et al., Investigation of the antibacterial activity and efflux pump inhibitory effect of co-loaded piperine and gentamicin nanoliposomes in methicillin-resistant Staphylococcus aureus. *Drug Development and Industrial Pharmacy*, 2015. **41**(6): p. 989–994.

78. Mabhiza, D., T. Chitemerere, and S. Mukanganyama, Antibacterial properties of alkaloid extracts from Callistemon citrinus and Vernonia adoensis against Staphylococcus aureus and Pseudomonas aeruginosa. *International Journal of Medicinal Chemistry*, 2016. **2016**(1): p. 6304163.
79. Huang, W., et al., Biosynthesis investigations of terpenoid, alkaloid, and flavonoid antimicrobial agents derived from medicinal plants. *Antibiotics*, 2022. **11**(10): p. 1380.
80. Cavalcante, G.M., A.E.d.S. Cabral, and C.C. Silva. *Leishmanicidal Activity of Flavonoids Natural and Synthetic: A Minireview*. Editora Científica Digital, 2022. 2: p. 266–282.
81. Feliciano, R.P., et al., Flavonoid intake and cardiovascular disease risk. *Current Opinion in Food Science*, 2015. **2**: p. 92–99.
82. Thilakarathna, S.H. and H.P.V. Rupasinghe, Flavonoid bioavailability and attempts for bioavailability enhancement. *Nutrients*, 2013. **5**(9): p. 3367–3387.
83. Kozlowska, A. and D. Szostak-Wegierek, Flavonoids-food sources and health benefits. *Roczniki Państwowego Zakładu Higieny*, 2014. **65**(2): p. 79–85.
84. Shkondrov, A., et al., Determination of flavonoids in related Astragalus species (Sect. Incani) occurring in Bulgaria. *Comptes rendus de l'Académie Bulg. des Sciences,* 2017. **70**: p. 363–366.
85. Pandey, K.B. and S.I. Rizvi, Plant polyphenols as dietary antioxidants in human health and disease. *Oxidative Medicine and Cellular Longevity*, 2009. **2**(5): p. 270–278.
86. Villela, A., et al., Photo-stability of a flavonoid dye in presence of aluminium ions. *Dyes and Pigments*, 2019. **162**: p. 222–231.
87. Ullah, A., et al., Important flavonoids and their role as a therapeutic agent. *Molecules*, 2020. **25**(22): p. 5243.
88. Danihelova, M., J. Viskupicová, and E. Sturdik, Lipophilization of flavonoids for their food, therapeutic and cosmetic applications. *Acta Chimica Slovaca*, 2012. **5**(1): p. 59.
89. Chuarienthong, P., N. Lourith, and P. Leelapornpisid, Clinical efficacy comparison of anti-wrinkle cosmetics containing herbal flavonoids. *International Journal of Cosmetic Science*, 2010. **32**(2): p. 99–106.
90. Verma, A., et al., Assessment of biological activities of fungal endophytes derived bioactive compounds isolated from Amoora rohituka. *Journal of Fungi*, 2022. **8**(3): p. 285.
91. Rai, N., et al., Bioprospecting of fungal endophytes from Oroxylum indicum (L.) Kurz with antioxidant and cytotoxic activity. *PloS One*, 2022. **17**(3): p. e0264673.
92. Gupta, P., A. Chattopadhaya, and V. Gautam, Myco-nanotechnological approach to synthesize gold nanoparticles using a fungal endophyte, Penicillium oxalicum, and unravelling its antibacterial activity and anti-breast cancer role via metabolic reprogramming. *Biomedical Materials*, 2024. **19**(6): p. 065030.
93. Zhao, L., et al., A review on flavones targeting serine/threonine protein kinases for potential anticancer drugs. *Bioorganic & Medicinal Chemistry*, 2019. **27**(5): p. 677–685.
94. Zhao, K., et al., LW-215, a newly synthesized flavonoid, exhibits potent anti-angiogenic activity in vitro and in vivo. *Gene*, 2018. **642**: p. 533–541.
95. Camero, C.M., et al., Anti-angiogenic activity of iridoids from G alium tunetanum. *Revista Brasileira de Farmacognosia*, 2018. **28**(3): p. 374–377.

96. Balasuriya, N. and H.P.V. Rupasinghe, Antihypertensive properties of flavonoid-rich apple peel extract. *Food Chemistry*, 2012. **135**(4): p. 2320–2325.
97. Patel, K., et al., New insights into the medicinal importance, physiological functions and bioanalytical aspects of an important bioactive compound of foods 'Hyperin': Health benefits of the past, the present, the future. *Beni-Suef University Journal of Basic and Applied Sciences*, 2018. 7(1): p. 31–42.
98. Xue, Z., et al., Antioxidant, antihypertensive, and anticancer activities of the flavonoid fractions from green, oolong, and black tea infusion waste. *Journal of Food Biochemistry*, 2018. **42**(6): p. e12690.
99. Khan, S., T. Khan, and A.J. Shah, Total phenolic and flavonoid contents and antihypertensive effect of the crude extract and fractions of Calamintha vulgaris. *Phytomedicine*, 2018. **47**: p. 174–183.
100. Lagunas-Herrera, H., et al., Acute and chronic antihypertensive effect of fractions, tiliroside and scopoletin from Malva parviflora. *Biological and Pharmaceutical Bulletin*, 2019. **42**(1): p. 18–25.
101. Mazidi, M., N. Katsiki, and M. Banach, A higher flavonoid intake is associated with less likelihood of nonalcoholic fatty liver disease: results from a multiethnic study. *The Journal of Nutritional Biochemistry*, 2019. **65**: p. 66–71.
102. Aguiar, L.M., et al., Functional food consumption and its physiological effects. In *Bioactive Compounds*, Maira Rubi Segura Campos, Editor. 2019, Elsevier. p. 205–225.
103. Masyita, A., et al., Terpenes and terpenoids as main bioactive compounds of essential oils, their roles in human health and potential application as natural food preservatives. *Food Chemistry: X*, 2022. **13**: p. 100217.
104. Keita, K., C. Darkoh, and F. Okafor, Secondary plant metabolites as potent drug candidates against antimicrobial-resistant pathogens. *SN Applied Sciences*, 2022. **4**(8): p. 209.
105. Haque, N., et al., Marine natural products in clinical use. *Marine Drugs*, 2022. **20**(8): p. 528.
106. Calder, P.C., Marine omega-3 fatty acids and inflammatory processes: Effects, mechanisms and clinical relevance. *Biochimica et Biophysica Acta (BBA)-Molecular and Cell Biology of Lipids*, 2015. **1851**(4): p. 469–484.
107. Malve, H., Exploring the ocean for new drug developments: Marine pharmacology. *Journal of Pharmacy and Bioallied Sciences*, 2016. **8**(2): p. 83–91.
108. Costa-Neto, E.M., Animal-based medicines: biological prospection and the sustainable use of zootherapeutic resources. *Anais da Academia Brasileira de ciências*, 2005. 77: p. 33–43.
109. King, G.F., Venoms as a platform for human drugs: translating toxins into therapeutics. *Expert Opinion on Biological Therapy*, 2011. **11**(11): p. 1469–1484.
110. Singh, A.P., Medicinal leech therapy (hirudotherapy): a brief overview. *Complementary Therapies in Clinical Practice*, 2010. **16**(4): p. 213–215.
111. Haefner, B., Drugs from the deep: marine natural products as drug candidates. *Drug Discovery Today*, 2003. **8**(12): p. 536–544.
112. Huang, G.S., et al., Potentiation of taxol efficacy by discodermolide in ovarian carcinoma xenograft-bearing mice. *Clinical Cancer Research*, 2006. **12**(1): p. 298–304.

113. Torres, F.A.E., et al., New drugs with antiprotozoal activity from marine algae: a review. *Revista Brasileira de Farmacognosia*, 2014. **24**: p. 265–276.
114. Lee, J.-C., et al., Marine algal natural products with anti-oxidative, anti-inflammatory, and anti-cancer properties. *Cancer Cell International*, 2013. **13**: p. 1–7.
115. Meenakshi, S., et al., Total flavanoid and in vitro antioxidant activity of two seaweeds of Rameshwaram coast. *Global Journal of Pharmacology*, 2009. **3**(2): p. 59–62.
116. Balaji Raghavendra Rao, H., A. Sathivel, and T. Devaki, Antihepatotoxic nature of Ulva reticulata (Chlorophyceae) on acetaminophen-induced hepatoxicity in experimental rats. *Journal of Medicinal Food*, 2004. **7**(4): p. 495–497.
117. Rai, N., et al., Plant associated fungal endophytes as a source of natural bioactive compounds. *Mycology*, 2021. **12**(3): p. 139–159.
118. Keshri, P.K., et al., Biological potential of bioactive metabolites derived from fungal endophytes associated with medicinal plants. *Mycological Progress*, 2021. **20**(5): p. 577–594.
119. Newman, D.J. and G.M. Cragg, Natural products as sources of new drugs over the last 25 years. *Journal of Natural Products*, 2007. **70**(3): p. 461–477.
120. Buriani, A., et al., Omic techniques in systems biology approaches to traditional Chinese medicine research: present and future. *Journal of Ethnopharmacology*, 2012. **140**(3): p. 535–544.
121. Liu, J., et al., Integrating a comprehensive DNA barcode reference library with a global map of yews (Taxus L.) for forensic identification. *Molecular Ecology Resources*, 2018. **18**(5): p. 1115–1131.
122. Chen, S., et al., A renaissance in herbal medicine identification: from morphology to DNA. *Biotechnology Advances*, 2014. **32**(7): p. 1237–1244.
123. Tnah, L.H., et al., DNA barcode database of common herbal plants in the tropics: a resource for herbal product authentication. *Food Control*, 2019. **95**: p. 318–326.
124. Kress, W.J., Plant DNA barcodes: applications today and in the future. *Journal of Systematics and Evolution*, 2017. **55**(4): p. 291–307.
125. Hussain, A., A phylogenetic perspective of antiviral species of the genus Artemisia (Asteraceae-Anthemideae): a proposal of anti SARS-CoV-2 (COVID-19) candidate taxa. *Journal of Herbal Medicine*, 2022. **36**: p. 100601.
126. *Baker, D.A., D.W.* Stevenson, and D.P. Little, DNA barcode identification of black cohosh herbal dietary supplements. *Journal of AOAC International*, 2012. **95**(4): p. 1023–1034.
127. Pulice, G., S. Pelaz, and L. Matías-Hernández, Molecular farming in Artemisia annua, a promising approach to improve anti-malarial drug production. *Frontiers in Plant Science*, 2016. **7**: p. 329.
128. Gantait, S., S. Debnath, and M. Nasim Ali, Genomic profile of the plants with pharmaceutical value. *3 Biotech*, 2014. **4**(6): p. 563–578.
129. Bumpus, S.B., et al., A proteomics approach to discovering natural products and their biosynthetic pathways. *Nature Biotechnology*, 2009. **27**(10): p. 951–956.
130. Ho, C.H., et al., A molecular barcoded yeast ORF library enables mode-of-action analysis of bioactive compounds. *Nature Biotechnology*, 2009. **27**(4): p. 369–377.

131. Martínez-Esteso, M.J., et al., The role of proteomics in progressing insights into plant secondary metabolism. *Frontiers in Plant Science*, 2015. **6**: p. 504.
132. Jung, H.J. and H.J. Kwon, Target deconvolution of bioactive small molecules: the heart of chemical biology and drug discovery. *Archives of Pharmacal Research*, 2015. **38**: p. 1627–1641.
133. Bassik, M.C., et al., A systematic mammalian genetic interaction map reveals pathways underlying ricin susceptibility. *Cell*, 2013. **152**(4): p. 909–922.
134. Shalem, O., et al., Genome-scale CRISPR-Cas9 knockout screening in human cells. *Science*, 2014. **343**(6166): p. 84–87.
135. Kasap, C., O. Elemento, and T.M. Kapoor, DrugTargetSeqR: a genomics-and CRISPR-Cas9–based method to analyze drug targets. *Nature Chemical Biology*, 2014. **10**(8): p. 626–628.
136. Neggers, J.E., et al., Identifying drug-target selectivity of small-molecule CRM1/XPO1 inhibitors by CRISPR/Cas9 genome editing. *Chemistry & Biology*, 2015. **22**(1): p. 107–116.
137. Clish, C.B., Metabolomics: an emerging but powerful tool for precision medicine. *Molecular Case Studies*, 2015. **1**(1): p. a000588.
138. Liu, X. and J.W. Locasale, Metabolomics: a primer. *Trends in Biochemical Sciences*, 2017. **42**(4): p. 274–284.
139. Thomford, N.E., et al., Natural products for drug discovery in the 21st century: innovations for novel drug discovery. *International Journal of Molecular Sciences*, 2018. **19**(6): p. 1578.
140. Thomford, N.E., et al., In vitro reversible and time-dependent CYP450 inhibition profiles of medicinal herbal plant extracts Newbouldia laevis and Cassia abbreviata: implications for herb-drug interactions. *Molecules*, 2016. **21**(7): p. 891.
141. Xie, G., et al., Ultra-performance LC/TOF MS analysis of medicinal Panax herbs for metabolomic research. *Journal of Separation Science*, 2008. **31**(6–7): p. 1015–1026.
142. Park, H.-W., et al., Metabolomic approach for discrimination of processed ginseng genus (Panax ginseng and Panax quinquefolius) using UPLC-QTOF MS. *Journal of Ginseng Research*, 2014. **38**(1): p. 59–65.
143. Chapman, T., Lab automation and robotics: automation on the move. *Nature*, 2003. **421**(6923): p. 661–663.
144. King, R.D., et al., The automation of science. *Science*, 2009. **324**(5923): p. 85–89.
145. Sparkes, A., et al., Towards robot scientists for autonomous scientific discovery. *Automated Experimentation*, 2010. **2**: p. 1–11.
146. Meanwell, N.A., Improving drug design: an update on recent applications of efficiency metrics, strategies for replacing problematic elements, and compounds in nontraditional drug space. *Chemical Research in Toxicology*, 2016. **29**(4): p. 564–616.
147. Schneider, G., Automating drug discovery. *Nature Reviews Drug Discovery*, 2018. **17**(2): p. 97–113.
148. Baranczak, A., et al., Integrated platform for expedited synthesis–purification–testing of small molecule libraries. *ACS Medicinal Chemistry Letters*, 2017. **8**(4): p. 461–465.
149. Eglen, R.M. and D.H. Randle, Drug discovery goes three-dimensional: goodbye to flat high-throughput screening? *Assay and Drug Development Technologies*, 2015. **13**(5): p. 262–265.

150. Esch, E.W., A. Bahinski, and D. Huh, Organs-on-chips at the frontiers of drug discovery. *Nature Reviews Drug Discovery*, 2015. **14**(4): p. 248–260.
151. Yang, X., et al., Concepts of artificial intelligence for computer-assisted drug discovery. *Chemical Reviews*, 2019. **119**(18): p. 10520–10594.
152. Li, G., et al., Artificial intelligence-guided discovery of anticancer lead compounds from plants and associated microorganisms. *Trends in Cancer*, 2022. **8**(1): p. 65–80.
153. Verma, A., et al., Integration of hyphenated techniques for characterizing and chemical profiling of natural products. *Chemistry & Biodiversity*, 2025. **22**(9): p. e202500234.
154. Das, K.P., Nanoparticles and convergence of artificial intelligence for targeted drug delivery for cancer therapy: current progress and challenges. *Frontiers in Medical Technology*, 2023. **4**: p. 1067144.
155. Adir, O., et al., Integrating artificial intelligence and nanotechnology for precision cancer medicine. *Advanced Materials*, 2020. **32**(13): p. 1901989.
156. Kumar, S. and S. Kumar, Molecular docking: A structure-based approach for drug repurposing. In *In Silico Drug Design*, K. Roy, Editor. 2019, Academic Press, Elsevier: Cambridge, MA. p. 161–189.
157. Sahoo, R.N., et al., Review on the use of molecular docking as the first line tool in drug discovery and development. *Indian Journal of Pharmaceutical Sciences*, 2022. **84**(5): p. 1334–1337.
158. Alharbi, Y.T.M., et al., Investigation of phytochemicals isolated from selected Saudi medicinal plants as natural inhibitors of SARS CoV-2 main protease: in vitro, molecular docking and simulation analysis. *Saudi Pharmaceutical Journal*, 2024. **32**(5): p. 102023.
159. Silvestrini, L., et al., The dimer-monomer equilibrium of SARS-CoV-2 main protease is affected by small molecule inhibitors. *Scientific Reports*, 2021. **11**(1): p. 9283.
160. Chouhan, M., et al., Inhibition of Mycobacterium tuberculosis resuscitation-promoting factor B (RpfB) by microbially derived natural compounds: a computational study. *Journal of Biomolecular Structure and Dynamics*, 2024. **42**(2): p. 948–959.
161. Patra, J.K., et al., Nano based drug delivery systems: recent developments and future prospects. *Journal of Nanobiotechnology*, 2018. **16**: p. 1–33.
162. Gupta, P., et al., Unveiling the cytotoxic and anti-proliferative potential of green-synthesized silver nanoparticles mediated by Colletotrichum gloeosporioides. *RSC Advances*, 2024. **14**(6): p. 4074–4088.
163. Singh, S., et al., Sustainable synthesis of novel green-based nanoparticles for therapeutic interventions and environmental remediation. *ACS Synthetic Biology*, 2024. **13**(7): p. 1994–2007.
164. Din, F.U., et al., Effective use of nanocarriers as drug delivery systems for the treatment of selected tumors. *International Journal of Nanomedicine*, 2017: p. 7291–7309.
165. Rahim, M.A., et al., Recent advancements in stimuli responsive drug delivery platforms for active and passive cancer targeting. *Cancers*, 2021. **13**(4): p. 670.
166. Rai, N., et al., Fungal endophyte-mediated green synthesis of silver nanoparticles as potential anticancer agent: current perspective and challenges. In *Handbook of Oncobiology: From Basic to Clinical Sciences*, R.C. Sobti, N.K. Ganguly, and R. Kumar, Editors. 2024, Springer *Nature Singapore*: Singapore. p. 1385–1408.

167. Ali Abdalla, Y.O., et al., Natural products for cancer therapy: a review of their mechanism of actions and toxicity in the past decade. *Journal of Tropical Medicine*, 2022. **2022**(1): p. 5794350.
168. Bajracharya, R., et al., Functional ligands for improving anticancer drug therapy: current status and applications to drug delivery systems. *Drug Delivery*, 2022. **29**(1): p. 1959–1970.
169. Kumari, A., et al., Nanoencapsulation for drug delivery. *EXCLI Journal*, 2014. **13**: p. 265.
170. Bano, S., et al., Microwave-assisted green synthesis of superparamagnetic nanoparticles using fruit peel extracts: surface engineering, T2 relaxometry, and photodynamic treatment potential. *International Journal of Nanomedicine*, 2016: p. 3833–3848.
171. Kharey, P., et al., Green synthesis of near-infrared absorbing eugenate capped iron oxide nanoparticles for photothermal application. *Nanotechnology*, 2019. **31**(9): p. 095705.
172. Idham, Z., et al., Improvement of extraction and stability of anthocyanins, the natural red pigment from roselle calyces using supercritical carbon dioxide extraction. *Journal of CO_2 Utilization*, 2022. **56**: p. 101839.
173. Urias-Lugo, D.A., et al., Anthocyanins and phenolic acids of hybrid and native blue maize (Zea mays L.) extracts and their antiproliferative activity in mammary (MCF7), liver (HepG2), colon (Caco2 and HT29) and prostate (PC3) cancer cells. *Plant Foods for Human Nutrition*, 2015. **70**: p. 193–199.
174. Bars-Cortina, D., et al. Chemopreventive effects of anthocyanins on colorectal and breast cancer: A review. *Seminars in Cancer Biology*, 2022. *81*: p. 241–258.
175. Mottaghipisheh, J., et al., The promising therapeutic and preventive properties of anthocyanidins/anthocyanins on prostate cancer. *Cells*, 2022. **11**(7): p. 1070.
176. Alsharairi, N.A., Insights into the mechanisms of action of proanthocyanidins and anthocyanins in the treatment of nicotine-induced non-small cell lung cancer. *International Journal of Molecular Sciences*, 2022. **23**(14): p. 7905.
177. Diaconeasa, Z., et al., Anthocyanins, vibrant color pigments, and their role in skin cancer prevention. *Biomedicines*, 2020. **8**(9): p. 336.
178. Munagala, R., et al., Exosomal formulation of anthocyanidins against multiple cancer types. *Cancer Letters*, 2017. **393**: p. 94–102.
179. Hamedi, S. and M. Koosha, Designing a pH-responsive drug delivery system for the release of black-carrot anthocyanins loaded in halloysite nanotubes for cancer treatment. *Applied Clay Science*, 2020. **197**: p. 105770.
180. Jafari, A., et al., Exosomes and cancer: from molecular mechanisms to clinical applications. *Medical Oncology*, 2021. **38**(4): p. 45.
181. Muddineti, O.S., et al., Cholesterol and vitamin E-conjugated PEGylated polymeric micelles for efficient delivery and enhanced anticancer activity of curcumin: evaluation in 2D monolayers and 3D spheroids. *Artificial Cells, Nanomedicine, and Biotechnology*, 2018. **46**(supl): p. 773–786.
182. Jafarinejad-Farsangi, S., et al., Curcumin loaded on graphene nanosheets induced cell death in mammospheres from MCF-7 and primary breast tumor cells. *Biomedical Materials*, 2021. **16**(4): p. 045040.
183. Zoghi, M., et al., Synthesis and characterization of chitosan/carbon quantum dots/Fe_2O_3 nanocomposite comprising curcumin for targeted drug delivery in breast cancer therapy. *International Journal of Biological Macromolecules*, 2023. **249**: p. 125788.

184. Kundu, M., et al., pH-responsive and targeted delivery of curcumin via phenylboronic acid-functionalized ZnO nanoparticles for breast cancer therapy. *Journal of Advanced Research*, 2019. **18**: p. 161–172.
185. Huang, W., et al., Synergistic antifungal activity of green synthesized silver nanoparticles and epoxiconazole against Setosphaeria turcica. *Journal of Nanomaterials*, 2020. **2020**(1): p. 9535432.
186. Srihasam, S., et al., Phytogenic generation of NiO nanoparticles using Stevia leaf extract and evaluation of their in-vitro antioxidant and antimicrobial properties. *Biomolecules*, 2020. **10**(1): p. 89.
187. Elangovan, D., et al., Coating of wallpaper with green synthesized silver nanoparticles from Passiflora foetida fruit and its illustrated antifungal mechanism. *Process Biochemistry*, 2022. **112**: p. 177–182.
188. Sathiya, P. and K. Geetha, Fruit extract mediated synthesis of silver oxide nanoparticles using Dimocarpus longan fruit and their assesment of catalytic, antifungal, antioxidant and cytotoxic potentials. *Inorganic and Nano-Metal Chemistry*, 2023. **53**(10): p. 1173–1183.
189. Swetha, M., et al., Cancer chemoprevention: a strategic approach using phytochemicals. *Frontiers in Pharmacology*, 2022. **12**: p. 809308.
190. Olayiwola, Y. and L. Gollahon, Natural compounds and breast cancer: chemo-preventive and therapeutic capabilities of chlorogenic acid and cinnamaldehyde. *Pharmaceuticals*, 2024. **17**(3): p. 361.
191. Chen, C. and A.-N.T. Kong, Dietary cancer-chemopreventive compounds: from signaling and gene expression to pharmacological effects. *Trends in Pharmacological Sciences*, 2005. **26**(6): p. 318–326.
192. Mocanu, M.-M., P. Nagy, and J. Szöllősi, Chemoprevention of breast cancer by dietary polyphenols. *Molecules*, 2015. **20**(12): p. 22578–22620.
193. Hussain, S.S., A.P. Kumar, and R. Ghosh. *Food-Based Natural Products for Cancer Management: Is the Whole Greater Than the Sum of the Parts?* in *Seminars in Cancer Biology*, 2016. **40**: p. 233–246. Academic Press..
194. Orlikova, B., M. Dicato, and M. Diederich, 1,000 Ways to die: natural compounds modulate non-canonical cell death pathways in cancer cells. *Phytochemistry Reviews*, 2014. **13**: p. 277–293.
195. Israel, B.e.B., et al., Phytochemicals: current strategies for treating breast cancer. *Oncology Letters*, 2018. **15**(5): p. 7471–7478.
196. Huang, L., et al., Berbamine inhibits SARS-CoV-2 infection by compromising TRPMLs-mediated endolysosomal trafficking of ACE2. *Signal Transduction and Targeted Therapy*, 2021. **6**(1): p. 168.
197. Xia, B., et al., SARS-CoV-2 envelope protein causes acute respiratory distress syndrome (ARDS)-like pathological damages and constitutes an antiviral target. *Cell Research*, 2021. **31**(8): p. 847–860.
198. Pizzorno, A., et al., In vitro evaluation of antiviral activity of single and combined repurposable drugs against SARS-CoV-2. *Antiviral Research*, 2020. **181**: p. 104878.
199. Varghese, F.S., et al., Berberine and obatoclax inhibit SARS-Cov-2 replication in primary human nasal epithelial cells in vitro. *Viruses*, 2021. **13**(2): p. 282.
200. Ku, K.B., et al., Repurposing screens of FDA-approved drugs identify 29 inhibitors of SARS-CoV-2. *Journal of Microbiology and Biotechnology*, 2020. **30**(12): p. 1843.
201. Grosse, M., et al., Quinine inhibits infection of human cell lines with SARS-CoV-2. *Viruses*, 2021. **13**(4): p. 647.

202. Salas Rojas, M., et al., Quinacrine, an antimalarial drug with strong activity inhibiting SARS-CoV-2 viral replication in vitro. *Viruses*, 2021. **13**(1): p. 121.
203. Verch, T., V. Yusibov, and H. Koprowski, Expression and assembly of a full-length monoclonal antibody in plants using a plant virus vector. *Journal of Immunological Methods*, 1998. **220**(1–2): p. 69–75.
204. McCormick, A.A., et al., Rapid production of specific vaccines for lymphoma by expression of the tumor-derived single-chain Fv epitopes in tobacco plants. *Proceedings of the National Academy of Sciences*, 1999. **96**(2): p. 703–708.
205. Nessa, M.U., M.A. Rahman, and Y. Kabir, Plant-produced monoclonal antibody as immunotherapy for cancer. *BioMed Research International*, 2020. **2020**(1): p. 3038564.
206. Bulaon, C.J.I., et al., Development of plant-derived bispecific monoclonal antibody targeting PD-L1 and CTLA-4 against mouse colorectal cancer. *Planta Medica*, 2024. **90**(4): p. 305–315.
207. Balammal, G., B.M. Sekar, and J.P. Reddy, Analysis of herbal medicines by modern chromatographic techniques. *International Journal of Preclinical and Pharmaceutical Research*, 2012. **3**(1): p. 50–63.
208. Pal, S.K. and Y. Shukla, Herbal medicine: current status and the future. *Asian Pacific Journal of Cancer Prevention*, 2003. **4**(4): p. 281–288.
209. Nema, A.K., A. Agarwal, and V. Kashaw, Hepatoprotective activity of Leptadenia reticulata stems against carbon tetrachloride-induced hepatotoxicity in rats. *Indian Journal of Pharmacology*, 2011. **43**(3): p. 254–257.
210. Gholijani, N., et al., Modulation of cytokine production and transcription factors activities in human Jurkat T cells by thymol and carvacrol. *Advanced Pharmaceutical Bulletin*, 2015. **5**(Suppl 1): p. 653.
211. Nooreen, Z., V.K. Rai, and N.P. Yadav, Phytopharmaceuticals: a new class of drug in India. *Annals of Phytomedicine*, 2018. **7**(1): p. 27–37.

Chapter 3

Epigenetic modifiers in healthcare mechanisms and applications

Ankita Saha
Department of Cell Biology, Albert Einstein College of Medicine, New York

3.1 INTRODUCTION

The epigenetics field has transformed modern medicine by showing that inheritable gene expression changes happen without DNA sequence modifications [1]. Molecular switches, known as epigenetic mechanisms including DNA methylation and histone modifications along with non-coding RNAs, regulate gene activity through responses to environmental and cellular cues [2,3,4]. The expanding knowledge in epigenetics has transformed disease management strategies and therapeutic advancements for complex life-threatening conditions like cancer, neurodegenerative disorders, and autoimmune diseases [5,6]. Basic epigenetic research breakthroughs enable earlier disease diagnosis and create targeted treatment strategies that boost survival rates and patient results.

Until epigenetics emerged as a significant field of study, the medical community primarily investigated genetic mutations as the leading cause of diseases. Most treatments focused on managing symptoms instead of addressing root causes and showed minimal effectiveness in stopping disease advancement [7]. Cancer treatment primarily utilized imprecise cytotoxic chemotherapies which often resulted in serious side effects and unfavorable prognoses [8,9]. Scientists have discovered that abnormal epigenetic changes play a major role in disease development, which opens new possibilities for treatment strategies. The focus on reversible epigenetic modifications has allowed scientists to create epigenetic drugs that act more precisely and reduce toxicity [10].

The latest research case studies demonstrate how epigenetic studies have significantly transformed clinical healthcare approaches. In the case of acute myeloid leukemia (AML), which traditionally showed poor survival statistics, the development of DNMT and HDAC enzyme inhibitors has led to better remission outcomes [11,12,13]. The application of epigenetic biomarkers in specific breast cancer types has enabled medical professionals to develop tailored treatment approaches that precisely attack therapy-resistant tumors [14,15]. Recent clinical trials investigate medications that adjust histone acetylation levels to improve cognitive function in Alzheimer's disease patients [16,17].

DOI: 10.1201/9781003733010-3

Healthcare has entered a new precision medicine phase due to epigenetic research, which enables personalized treatment and diagnosis based on specific epigenetic patterns of patients. This chapter thoroughly examines the dual functionality of epigenetic modifiers as diagnostic instruments and therapeutic targets while highlighting their impact on the evolution of contemporary healthcare practices.

3.2 OVERVIEW OF EPIGENETICS: CONCEPTS AND KEY MECHANISMS

Epigenetics examines reversible gene expression changes passed through generations, which occur without DNA sequence mutations [18]. These molecular changes serve as control mechanisms that determine gene activation or repression within specific cells at any moment. DNA methylation together with histone modifications and non-coding RNAs stands as essential epigenetic regulatory mechanisms that sustain cellular identity and functional state (Figure 3.1) [19].

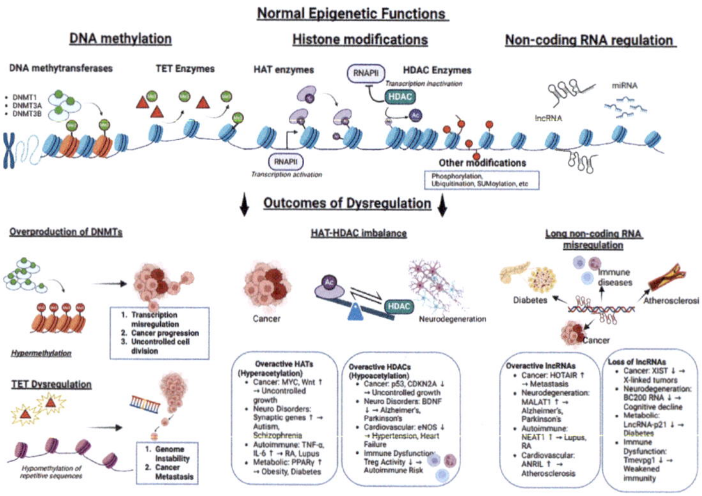

Figure 3.1 Epigenetic regulation and its dysregulation: mechanisms, consequences, and disease associations. This figure illustrates the normal functions of epigenetic mechanisms—DNA methylation, histone modifications, and non-coding RNA regulation—and their dysregulation in disease. At the top, normal epigenetic functions are shown. DNA methylation, mediated by DNA methyltransferases (DNMTs), silences gene expression, while TET enzymes promote demethylation. Histone modifications involve histone acetyltransferases (HATs) activating transcription and histone deacetylases (HDACs) repressing it. Other modifications, such as phosphorylation, ubiquitination, and SUMOylation, contribute to chromatin regulation. Non-coding RNAs (miRNAs and lncRNAs) regulate gene expression at the post-transcriptional level.

Below, the figure highlights the outcomes of epigenetic dysregulation. Overactive DNMTs cause hypermethylation, leading to transcriptional misregulation, cancer progression, and uncontrolled cell division. Ten-eleven translocation (TET) enzyme dysregulation contributes to genomic instability and metastasis. The HAT-HDAC imbalance impacts gene expression. Overactive HATs (hyperacetylation) lead to cancer (MYC, Wnt overactivation), neurodevelopmental disorders (synaptic gene dysregulation), autoimmune diseases (TNF-α, IL-6 upregulation), and metabolic disorders (PPARγ overactivation). In contrast, overactive HDACs (hypoacetylation) silence tumor suppressor genes (p53, CDKN2A), contribute to neurodegeneration (BDNF repression), cardiovascular dysfunction (eNOS silencing), and immune dysregulation (Treg suppression). Long non-coding RNA misregulation affects disease states. lncRNA overexpression (HOTAIR, MALAT1, NEAT1, and ANRIL) drives cancer, neurodegeneration, immune diseases, and atherosclerosis, while lncRNA loss (XIST, BC200, lncRNA-p21, and Tmevpg1) contributes to X-linked cancers, neurodegenerative decline, diabetes, and immune dysfunction.

3.2.1 DNA methylation

DNA methylation functions as an essential epigenetic mechanism which controls gene activity, without changing the DNA sequence. The process attaches a methyl group (-CH_3) to the cytosine base within DNA and occurs mainly at cytosine-guanine (CpG) dinucleotides. The modification serves as a fundamental mechanism in cellular differentiation as well as genomic imprinting, X-chromosome inactivation, and genome stability maintenance. During cell division DNA methylation patterns transmit heritable information, which maintains cellular identity yet allows dynamical adjustments according to environmental signals and developmental stages [2,20].

DNA methyltransferases (DNMTs) facilitate DNA methylation by transferring a methyl group from *S*-adenosylmethionine (SAM) to the fifth carbon of cytosine to produce 5-methylcytosine (5mC). DNMT1 maintains DNA methylation patterns through replication by identifying hemimethylated DNA strands and transferring methylation marks to the new strand. The process maintains accurate epigenetic information transfer through multiple generations of cells. DNMT3A together with DNMT3B performs *de novo* methylation to create new methylation marks during embryonic development and cell differentiation stages. The enzymes responsible for establishing tissue-specific methylation patterns, which determine cell fate, are essential [21,22].

Gene activity is controlled through the regulatory mechanism of DNA methylation. Genes linked to unmethylated CpG islands in their promoter regions maintain transcriptional activity. Gene expression becomes repressed when these regions undergo methylation because it blocks transcription factor access and attracts chromatin-modifying proteins that establish a repressive chromatin state [23]. Methylated DNA serves as an anchor for methyl-CpG-binding domain (MBD) proteins, including MeCP2, MBD1, and MBD2, which partner with histone deacetylases (HDACs) and chromatin remodeling enzymes to create a compact chromatin structure that restricts transcriptional machinery access [24]. X-chromosome inactivation in females depends heavily on this type of gene silencing because it involves heavy methylation and transcriptional silencing of one X-chromosome [25].

DNA methylation maintains cell stability but can be reversed for gene expression control. The demethylation process utilizes two primary mechanisms, which are passive and active demethylation. Passive demethylation takes place when methylation marks become diluted during DNA replication due to the lack of DNMT activity. TET enzymes facilitate active demethylation by converting 5-methylcytosine into 5-hydroxymethylcytosine and further intermediates, which result in base excision repair that restores unmethylated cytosine [26].

Many diseases show abnormal DNA methylation patterns, with cancer being the most noted example. The promoter regions of tumor suppressor genes frequently become hypermethylated, which causes transcriptional silencing followed by functional loss. It results in tumor progression through unregulated cell division. Global DNA hypomethylation in repetitive elements and oncogenes results in genomic instability and promotes mutation rates and metastasis [27]. Therapeutic interventions now target DNA methylation due to its important role in disease development. 5-Azacytidine and decitabine function as DNA methylation inhibitors for myelodysplastic syndromes and specific leukemias through demethylation processes that activate previously silenced genes [28,29].

DNA methylation functions as a fundamental epigenetic regulatory system that maintains correct gene expression patterns while preserving cellular identity and genome integrity. It coordinates essential gene expression patterns needed for normal development and differentiation by interacting with histone modifications and non-coding RNAs. Research into the regulatory functions and enzymes of epigenetic processes has revealed essential knowledge about their impact on health and disease while creating new opportunities for therapeutic development.

3.2.2 Histone modifications

Histones serve as essential structural proteins that organize DNA into chromatin. The dynamic structure of chromatin undergoes regulation by histone PTMs, which affect gene expression by controlling how accessible DNA is to the transcriptional machinery. The N-terminal tails of histones undergo key modifications such as acetylation, methylation, phosphorylation, ubiquitination, and SUMOylation [3].

3.2.2.1 Histone acetylation and deacetylation

The epigenetic modifications of histone acetylation and deacetylation function as key regulators of gene expression through their effects on chromatin structure. These processes occur on the lysine residues of histone proteins and are controlled by two opposing enzyme families: Histone acetyltransferases (HATs) add acetyl groups while histone deacetylases (HDACs) serve to remove these groups [30]. The interaction between histone acetyltransferases and histone deacetylases regulates chromatin structure by controlling its state between an open transcriptionally active form and a condensed repressed form. The regulation of these biological mechanisms depends on this balance, which affects gene regulation, DNA replication, and cellular differentiation.

Histone acetylation represents the post-translational modification where HAT enzymes attach an acetyl group from acetyl-CoA to the ε-amino group located on lysine residues of histone tails. The acetylation process removes the positive charge from lysine residues, which reduces the electrostatic bond between histones and negatively charged DNA. The chromatin structure loosens its configuration, which enables transcription factors and RNA polymerase to reach gene promoters with greater ease. Euchromatin describes the open chromatin state that facilitates active gene transcription. HATs not only modify histones, but they also acetylate proteins outside of histones, such as transcription factors and chromatin remodelers, resulting in enhanced gene activation [3,31].

Scientific classification of histone acetyltransferases (HATs) divides them into multiple families according to sequence similarity and their enzymatic activities [32]. The GNAT family of HAT enzymes, which contains GCN5 and PCAF, targets histone acetylation at active promoter regions, thereby serving as key regulators of transcription. The p300/CBP family members, including p300 and CBP, function as transcriptional co-activators that enable histone acetylation through various signaling pathways [33]. The MYST family members Tip60 and MOF participate in chromatin remodeling activities and the cellular response to DNA damage [34]. All of these HAT families serve specialized regulatory roles, which influence chromatin accessibility while supporting cellular differentiation and stress response processes [35].

Histone deacetylation eliminates acetyl groups from lysine residues, which reverses the effects of acetylation. HDACs drive this process, which restores the positive charge to histones, thus tightening DNA-histone interactions and producing a condensed chromatin structure referred to as heterochromatin. The closed chromatin arrangement limits the ability of transcription factors and RNA polymerase to reach gene promoters thereby causing gene repression. HDACs work within multiprotein complexes to recruit other chromatin-modifying enzymes, which allow them to create transcriptionally silent chromatin domains.

Four groups encompass the HDAC family classification through sequence homology and functional characteristics. The principal location of Class I HDACs, including HDAC1 and HDAC2, is within the nucleus, where they perform essential functions in transcriptional repression [36]. The Class II HDAC enzymes HDAC4 and HDAC5 move between the nucleus and cytoplasm to regulate gene expression based on external stimuli. Sirtuins, which belong to Class III HDACs, need NAD^+ to function and play roles in metabolic control, stress reaction, and life span extension. Class IV HDACs like HDAC11 share structural features with Class I and II enzymes yet perform different biological roles [37].

Normal gene expression patterns rely on the balance between histone acetylation and deacetylation, which also supports cellular homeostasis and environmental response mechanisms. Multiple diseases, including cancer and neurodegenerative disorders, have been linked to disruptions in this balance which also affects inflammatory conditions [33]. Unregulated HAT activity triggers excessive gene activation, and HDAC overactivity leads to incorrect gene silencing, which results in pathological conditions. Because of their critical role in disease pathology, both HATs and HDACs represent important targets for new medical therapies. Cancer treatments now include small-molecule HDAC inhibitors like vorinostat and romidepsin because they enable tumor suppressor genes to regain their expression [5]. Researchers are investigating HAT activators as potential treatments for neurodegenerative and inflammatory disorders because they help reverse transcriptional repression linked to disease advancement [38,39].

Research on histone acetylation and deacetylation regulation reveals gene control mechanisms and opens pathways to develop therapeutic interventions. Targeted epigenetic therapies will become more important in precision medicine as research reveals more about these modifications, which affect conditions caused by epigenetic dysregulation.

3.2.2.2 Histone methylation

Histone methylation stands as an essential epigenetic modification that controls chromatin structure while regulating gene expression. While histone acetylation typically leads to transcriptional activation, histone methylation can function to either activate or repress transcription based on which histone residue is modified along with the level of methylation. Histone proteins undergo this modification at lysine and arginine residues through dynamic regulation by histone methyltransferases that add methyl groups and histone demethylases that remove them. The exact regulation of these enzymes establishes both chromatin accessibility levels and gene transcription activity [40,41].

The trimethylation of lysine 4 on histone H3 (H3K4me3) serves as an indicator of active gene promoters, which correlates with transcriptional activation. Trimethylation at histone H3 lysine 9 (H3K9me3) and lysine 27

(H3K27me3) functions as a marker for transcriptionally inactive genomic areas, which leads to gene silencing and heterochromatin development. Specific chromatin-binding proteins recognize and interpret histone methylation marks to determine their functional outcome [42,43].

The histone methyltransferase family consists of two major groups based on their catalytic domains: lysine methyltransferases (KMTs) and arginine methyltransferases (PRMTs) [44,45]. KMTs target lysine residues and contain enzymes like SET1 for H3K4 methylation and SUV39H1 for H3K9 methylation to form repressive chromatin domains. PRMTs target arginine residues for modification while playing dual roles in transcriptional regulation through activation or repression, which depends on the methylation pattern [45]. These enzymes operate within substantial protein assemblies and synchronize their functions with additional chromatin modifiers to adjust gene expression based on developmental and environmental signals.

Histone demethylases (HDMs) perform histone demethylation by removing methyl groups from lysine and arginine residues, which results in changes to chromatin structure and gene expression patterns. The identification of histone demethylases proved that histone methylation could be reversed rather than being permanent. HDMs are divided into two major classes: The two major categories of histone demethylases are lysine-specific demethylases (LSDs) and Jumonji domain-containing demethylases (JmjC). LSD1 functions as an established HDM that targets H3K4me1 and H3K4me2 for demethylation to achieve gene repression, while JmjC domain-containing enzymes exert their function by removing methyl groups from diverse histone lysine residues to modulate gene activity dynamically [46,47].

Proper gene expression and cellular identity depend on maintaining the equilibrium between histone methylation and demethylation. A disruption of this molecular balance has been connected to multiple health conditions such as cancer and developmental anomalies while also affecting neurological function. The abnormal methylation of histones can cause excessive gene silencing or the inappropriate activation of genes, which contributes to tumor development and spread [48]. Mutations in EZH2, which functions as a histone methyltransferase responsible for H3K27 methylation, have been connected to aggressive cancer forms because they excessively silence tumor suppressor genes [49]. When demethylases like KDM6A that remove H3K27me3 marks lose their function, tumors may develop through unchecked activation of oncogenes [50].

The essential function of histone methylation in disease mechanisms has led to substantial interest in therapeutic approaches that target both methyltransferases and demethylases. Researchers are developing small-molecule blockers for methyltransferases, including EZH2 inhibitors, to treat cancers that result from excessive histone methylation [51]. Demethylase inhibitors display potential to reactivate genes that are turned off in specific diseases. Epigenetic therapies enable precise gene expression modification without changing genetic sequences, thereby creating innovative personalized medical approaches [52].

Research into histone methylation shows that it remains a complex and essential aspect of cellular regulation. The process of how chromatin-binding proteins read these modifications and their interactions with other histone changes provides key insights into unraveling the "histone code" [53]. The advancement of scientific research shows that focusing on histone methylation pathways could provide significant treatment options for numerous diseases, which demonstrates the crucial role of epigenetic regulation in human health and disease.

3.2.2.3 Phosphorylation

Histone phosphorylation serves as an essential post-translational modification responsible for chromatin remodeling and gene regulation, DNA repair mechanisms, and cell cycle progression. Phosphorylation acts as a molecular signal directing chromatin transitions in response to cellular cues such as stress and DNA damage, while acetylation and methylation primarily affect transcription. The attachment of phosphate groups to serine, threonine, or tyrosine residues on histone proteins results in structural and functional changes to chromatin. Histone phosphorylation enables cells to swiftly adjust to environmental shifts, which results in precise gene regulation and proper genome maintenance along with chromatin organization [54,55].

Phosphate group addition creates negative charges that disrupt histone-DNA interactions, resulting in either chromatin relaxation or compaction based on the situation [3]. The phosphorylation of serine 139 on histone H2AX to create γH2AX represents a highly researched post-translational modification that acts as an initial indicator of DNA double-strand breaks [56]. This modification enables DNA damage repair proteins to attach and triggers the DNA damage response (DDR), which leads to increased genomic stability. Histone H3 undergoes essential phosphorylation at serine 10 (H3S10ph), which plays a role in both the condensation of mitotic chromosomes and the activation of transcription. H3S10ph facilitates chromatin compaction throughout mitosis to ensure chromosomes segregate properly [57]. Non-mitotic cells experience this modification as a mechanism for activating transcription, especially when reacting to external stimulation like stress or growth factors [58].

Linker histone H1 experiences phosphorylation, which functions as a modification associated with transcriptional activation [59]. The process of H1 phosphorylation lowers its DNA binding affinity, which causes chromatin decondensation and allows transcription factors to access gene promoters more easily [60]. Research indicates that phosphorylation of H1 demonstrates a consistent relationship with active gene transcription across multiple cell types, which implies its function in adjusting chromatin structure to support gene expression [61]. Histone phosphorylation shows functional complexity because its impact on transcription can vary between activation and repression based on which histone and residue undergo modification.

Multiple kinases that react to distinct cellular signals facilitate histone phosphorylation. The ATM and ATR kinases phosphorylate H2AX which occurs following DNA damage and results in the recruitment of DNA repair proteins to damaged chromatin [62]. During mitosis, Aurora kinases together with cyclin-dependent kinases (CDKs) direct histone phosphorylation to maintain correct chromosome structure and advance the cell cycle [63]. CDK's phosphorylation of histone H1 triggers transcriptional activation which confirms its function in supporting gene expression [64,65]. Kinases operate through precise regulatory mechanisms to synchronize histone modifications with essential cellular activities including DNA replication and chromatin structure alteration.

Histone dephosphorylation happens through protein phosphatases, which eliminate phosphate groups from histones, thereby reversing phosphorylation and restoring chromatin structure. Protein phosphatase 1 (PP1) and protein phosphatase 2A (PP2A) serve as critical enzymes for histone dephosphorylation, which maintains chromatin flexibility to meet cellular requirements [66]. Histone kinases and phosphatases interact to balance chromatin condensation with relaxation, which enables cells to control transcription, DNA repair, and cell division efficiently [67].

Histone phosphorylation performs its well-known function in transcriptional regulation while playing a vital role in sustaining genome stability. When DNA gets damaged, cells use γH2AX phosphorylation to attract chromatin remodelers and repair factors that help fix the damaged DNA [68]. Research has linked histone H2B phosphorylation at serine 14 (H2BS14ph) to apoptosis because it encourages chromatin condensation and triggers programmed cell death [69,70]. Histone phosphorylation functions as a key mechanism for cellular stress responses because environmental shifts and inflammatory signals alongside oxidative stress activate specific phosphorylation events.

Numerous diseases such as cancer and neurodegenerative disorders manifest because of abnormal histone phosphorylation regulation, which also affects inflammatory conditions. The irregular activation of histone kinases, including Aurora kinases, results in improper chromosome condensation and unrestrained cell growth which promotes tumor development [63]. Mutations affecting ATM and ATR kinases that control H2AX phosphorylation cause defects in DNA repair systems while promoting genomic instability. Cancer cells display disrupted H1 phosphorylation patterns which implies that altered linker histone modifications could drive oncogenic transcriptional activity. Research into therapeutic methods specifically targets histone kinases and phosphatases because of the significant role histone phosphorylation plays in disease progression [71]. The inhibition of Aurora kinases and CDKs through kinase inhibitors appears to be an effective cancer treatment because they help reinstate normal chromatin regulation alongside proper cell cycle management [72].

The expanding research into histone phosphorylation reveals its growing significance in chromatin regulation and gene expression control. This modification acts as an essential part of the histone code by linking signaling pathways to chromatin functions, including transcription as well as DNA repair and mitosis. The identification of H1 phosphorylation as a transcriptional regulator highlights both the complex nature of this modification and its significant role in controlling gene expression. Research into histone phosphorylation regulation will enhance our knowledge of epigenetic mechanisms while expanding potential treatment options for diseases caused by abnormal chromatin alterations.

3.2.2.4 Ubiquitination

Histone ubiquitination serves as an essential post-translational modification that regulates chromatin through its ability to activate or repress transcription based on specific histones and their modification sites as well as the cellular environment. Ubiquitination differs from acetylation and methylation because it involves attaching ubiquitin, which is a regulatory protein of 76 amino acids, to lysine residues on histones through covalent bonds [73]. Histones H2A and H2B experience this modification, which functions as a dynamic signal controlling chromatin accessibility alongside transcription and DNA repair processes. Histone ubiquitination drives changes in gene expression and promotes chromatin remodeling through histone removal or exchange which modifies nucleosome structure and stability [74].

A three-step enzymatic cascade consisting of ubiquitin-activating enzymes (E1), ubiquitin-conjugating enzymes (E2), and ubiquitin ligases (E3) jointly facilitate the covalent attachment of ubiquitin to histone lysine residues during histone ubiquitination [75]. Histone ubiquitination produces different functional effects based on whether it involves the attachment of just one ubiquitin molecule (monoubiquitination) or several ubiquitin molecules (polyubiquitination). Histone monoubiquitination primarily functions in transcriptional regulation, whereas polyubiquitination acts as a signal for protein degradation or chromatin remodeling [76].

The ubiquitination of histone H2A usually results in transcriptional repression along with the creation of heterochromatin. The Polycomb Repressive Complex 1 (PRC1) mediates H2A ubiquitination at lysine 119 (H2AK119ub), which serves as a vital mechanism for gene repression and maintenance of repressive chromatin states through epigenetic means [77]. The modification functions as a crucial mechanism to regulate genes during development and to create stable transcriptional networks [78].

Transcriptional activation and cell differentiation require the removal of H2A ubiquitination by deubiquitinating enzymes, including USP16. H2A ubiquitination serves as a recruitment signal for repair factors that target DNA lesions, thereby maintaining genome integrity during DNA damage repair [79].

Histone H2B ubiquitination mainly functions to activate transcription. The monoubiquitination at lysine 120 of H2B (H2Bub1), which the E3 ligase complex RNF20-RNF40 performs, serves as a signature modification for genes that undergo active transcription. This modification increases chromatin accessibility, which enables the addition of other histone modifications like H3K4 and H3K79 methylation that trigger gene activation. H2B ubiquitination facilitates both nucleosome remodeling and transcription elongation, which enables efficient RNA polymerase II movement throughout the gene body [80]. H2B ubiquitination plays a dual role in both transcription regulation and DDR pathways, which help maintain chromatin organization during repair functions [79].

Deubiquitinating enzymes (DUBs) control the removal of ubiquitin from histones, which reverses ubiquitination and adjusts chromatin structure. Histone function restoration and transcriptional state reversibility depend on deubiquitination processes. The imbalance of ubiquitination and deubiquitination processes has been linked to multiple diseases such as cancer along with neurodegenerative disorders and inflammatory diseases. Tumors frequently show abnormal H2A ubiquitination, which leads to excessive gene silencing and subsequent suppression of tumor suppressor genes. H2B ubiquitination disruption leads to impaired transcriptional activation and results in defects within cellular differentiation and proliferation processes [81].

Because histone ubiquitination plays a critical role in both gene regulation and disease progression, targeting ubiquitin ligases and deubiquitinating enzymes represents a promising therapeutic approach. Scientists are investigating small-molecule inhibitors that alter ubiquitination enzyme functions to reestablish normal chromatin dynamics alongside gene expression patterns [82]. Progress in research will enable scientists to gain a more thorough understanding of histone ubiquitination interactions with chromatin modifications, which will reveal new information about genome function control and epigenetic inheritance mechanisms.

3.2.3 Non-coding RNAs

Non-coding RNAs (ncRNAs) serve important cellular functions that do not involve protein encoding by participating in gene regulation along with chromatin organization and RNA processing. Through their regulation of transcriptional and post-transcriptional networks, these RNAs impact epigenetic mechanisms, which then alter chromatin structure and cellular identity. MicroRNAs (miRNAs) and long non-coding RNAs (lncRNAs) function as key regulators of gene expression and control chromatin accessibility within the broad spectrum of ncRNA classes. The significance of these components in cellular functionality and disease development becomes clear through their interactions with transcriptional machinery and epigenetic modifiers like histone-modifying complexes and DNA methylation regulators [4,83].

MicroRNAs (miRNAs) consist of short single-stranded RNA molecules with about 22 nucleotides that control gene expression after transcription through binding to complementary target messenger RNA (mRNA) sequences [84]. The binding interaction between miRNA and its target mRNA usually results in translational repression or mRNA degradation based on how well they complement each other [85]. Primary miRNAs (pri-miRNAs) emerge from transcription and undergo processing in the nucleus to become precursor miRNAs (pre-miRNAs), which then move to the cytoplasm for additional maturation by the Dicer enzyme. After reaching maturity, miRNAs join the RNA-induced silencing complex (RISC), which enables them to bind target mRNAs and silence genes. miRNAs perform regulatory functions that affect multiple biological processes such as cell proliferation, differentiation, apoptosis, and immune responses [86]. These precise gene network controls enable tissues to sustain homeostasis and respond to environmental changes. Multiple diseases including cancer involve miRNA expression dysregulation which allows miRNAs to act either as oncogenes or tumor suppressors [87]. OncomiRs such as miR-21 drive tumor formation by inhibiting tumor suppressor genes which regulate apoptosis and cell cycle regulation [88]. The let-7 family of tumor-suppressive miRNAs inhibits uncontrolled cell growth by targeting oncogenes such as RAS and MYC. The abnormal expression of miRNAs occurs not only in cancer but is also present in cardiovascular diseases, neurodegenerative disorders, and autoimmune conditions, which leads to pathological gene expression regulation [89].

Long non-coding RNAs (lncRNAs) are transcripts that exceed 200 nucleotides in length and serve as major regulators of chromatin dynamics alongside transcriptional activation and repression mechanisms as well as RNA processing [90]. In contrast to miRNAs that function after transcription, lncRNAs usually influence gene expression at the transcriptional stage by associating with chromatin-modifying complexes and transcription factors along with other regulatory proteins [4]. Through their interactions with specific binding partners and their functional roles, lncRNAs can either boost or inhibit gene expression. The most extensively studied lncRNA is Xist, which performs a crucial function in the X-chromosome inactivation process of female mammals [91]. In the early stages of development, Xist RNA transcripts emerge from one X-chromosome's X-inactivation center to cover that X-chromosome and draw in Polycomb Repressive Complex 2 (PRC2), which places repressive histone marks including H3K27me3.

The coated X-chromosome becomes inactive through this process, which maintains equal gene dosage between male and female organisms [92]. LncRNAs like HOTAIR and MALAT1 regulate gene expression through the recruitment of histone-modifying enzymes to specific genomic positions, which changes chromatin structure and affects transcriptional activity. HOTAIR interacts with PRC2 to silence tumor suppressor genes which supports cancer metastasis and disease progression [93].

LncRNAs help create chromatin loops, which move distant parts of the genome together, allowing enhancer-promoter interactions that control gene regulation. LncRNAs function as molecular scaffolds, which recruit chromatin remodelers alongside transcription factors and RNA-binding proteins to build complex regulatory networks [94]. The involvement of these molecules in alternative splicing as well as RNA stabilization and nuclear-cytoplasmic transport highlights their functional diversity. LncRNAs control chromatin accessibility and transcription to play essential roles in stem cell differentiation as well as immune cell activation and metabolic adaptation [95].

ncRNAs perform regulatory functions that interact closely with traditional epigenetic changes like DNA methylation and histone modifications to create and sustain patterns of gene expression. miRNAs and lncRNAs work together with DNA methyltransferases (DNMTs), histone acetyltransferases (HATs), and histone deacetylases (HDACs) to regulate chromatin states. Some miRNAs directly interact with DNMTs to modify DNA methylation patterns which then affects gene expression levels. DNA methylation functions to control the expression of miRNA and lncRNA, which form feedback loops that enable precise transcriptional control [96,97]. The interaction between histone modifications and ncRNAs dictates chromatin accessibility levels. LncRNAs like Xist and HOTAIR recruit H3K27me3, which is a repressive histone mark produced by PRC2, to silence gene expression. Research has demonstrated that miRNAs control histone-modifying enzymes to shift chromatin states between active and repressive conditions. The complex interactions between ncRNAs and conventional epigenetic mechanisms enable precise gene expression regulation according to developmental signals and external environmental factors [98].

Various diseases such as cancer and cardiovascular disorders alongside autoimmune diseases and neurodegenerative conditions commonly display disruption in ncRNA-mediated epigenetic regulation. Cancer cells often show a dual pattern of DNA methylation changes where they experience widespread DNA hypomethylation which causes genomic instability, while simultaneously showing hypermethylation at tumor suppressor gene promoters, which silence those genes [99]. The abnormal expression patterns of miRNAs and lncRNAs contribute to cancer progression by strengthening both oncogenic and suppressive epigenetic states. ncRNAs control endothelial behavior while directing inflammation and fibrosis which leads to the advancement of heart disease [100]. Research shows that miRNAs, which are not properly regulated, play a role in heart failure, hypertension, and atherosclerosis by controlling the expression of important cardiovascular genes. LncRNAs, including ANRIL and SENCR, contribute to vascular homeostasis regulation and disease susceptibility [101,102].

Epigenetic modifications mediated by ncRNAs affect the progression of neurodegenerative disorders like Alzheimer's and Parkinson's disease [103]. miRNAs play a role in controlling neuronal differentiation as well as synaptic plasticity and neuroinflammation, yet the dysregulation of selected miRNAs correlates with amyloid-beta deposition and tau abnormalities alongside neuronal cell death [104]. The lncRNA BACE1-AS controls gene expression related to neurodegeneration thereby impacting disease progression [105]. Multiple sclerosis, along with rheumatoid arthritis and systemic lupus erythematosus, display changed ncRNA expression patterns which influence immune cell differentiation and inflammatory responses. Specific miRNAs control the production levels of cytokines, immune receptors, and transcription factors which drives immune system irregularities [106]. The lncRNA NEAT1 contributes to inflammasome activation and shows a connection to chronic inflammatory diseases [107].

Because ncRNAs play vital parts in both gene regulation and disease development, they stand out as promising targets for therapeutic intervention. Researchers have started developing miRNA-based therapies that use miRNA mimics [108] and antagomiRs [109] to restore normal gene expression in cancer and various diseases. Scientists investigate lncRNAs to develop new ways to control chromatin states and transcriptional activities. New RNA-based delivery methods like antisense oligonucleotides (ASOs) and small interfering RNAs (siRNAs) enable precise targeting of ncRNAs through enhanced specificity [110]. The role of ncRNAs is becoming clearer as research advances and their importance for precision medicine applications becomes more visible. The use of ncRNAs in diagnostic and therapeutic contexts enables the creation of innovative treatment methods for diseases resulting from epigenetic dysregulation. Combining ncRNA-based treatments with established epigenetic medications offers significant potential to create personalized therapies that address the fundamental causes of gene misregulation.

3.3 EPIGENETIC MODIFICATIONS IN DISEASE PATHOGENESIS

The disruption of epigenetic regulatory mechanisms results in abnormal gene expression profiles, which play a role in the development of diseases. Recent studies show epigenetic dysregulation plays a critical role in the development of cancer along with neurodegenerative disorders, autoimmune diseases, metabolic disorders, and cardiovascular diseases. Insights into disease mechanisms and potential therapeutic approaches become possible through the study of epigenetic changes (Tables 3.1–3.3).

Table 3.1 Epigenetic mechanisms and disease associations

Epigenetic mechanism	*Aberrations*	*Key genes affected*	*Diseases*	*Therapeutics*	*Biomarkers*
DNA methylation	Hypermethylation of tumor suppressor genes, global hypomethylation	TP53, BRCA1, MGMT, CDKN2A	Cancer, Alzheimer's, autoimmune diseases	DNMT inhibitors (5-azacytidine, decitabine)	MGMT methylation (glioblastoma), SEPT9 (colorectal cancer)
Histone acetylation	Reduced acetylation (gene silencing), hyperacetylation (aberrant activation)	CREB, P300, HDAC1, SIRT1	Neurodegeneration, cancer, inflammation	HDAC inhibitors (vorinostat, romidepsin)	H3K9ac, H4K16ac (neurodegeneration), CBP/p300 mutations
Histone methylation	H3K27/H3K9 methylation (repressive), H3K4 methylation loss (activating)	EZH2, SUV39H1, KMT2A	Cancer, psychiatric disorders, developmental diseases	EZH2 inhibitors (tazemetostat), LSD1 inhibitors	H3K27me3 (cancer prognosis), EZH2 overexpression
Histone phosphorylation	Increased γH2AX in DNA damage, altered mitotic phosphorylation	H2AX, ATM, ATR	Neurodegeneration, cancer, DNA repair defects	ATM/ATR inhibitors, kinase modulators	γH2AX foci (DNA damage marker), ATM/ATR phosphorylation
Histone ubiquitination	Altered H2A/H2B ubiquitination affecting transcription and histone turnover	RNF20, USP22, BRCA1	Cancer, neurodevelopmental disorders	Deubiquitinase inhibitors, CRISPR-based ubiquitination control	H2Bub1 (cancer regulation), USP22 overexpression
Non-coding RNA regulation	Dysregulated miRNAs/lncRNAs, impaired RNA-mediated gene regulation	miR-21, HOTAIR, MALAT1, XIST	Cancer, cardiovascular diseases, metabolic disorders	miRNA mimics/inhibitors, lncRNA therapies	miR-21 (cancer biomarker), HOTAIR (metastasis predictor)

Table 3.2 Enhanced epigenetic biomarkers and diagnostic applications

Biomarker	*Biomarker type*	*Detection method*	*Associated disease*	*Clinical significance*	*FDA-approved status*
MGMT methylation	DNA methylation	Bisulfite sequencing, qPCR	Glioblastoma, colorectal cancer	Predicts chemotherapy response in glioblastoma	Yes (Glioblastoma)
SEPT9 methylation	DNA methylation	Methylation-specific PCR	Colorectal cancer	Non-invasive early colorectal cancer screening	Yes (Colorectal cancer)
H3K27me3 levels	Histone modification	ChIP-seq, immunohistochemistry	Cancer prognosis, tumor suppression	Prognostic indicator in aggressive tumors	No (under research)
γH2AX foci	Histone phosphorylation	Immunofluorescence, Western blot	DNA damage response, cancer	DNA damage marker, useful in radiation therapy response	No (under research)
miR-21 expression	MicroRNA	RT-qPCR, RNA-seq	Various cancers, cardiovascular diseases	Cancer biomarker for diagnosis and progression	No (research and clinical trials)
HOTAIR lncRNA expression	Long non-coding RNA	RNA hybridization, RNA-seq	Cancer metastasis predictor	Predicts tumor metastasis potential in multiple cancers	No (under research)

Table 3.3 Epigenetic therapies and their applications

Therapeutic strategy	*Mechanism of action*	*Example drugs/methods*	*Clinical applications*	*FDA-approved status*
DNMT inhibitors	Inhibit DNA methyltransferases to reverse abnormal gene silencing	5-Azacytidine, decitabine, guadecitabine	Leukemia, myelodysplastic syndromes, colorectal and lung cancer	Yes (5-azacytidine, decitabine)
HDAC inhibitors	Block histone deacetylases to promote chromatin relaxation and transcription activation	Vorinostat, romidepsin, panobinostat	Lymphomas, neurodegenerative diseases, inflammatory disorders	Yes (vorinostat, romidepsin, panobinostat)
Histone methyltransferase inhibitors	Target histone methyltransferases to modulate transcriptional activation/repression	Tazemetostat (EZH2 inhibitor), LSD1 inhibitors, DOT1L inhibitors	Lymphoma, prostate cancer, glioblastoma	Yes (Tazemetostat for EZH2-mutant follicular lymphoma)
CRISPR-based epigenetic editing	Use CRISPR-dCas9 to add/remove epigenetic marks without altering DNA sequence	dCas9-TET1 (demethylation), dCas9-KRAB (gene repression)	Targeted therapy for cancer, neurodevelopmental, and metabolic disorders	No (experimental, preclinical and clinical trials)
miRNA-based therapies	Restore or inhibit specific miRNAs to regulate gene expression post-transcriptionally	miRNA mimics, miRNA inhibitors, lncRNA modulators	Cancer, cardiovascular diseases, metabolic disorders	No (clinical trials ongoing for miRNA-based therapies)
Dietary epigenetic modulation	Utilize dietary compounds to influence DNA methylation and histone modifications	Resveratrol, curcumin, folate, green tea polyphenols	Cancer prevention, neuroprotection, anti-inflammatory effects	No (Dietary compounds studied for epigenetic influence)

3.3.1 Epigenetics in cancer: tumor suppressor silencing and oncogene activation

Cancer represents the disease with the most extensive research on epigenetic changes. Genetic mutations modify DNA sequences directly, while epigenetic modifications contribute to cancer development by turning off tumor suppressor genes and turning on oncogenes [111]. Cancer features DNA methylation changes as a key characteristic because hypermethylation of tumor suppressor gene promoters causes transcriptional repression and loss of functional activity [112]. The tumor suppressor genes, CDKN2A, that produce p16 and p14ARF proteins undergo frequent silencing across multiple cancer types because of promoter hypermethylation [113]. BRCA1 functions as an important DNA repair gene but becomes epigenetically silenced in breast and ovarian cancers, which leads to genomic instability and malignant transformation [114].

Cancer cells often show widespread hypomethylation which activates oncogenes and results in genomic instability. Proto-oncogenes like MYC and RAS experience increased expression when hypomethylated which triggers unregulated cellular multiplication [115,116]. The loss of DNA methylation throughout repetitive sequences causes chromosomal instability which leads to genetic changes that support tumor progression [117]. Histone modifications serve as crucial components in the development of cancer diseases. Histone acetylation and methylation changes can affect chromatin structure and gene activity levels. Cancer cells showing excessive histone deacetylase (HDAC) activity close chromatin structure and suppress tumor suppressor genes [118]. The histone methyltransferase EZH2 induces H3K27 trimethylation which results in transcriptional silencing of genes controlling cell differentiation and apoptosis [119].

New anticancer treatments work by attacking these changes in epigenetic functions. Cancer treatment uses DNMT inhibitors like 5-azacytidine and decitabine to eliminate hypermethylation and restore the function of silenced tumor suppressor genes. The drugs vorinostat and romidepsin function as HDAC inhibitors which enable chromatin to return to normal accessibility while restoring gene expression levels. CRISPR-based epigenetic editing advancements create innovative methods to accurately adjust epigenetic marks, which promise fresh therapeutic approaches for cancer [120].

3.3.2 Epigenetic changes in neurodegenerative diseases affecting Alzheimer's and Parkinson's patients

Patients with neurodegenerative diseases like Alzheimer's and Parkinson's disease display progressive brain cell deterioration together with declining mental abilities. The regulation of neuronal gene expression and synaptic plasticity depends heavily on epigenetic modifications which, when disrupted, become linked to neurodegeneration [121]. As people age, their DNA methylation patterns alter, and abnormal methylation processes play a role in developing neurodegenerative disease pathologies. Alzheimer's disease triggers hypermethylation of key genes, including BDNF, which results in diminished gene expression and compromised neuronal survival [122]. Global hypomethylation has emerged as a characteristic feature in AD, which triggers enhanced gene expression related to neuroinflammation and amyloid-beta buildup that define this disease [123].

Histone modifications also influence neurodegenerative disease progression. The transcription of genes related to dopaminergic neuron survival in Parkinson's disease (PD) is influenced by alterations in histone acetylation levels. When histone acetylation decreases at neuroprotective genes, gene expression is repressed, which leads to impaired neuronal function [124]. Scientists investigate HDAC inhibitors as neurodegenerative disease treatments because these compounds increase histone acetylation and deliver neuroprotective benefits [125]. Scientific research shows that blocking HDAC activity enhances brain function and preserves neurons through the activation of genes crucial for synaptic plasticity and neuron protection [126].

MicroRNAs (miRNAs), among other non-coding RNAs, influence neurodegenerative diseases through their regulatory functions on genes that control neuronal activity and inflammatory responses [127]. In AD [128] and PD [129], the aberrant miRNA expression modifies pathways that control neuroinflammation and protein aggregation alongside mitochondrial dysfunction. Research efforts are underway to develop therapeutic strategies that target miRNAs to reestablish regular gene control mechanisms while decelerating disease advancement.

Epigenetics plays a vital role in modulating immune system activity, which directly affects autoimmune diseases and chronic inflammatory processes. Epigenetic changes serve as key regulators of immune function by directing immune cell differentiation as well as activation and tolerance processes [130]. Autoimmune diseases occur when epigenetic mechanism dysregulation causes the immune system to attack the body's own tissues. Autoimmune conditions including systemic lupus erythematosus (SLE), rheumatoid arthritis (RA), and multiple sclerosis (MS) exhibit common alterations to DNA methylation patterns. The reduction of methylation levels in immune-related genes of T cells triggers atypical immune activation in SLE patients, which results in heightened inflammation and tissue deterioration [131,132].

Histone modifications regulate inflammatory gene expression patterns in autoimmune diseases. Variations in histone acetylation and methylation patterns regulate the transcription activity of immune signaling molecules and cytokines [133]. The expression of pro-inflammatory cytokine genes goes up because of increased H3K4 methylation and histone acetylation at their promoters, leading to worsened inflammation in RA and MS [134].

Altered miRNA expression disrupts important immune signaling pathways which leads to continuous inflammation as well as autoimmune disorders. The potential use of DNMT and HDAC inhibitors to influence immune responses and achieve immune tolerance represents a research focus in therapeutic epigenetic regulation [135].

The disruption of epigenetic pathways plays a critical role in the development of metabolic disorders such as diabetes and obesity. Epigenetic modifications perform a pivotal role in metabolic processes, which govern glucose metabolism alongside insulin signaling and the development of fat cells. Metabolic disorders such as type 2 diabetes (T2D) and obesity occur due to disruptions in epigenetic mechanisms [136]. Modifications to DNA methylation patterns within essential metabolic genes directly influence insulin secretion and glucose homeostasis. Hypermethylation of insulin production-related genes in pancreatic beta cells results in compromised insulin secretion and glucose intolerance [137]. Research of DNA methylation patterns across the genome revealed specific differences between obese individuals and those with type 2 diabetes, which demonstrates epigenetic factors' effect on disease risk [138].

Histone modifications also regulate metabolic pathways. Changes in histone acetylation and methylation at energy metabolism genes lead to insulin resistance and lipid accumulation. Genes that regulate fat cell formation experience enhanced H3K27 acetylation which leads to increased fat storage and obesity [139]. Metabolic processes, including insulin signaling and lipid metabolism regulation, are influenced by non-coding RNAs such as miRNAs, which target specific genes. Obesity and diabetes lead to abnormal miRNA expression which disrupts pathways responsible for glucose absorption and adipocyte development along with inflammatory responses [140,141]. Researchers are examining HDAC inhibitors and miRNA mimics as potential epigenetic therapies to control metabolic gene expression and enhance insulin sensitivity [142].

3.3.3 Epigenetics in cardiovascular diseases

The development of cardiovascular diseases involves epigenetic mechanisms, which control both endothelial function and cardiac remodeling as well as vascular inflammation. Research shows that alterations in DNA methylation patterns correlate with the development of atherosclerosis, which leads to heart disease. Pro-inflammatory gene hypomethylation in endothelial cells triggers ongoing vascular inflammation while promoting plaque development. The hypermethylation of genes responsible for vascular healing and endothelial cell functions disrupts cardiovascular system stability [143,144].

Gene expression related to cardiac hypertrophy, heart failure, and vascular disease undergoes regulation through histone modifications [145]. The process of increased histone acetylation at genes that stimulate fibrosis and inflammation results in negative cardiac remodeling patterns [146]. Medical researchers examine HDAC inhibitors as possible heart disease treatments because these compounds modify chromatin structure while restoring normal gene expression patterns [147]. The cardiovascular gene network is regulated by non-coding RNAs, which include miRNAs that significantly affect cardiac function and vascular integrity. Ongoing research aims to develop epigenetic drugs as a promising strategy for cardiovascular disease prevention and treatment by restoring normal vascular and cardiac function.

3.4 THERAPEUTIC APPLICATIONS OF EPIGENETIC MODIFIERS

The dysregulation of epigenetic modifications, which control gene expression, leads to multiple diseases such as cancer and neurodegenerative disorders as well as metabolic syndromes (Figure 3.2). Therapeutic interventions now have new prospects because researchers can adjust epigenetic states. Researchers are developing multiple strategies such as pharmacological inhibitors and genome editing tools to address epigenetic abnormalities and bring gene function back to normal. The development of these therapeutic approaches faces hurdles concerning their effectiveness and precision along with ethical implications. Current investigations persist in advancing these therapeutic approaches through the adoption of cutting-edge technologies like single-cell epigenomics and artificial intelligence-based drug discovery along with sophisticated gene delivery systems which aim to improve treatment precision and effectiveness [148].

3.4.1 Epigenetic drugs: DNMT inhibitors and HDAC inhibitors

Disease treatments, which target epigenetic changes, have demonstrated potential for addressing conditions that involve abnormal gene expression patterns. Tumor suppressor genes that become silenced due to abnormal DNA methylation patterns in leukemia and myelodysplastic syndromes can be reactivated with the help of DNMT inhibitors like 5-azacytidine and decitabine. The drugs integrate into DNA to interfere with methylation processes and activate genes that were previously silenced. Next-generation DNMT inhibitors like guadecitabine show improved specificity and reduced toxicity while demonstrating longer half-life and increased effectiveness in clinical trials [5]. Scientists continue to develop new DNMT inhibitors, which selectively target precise methylation sites while preserving overall DNA methylation patterns, thereby reducing side effects and enhancing therapeutic results.

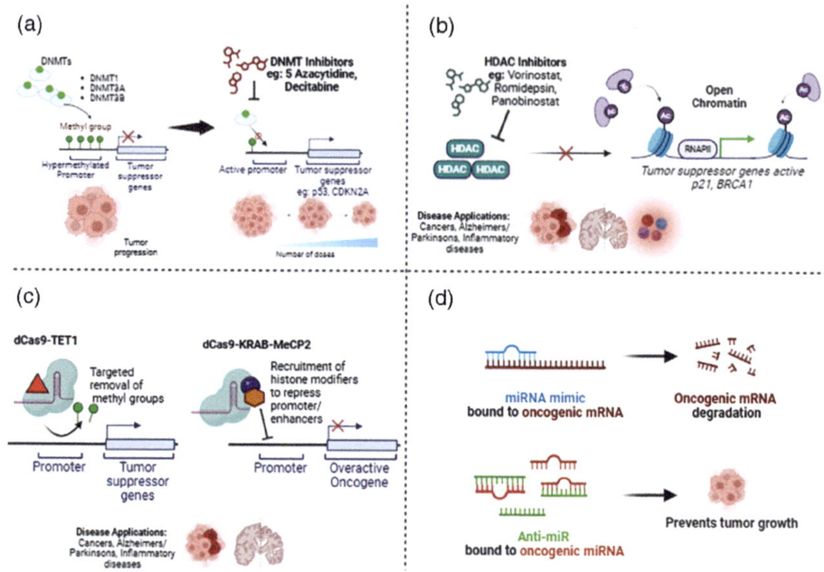

Figure 3.2 Epigenetic therapies and their mechanisms of action. (This figure illustrates four major epigenetic therapy strategies used to regulate gene expression by targeting DNA methylation, histone modifications, and non-coding RNA interactions. (a) DNMT Inhibitors (Demethylation Therapy): DNA methyltransferases (DNMTs) catalyze the addition of methyl groups to gene promoters, leading to tumor suppressor gene silencing and tumor progression. DNMT inhibitors, such as 5-azacytidine and decitabine, block methylation, allowing reactivation of tumor suppressor genes (e.g., p53, CDKN2A). (b) HDAC Inhibitors (Histone Acetylation Therapy): Histone deacetylases (HDACs) remove acetyl groups from histones, leading to chromatin compaction and gene repression. HDAC inhibitors, including vorinostat, romidepsin, and panobinostat, prevent deacetylation, maintaining an open chromatin state and reactivating genes like p21 and BRCA1. (c) CRISPR-Based Epigenetic Editing: dCas9-TET1 removes DNA methylation to restore tumor suppressor gene activity, while dCas9-KRAB-MeCP2 recruits histone modifiers to repress oncogene expression. This approach provides a targeted and reversible method for epigenetic regulation. (d) miRNA-Based Therapies: miRNA mimics bind oncogenic mRNAs, triggering degradation and reducing tumor growth. Anti-miRs block oncogenic miRNAs, preventing them from suppressing tumor suppressor genes. These strategies offer precise post-transcriptional gene regulation.)

Vorinostat, romidepsin, and panobinostat function as HDAC inhibitors that increase histone acetylation which ultimately causes chromatin to relax and elevates transcription levels of genes that manage cell cycle regulation and apoptosis [149,150]. Doctors have approved HDAC inhibitors to treat some types of lymphomas while research continues to explore their neuroprotective capabilities through enhanced neuronal survival and plasticity for neurodegenerative diseases [151]. Research into HDAC-targeted therapies has resulted in the creation of isoform-selective HDAC inhibitors, which help reduce off-target effects [152]. Researchers are examining combination treatments that use DNMT inhibitors together with HDAC inhibitors to achieve enhanced synergistic effects and address cancer resistance. Recent studies indicate that cancer immunotherapy effectiveness increases when HDAC inhibitors are combined with immune checkpoint inhibitors because this combination alters the tumor microenvironment and strengthens anti-tumor immune reactions [153].

3.4.2 CRISPR-based epigenetic editing for gene regulation

CRISPR-based epigenetic editing transformed researchers' ability to finely regulate gene expression without making permanent DNA sequence changes. Epigenetic CRISPR editing differentiates itself from traditional methods by deploying modified Cas proteins like dCas9 combined with histone-modifying enzymes to mark specific epigenetic targets [154,155]. By altering DNA methylation or histone acetylation, these systems control gene activation and repression. The application of CRISPR technology for epigenetic editing shows tremendous promise as a treatment for genetic disorders as well as cancer and neurological diseases through the restoration of correct gene regulation [156]. The latest scientific discoveries have shown that dCas9-TET1 fusion proteins can reverse hypermethylation of promoters and activate disease-linked silenced genes. Preclinical cancer models have shown that CRISPR-dCas9-KRAB systems effectively suppress oncogene expression, according to recent research [157,158].

The integration of epigenome-wide association studies (EWAS) with CRISPR epigenetic editing provides a potent method for identifying disease-specific epigenetic modifications and creating personalized treatment strategies [159]. To improve targeted gene regulation and reduce immune responses, scientists combine epigenetic editing techniques with nanoparticle delivery systems [160]. Before CRISPR epigenetic editing can reach full clinical application, researchers need to solve issues with delivery systems and operational efficiency while also preventing unanticipated effects. Research on lipid nanoparticles and viral vector systems aims to enhance stability and specificity outcomes for CRISPR-based epigenetic treatment methods [161].

3.4.3 Personalized epigenetic therapy: precision medicine approaches

By incorporating epigenetic profiling into personalized medicine, we seek to customize medical treatments according to each person's distinct epigenetic profile [162]. The development of high-throughput sequencing technologies has made it possible to identify epigenetic biomarkers, which serve as predictors for disease susceptibility as well as treatment response and prognosis [163]. Personalized epigenetic therapies utilize drugs or interventions that target specific epigenetic changes in patients to achieve therapeutic effects. Cancer patients with gene silencing from hypermethylation might find DNMT inhibitors beneficial, whereas HDAC inhibitors could help those with neurodegenerative conditions associated with histone deacetylation [164].

The latest research has revealed epigenetic markers that enable healthcare providers to group patients by risk levels, which leads to better treatment choices and patient outcomes. Medical professionals use liquid biopsy methods to track epigenetic markers in circulation which enables real-time evaluation of disease progression and treatment effectiveness. Treatment effectiveness improves when epigenetic drugs are used in combination with immunotherapy and chemotherapy because they enhance immune system responses while making cancer cells more susceptible to conventional treatments [165]. Machine learning algorithms applied to epigenetic data analysis enable the development of refined predictive models which improve patient selection for specific epigenetic treatments [166]. The development of personalized epigenetic medicine promises to revolutionize disease treatment through individualized precision medicine approaches.

3.4.4 Natural compounds and diet-based epigenetic modulation

Epigenetic modifications are heavily affected by lifestyle choices and dietary habits, which demonstrate natural compounds as effective tools for epigenetic treatment. Polyphenols together with flavonoids and vitamins as bioactive dietary components modify epigenetic marks to improve health outcomes [167–169]. As a turmeric compound, curcumin modifies DNA methylation along with histone acetylation, which underpins its anti-inflammatory and anti-tumor activities [170]. Folate, together with other B vitamins, drives DNA methylation processes and is vital for blocking developmental disorders as well as specific cancer types [171].

Recent studies reveal that the gut microbiome participates in epigenetic regulation through microbial metabolites like short-chain fatty acids, which affect histone modifications and DNA methylation patterns [172]. Targeting microbiomes with specific dietary interventions could enable more complex epigenetic regulation possibilities. Scientists are currently conducting clinical research on dietary epigenetic interventions to create effective dietary guidelines for preventing and managing diseases. Nutriepigenomics is currently examining how dietary epigenetic modifications can create effective preventive and therapeutic diets to treat chronic diseases [173,174].

Epigenetic therapies show promise but face challenges in precision, durability, and ethics. Off-target effects can unintentionally alter gene activity, causing side effects. Their reversibility demands ongoing monitoring, and targeted delivery to specific tissues remains a major hurdle. Careful ethical oversight is essential for epigenetic modifications. These technologies alter gene expression without changing DNA sequences, raising concerns about misuse in human enhancement and non-therapeutic applications. Potential inheritance of epigenetic changes adds to fears of unforeseen consequences. Strong regulatory frameworks and ethical guidelines are crucial for responsible development and use [175,176].

Current research aims to boost the precision, safety, and effectiveness of epigenetic therapies as advancements continue in the epigenetic field. Treatment of epigenetic diseases benefits from a multidimensional approach consisting of drug treatments combined with CRISPR technology and dietary changes. Epigenetic therapies promise to transform medical treatments while addressing existing challenges and ethical concerns because they offer advanced personalized solutions for numerous diseases.

3.5 EPIGENETICS IN DIAGNOSTICS AND BIOMARKER DEVELOPMENT

Epigenetic modifications serve as powerful biomarkers for disease detection, prognosis, and treatment response. Unlike permanent genetic mutations, epigenetic changes are dynamic and reversible, making them ideal for monitoring disease progression and therapeutic success. Advances in high-throughput sequencing, bioinformatics, and single-cell epigenomics have identified disease-associated epigenetic signatures, enabling non-invasive diagnostics. Biomarkers like DNA methylation patterns, histone modifications, and non-coding RNA profiles provide insights into disease mechanisms and early detection. Liquid biopsy-based epigenetic screening enhances real-time monitoring with minimal invasiveness, paving the way for personalized risk assessment, treatment optimization, and recurrence prediction (Table 3.2) [177,178].

3.5.1 DNA methylation-based biomarkers for early disease detection

DNA methylation stands as the most extensively studied epigenetic modification because of its significant role in disease development. Cancer development features abnormal DNA methylation patterns, which include tumor suppressor gene promoter hypermethylation and widespread hypomethylation that causes genomic instability. Methylation changes emerge early during tumor development, which positions them as perfect targets for cancer screening. Researchers have created effective DNA methylation-based biomarkers to identify lung cancer as well as colorectal, breast, and prostate cancer. The FDA-approved methylation-based test Epi proColon screens for SEPT9 gene methylation in circulating tumor DNA to detect colorectal cancer non-invasively. MGMT promoter methylation status acts as a predictive biomarker, which indicates glioblastoma patient responses to alkylating chemotherapy [179,180].

DNA methylation biomarkers demonstrate potential applications in treating neurodegenerative diseases as well as cardiovascular conditions and autoimmune illnesses. The association between hypermethylation of APP and MAPT genes and Alzheimer's disease creates potential early diagnostic targets [181]. Studies show that changes to methylation patterns in genes responsible for inflammation and vascular health, including FOXP3 and NOS3, correlate with higher risk for developing atherosclerosis and hypertension in cardiovascular disease patients [182]. DNA methylation profiling has revealed unique patterns linked to immune system dysfunction in autoimmune diseases such as lupus and RA, which help in both disease classification and prognosis [183]. The evolution of technology enhances the precision and dependability of methylation-based diagnostic tools through DNA methylation arrays, targeted bisulfite sequencing, and machine learning algorithms which boosts their clinical usefulness [184].

3.5.2 Histone modification signatures as diagnostic tools

Acetylation, methylation, phosphorylation, and ubiquitination represent key histone modifications, which regulate chromatin dynamics and control gene expression. The connection between histone modification patterns and various diseases has positioned them as potential targets for biomarker discovery. The analysis of histone modification signatures reveals transcriptional dysregulation patterns and chromatin remodeling events that track disease development. Increased levels of H3K27me3 and H3K9me2 have been correlated with transcriptional repression in cancer, while abnormal H3K4me3 enrichment drives tumor progression and metastasis [185,186].

Histone acetylation levels have shown alterations in both Alzheimer's disease and PD among neurodegenerative conditions. The correlation between decreased H3K9ac and H4K16ac levels and cognitive decline suggests the potential of histone acetylation profiles to act as early predictors for neurodegenerative disease risk [187]. The methylation of histones at synaptic plasticity and neurotransmission genes presents possible disease classification biomarkers and treatment response predictors in schizophrenia and bipolar disorder [188,189].

Technologies including chromatin immunoprecipitation followed by sequencing (ChIP-seq), mass spectrometry-based histone modification analysis, and high-throughput proteomics enable researchers to identify histone signatures in clinical samples [190,191]. By using these methods, researchers can identify disease-related histone marks precisely, which leads to advances in personalized diagnostic techniques. The integration of histone modification profiling with DNA methylation and non-coding RNA expression studies leads to improved diagnostic precision while offering a broader understanding of disease origins.

3.5.3 Non-coding RNA profiling in disease monitoring

A variety of non-coding RNAs such as microRNAs, long non-coding RNAs, and circular RNAs serve as crucial elements in controlling gene expression and epigenetic changes. Because non-coding RNAs demonstrate stable expression in bodily fluids such as blood, saliva, and urine, they provide strong potential as non-invasive biomarkers. The extensive research on miRNA expression patterns across cancer, cardiovascular diseases, and neurodegenerative disorders highlights their potential applications as diagnostic and prognostic tools.

The miRNAs miR-21, miR-155, and miR-34a show consistent upregulation across different cancer types and have been suggested as universal cancer biomarkers. The ability of these miRNAs to control oncogenic pathways and apoptotic mechanisms renders them useful markers for monitoring tumor advancement [192]. The connection between myocardial infarction and heart failure with altered miR-1, miR-133, and miR-208 expression in cardiovascular diseases enables early diagnosis and risk evaluation [193]. Alzheimer's pathology includes characteristic ncRNA signatures where miR-29 and miR-124 play a role through their regulation of tau protein aggregation and amyloid-beta production [194].

The discovery of lncRNAs and circRNAs adds additional dimensions to epigenetic biomarker research. The lncRNAs HOTAIR, MALAT1, and GAS5 function as essential regulatory elements for cancer metastasis and mechanisms of drug resistance and immune system evasion [195,196]. The correlation between their expression patterns and disease severity positions these molecules as strong candidates for clinical use. CircRNAs maintain high stability through their covalently closed-loop structure, which enables them to regulate miRNA activity and transcriptional networks. Research suggests CircHIPK3 and circCDR1as are innovative biomarkers for neurodegenerative and cardiovascular disease diagnosis and treatment monitoring [197].

The integration of next-generation sequencing (NGS) techniques with single-cell RNA sequencing (scRNA-seq) and bioinformatics-driven analysis of ncRNA expression profiles now facilitates precise classification of diseases and effective patient stratification strategies. Liquid biopsy-based diagnostics integrated with ncRNA profiling improves early disease detection and delivers ongoing updates about disease progression and treatment effectiveness along with insights into resistance mechanisms.

3.5.4 Liquid biopsy and epigenetic screening techniques

The emergence of liquid biopsy transformed disease diagnostics through its minimally invasive detection capabilities for epigenetic changes in ctDNA, cfDNA, exosomes, and circulating RNA [198]. Researchers can track disease progression and treatment effectiveness in real time while detecting minimal residual disease using this method. Early-stage cancer detection and therapeutic outcome prediction become feasible through the study of DNA methylation changes and histone modifications in ctDNA [199].

Liquid biopsy tests centered on methylation analysis have shown strong sensitivity and specificity for cancer detection. The Galleri test functions as a multicancer early detection assay, which uses DNA methylation patterns to identify more than 50 different cancer types from just one blood sample. The LUNAR-2 test identifies colorectal cancer through its analysis of methylation changes found in circulating DNA [200,201].

Studies are extending to liquid biopsy-based epigenetic screening to evaluate neurodegenerative diseases and prenatal diagnostics beyond the field of oncology. The examination of cfDNA methylation patterns demonstrates potential for diagnosing fetal abnormalities along with pregnancy complications [202]. Researchers are exploring blood-based epigenetic markers such as DNA methylation patterns and miRNA signatures in Alzheimer's disease to develop non-invasive diagnostic tools for early detection of cognitive decline [203].

Scientists are actively working to improve cfDNA enrichment methods while optimizing sequencing depth and using artificial intelligence for better biomarker identification [204]. Epigenetic diagnostics will evolve through multiomics approaches that merge DNA methylation analysis with histone modification studies and ncRNA profiling to establish thorough disease monitoring and early intervention methods.

Epigenetic biomarkers revolutionize disease diagnostics with their exceptional sensitivity and specificity for early detection and monitoring of prognosis and therapy. Multiomics integration approaches combined with liquid biopsy technologies and AI-supported epigenetic analysis will improve both precision and accessibility for epigenetic diagnostic methods [205]. Ongoing research developments in epigenetic screening technologies and biomarker validation will transform personalized medicine by enabling earlier and more precise disease detection which will improve patient outcomes.

3.6 FUTURE PERSPECTIVES IN EPIGENETICS AND HEALTHCARE

Epigenetics represents a transformative domain in biomedical research, offering broad applications in healthcare management and personalized medicine. By manipulating epigenetic states, researchers can develop targeted therapies and disease prevention strategies, particularly in conjunction with advancements in regenerative medicine. The rapid expansion of epigenetic research is driven by technological innovations that enable the use of sophisticated tools such as single-cell epigenomics, CRISPR-based epigenetic editing, and artificial intelligence for biomarker identification [206]. These advancements have revolutionized our understanding of gene regulation and paved the way for novel therapeutic approaches that extend beyond traditional genetic interventions. As epigenetic interventions become increasingly precise and accessible, their responsible application necessitates rigorous ethical and social evaluations. The future of epigenetics in healthcare will be shaped by the integration of translational research, regulatory frameworks, and the clinical implementation of epigenetic discoveries [207].

3.6.1 Emerging technologies in epigenetic research

The development of innovative technologies, which enable precise epigenome mapping and analysis, has driven significant progress in epigenetic research. Single-cell epigenomic techniques enable researchers to study cell-to-cell differences in epigenetic marks and reveal the intricate regulatory networks that determine cellular fate and disease progression. Single-cell ATAC-seq and single-cell bisulfite sequencing enable researchers to achieve highly resolved chromatin accessibility maps and DNA methylation patterns for cellular diversity research in cancer and neurodegenerative and developmental biological studies [208,209].

The combined use of multiomics methods in epigenomics with transcriptomics, proteomics, and metabolomics enables researchers to better understand complex gene regulatory networks. AI systems operate in conjunction with machine learning algorithms to examine large epigenetic datasets with the aim of identifying biomarkers and enhancing therapeutic strategies while evaluating disease risks. The application of AI in epigenetic analysis holds significant potential for early disease detection, as it enables scientists to spot small epigenetic alterations before clinical symptoms develop [210,211]. Scientists now systematically explore epigenetic changes through high-throughput CRISPR-based epigenetic screening which accelerates drug discovery and therapeutic development.

Recent breakthroughs in epigenetic research stem from the application of nanotechnology in targeted epigenetic modulation. Scientists create nanoparticle delivery systems that transport epigenetic drugs to diseased cells while minimizing side effects and improving therapeutic outcomes [212]. Nanomedicine together with epigenetic science establishes advanced treatment possibilities that demonstrate outstanding potential for both cancer treatment and management of neurological diseases.

3.6.2 Epigenetic reprogramming holds significant potential for advancing regenerative medicine treatment techniques

Within regenerative medicine, epigenetic reprogramming stands out as a strong approach for both restoring cell function and eliminating disease-related epigenetic alterations. Through epigenetic state modulation, scientists convert somatic cells into induced pluripotent stem cells (iPSCs) which become capable of developing into various cell types for therapeutic applications. The alteration and erasure of epigenetic marks open treatment possibilities for tissue regeneration and degenerative diseases as well as enhance wound healing mechanisms [213,214].

Studies have investigated epigenetic reprogramming techniques to restore neuron functionality while stopping neurodegenerative progress in neurological conditions [215]. Research reveals that HDAC inhibitors enhance neuroplasticity while they help maintain neuron survival in models of Alzheimer's and PD [216]. Scientists have utilized DNA methylation editing tools such as CRISPR-dCas9-TET1 to revive silenced genes responsible for neuroprotective functions and synaptic operations. Epigenetic treatments have shown positive outcomes which could help treat cognitive decline that occurs with aging and neurodegenerative conditions [217].

Cardiovascular researchers use epigenetic reprogramming methods to repair heart tissue following myocardial infarction [145]. Studies have found that modifications in histone acetylation and DNA methylation patterns improve cardiomyocyte production and help restore heart function following injury. Studies demonstrate that epigenetic reprogramming of mesenchymal stem cells enhances cartilage and bone regeneration and offers new treatment possibilities for osteoarthritis and fracture repair [146].

Partial cellular reprogramming to restore youthful epigenetic states shows cutting-edge progress in regenerative medical science. Animal studies show that temporary activation of Yamanaka factors (Oct4, Sox2, Klf4, and c-Myc) rejuvenates aging tissues without reaching full pluripotency, which suggests epigenetic reprogramming could extend health span by slowing down the aging process [218]. Clinical applications derived from these findings require further research to validate their safety and effectiveness while ensuring controlled reprogramming to prevent tumor growth.

3.7 ETHICAL CONSIDERATIONS AND FINAL PERSPECTIVE

The swift progression of epigenetic therapies demands thorough assessment of their ethical, legal, and societal consequences. New epigenetic therapies present potential disease treatments but currently face challenges such as unknown side effects, inheritable epigenetic alterations, data privacy matters, and fair treatment distribution [177].

The dynamic nature of epigenetic modifications, which depend on context, creates doubt about unexpected activation or silencing of genes. Due to the uncertain long-term effects of these treatments, there exists an urgent requirement for thorough safety evaluations to prevent negative consequences. The possibility of epigenetic alterations being passed to subsequent generations creates intricate ethical issues, notably in relation to germline modifications. Regulatory systems need to establish precise rules to make certain epigenetic modifications serve therapeutic purposes instead of genetic enhancement [219]. Incorporating epigenetic profiling into personalized medicine for disease prediction prompts major privacy challenges. Individuals could face discrimination because of their epigenetic data if employers or insurance companies misuse this information. The development and enforcement of rigorous data protection regulations by policymakers are essential to protect personal data and to prevent discriminatory practices [220].

Expensive epigenetic treatment development and deployment could restrict access to wealthy populations and deepen current healthcare inequality gaps. Guaranteeing fair access to these medical treatments helps prevent increasing socioeconomic differences in health outcomes [221,222]. The clinical application of epigenetic research depends on continuous financial support for translational studies and strong regulatory supervision along with teamwork between scientists, clinicians, and industry representatives. Research through thorough clinical trials must validate epigenetic therapies' safety and effectiveness throughout various patient groups. The FDA and EMA need to establish evaluation guidelines for epigenetic drugs and gene-editing technologies along with biomarker diagnostics to confirm their clinical effectiveness and ethical integrity.

The medical field will soon integrate epigenetic profiling into standard practice to allow clinicians to customize treatment strategies according to individual epigenetic data alongside lifestyle and environmental influences. Future medical practice will likely see increased use of liquid biopsy-based epigenetic screenings which will aid early disease detection while allowing ongoing treatment response monitoring. Progress in artificial intelligence technology will boost the precision of predictive models which will lead to more accurate and efficient epigenetic diagnostic methods. The integration of epigenetic therapies with immunotherapy techniques and gene therapy approaches along with regenerative medical strategies can lead to innovative treatment solutions for complex diseases. The effectiveness of modern medical interventions relies on continuous scientific progress alongside ethical monitoring and fair access under strong regulatory systems. Epigenetic therapies present revolutionary opportunities for personalized medicine and regenerative treatments but require careful consideration of ethical, legal, and social issues to achieve responsible healthcare integration.

REFERENCES

1. Waddington, C.H., The epigenotype. 1942. *Int J Epidemiol*, 2012. 41(1): p. 10–3.
2. Moore, L.D., T. Le, and G. Fan, DNA methylation and its basic function. *Neuropsychopharmacology*, 2013. 38(1): p. 23–38.
3. Bannister, A.J. and T. Kouzarides, Regulation of chromatin by histone modifications. *Cell Res*, 2011. 21(3): p. 381–95.
4. Statello, L., et al., Gene regulation by long non-coding RNAs and its biological functions. *Nat Rev Mol Cell Biol*, 2021. 22(2): p. 96–118.
5. Dai, W., et al., Epigenetics-targeted drugs: current paradigms and future challenges. *Signal Transduct Target Ther*, 2024. 9(1): p. 332.
6. Verma, A., et al., Epigenetic manipulation for secondary metabolite activation in endophytic fungi: current progress and future directions. *Mycology*, 2023. 14(4): p. 275–91.
7. Hamilton, J.P., Epigenetics: principles and practice. *Dig Dis*, 2011. 29(2): p. 130–5.
8. Kim, J.Y., et al., Retrospective analysis of the clinical characteristics of patients with breast cancer treated with telomerase peptide immunotherapy combined with cytotoxic chemotherapy. *Breast Cancer* (Dove Med Press), 2023. 15: p. 955–66.
9. Li, L.J., et al., Different treatment efficacies and side effects of cytotoxic chemotherapy. *J Thorac Dis*, 2020. 12(7): p. 3785–3795.
10. Yu, X., et al., Cancer epigenetics: from laboratory studies and clinical trials to precision medicine. *Cell Death Discov*, 2024. 10(1): p. 28.
11. Im, A.P., et al., DNMT3A and IDH mutations in acute myeloid leukemia and other myeloid malignancies: associations with prognosis and potential treatment strategies. *Leukemia*, 2014. 28(9): p. 1774–83.
12. Fredly, H., B.T. Gjertsen, and O. Bruserud, Histone deacetylase inhibition in the treatment of acute myeloid leukemia: the effects of valproic acid on leukemic cells, and the clinical and experimental evidence for combining valproic acid with other antileukemic agents. *Clin Epigenetics*, 2013. 5(1): p. 12.
13. Li, Y. and E. Seto, HDACs and HDAC inhibitors in cancer development and therapy. *Cold Spring Harb Perspect Med*, 2016. 6(10), Article a026831.
14. Brown, L.J., et al., Epigenetic therapies and biomarkers in breast cancer. *Cancers (Basel)*, 2022. **14**(3): p. 474.
15. Garcia-Martinez, L., et al., Epigenetic mechanisms in breast cancer therapy and resistance. *Nat Commun*, 2021. 12(1): p. 1786.

16. Ljubenkov, P.A., et al., Effect of the histone deacetylase inhibitor FRM-0334 on progranulin levels in patients with progranulin gene haploinsufficiency: a randomized clinical trial. *JAMA Netw Open*, 2021. 4(9): p. e2125584.
17. Liu, J., et al., A class I HDAC inhibitor BG45 alleviates cognitive impairment through the CaMKII/ITPKA/Ca($^{2+}$) signaling pathway. *Pharmaceuticals (Basel)*, 2022. **15**(12): p. 1481. https://doi.org/10.3390/ph15121481.
18. Allis, C.D. and T. Jenuwein, The molecular hallmarks of epigenetic control. *Nat Rev Genet*, 2016. 17(8): p. 487–500.
19. Bure, I.V., M.V. Nemtsova, and E.B. Kuznetsova, Histone modifications and non-coding RNAs: mutual epigenetic regulation and role in pathogenesis. *Int J Mol Sci*, 2022. **23**(10): p. 5801. doi:10.3390/ijms23105801
20. Mattei, A.L., N. Bailly, and A. Meissner, DNA methylation: a historical perspective. *Trends Genet*, 2022. 38(7): p. 676–707.
21. Jin, B. and K.D. Robertson, DNA methyltransferases, DNA damage repair, and cancer. *Adv Exp Med Biol*, 2013. 754: p. 3–29.
22. Lyko, F., The DNA methyltransferase family: a versatile toolkit for epigenetic regulation. *Nat Rev Genet*, 2018. 19(2): p. 81–92.
23. Lande-Diner, L., et al., Role of DNA methylation in stable gene repression. *J Biol Chem*, 2007. 282(16): p. 12194–200.
24. Jones, P.L., et al., Methylated DNA and MeCP2 recruit histone deacetylase to repress transcription. *Nat Genet*, 1998. 19(2): p. 187–91.
25. Pinheiro, I. and E. Heard, X chromosome inactivation: new players in the initiation of gene silencing. *F1000Res*, 2017. **6**, F1000 Faculty Rev-344. doi:10.12688/f1000research.10707.1.
26. Wu, H. and Y. Zhang, Reversing DNA methylation: mechanisms, genomics, and biological functions. *Cell*, 2014. 156(1–2): p. 45–68.
27. Nishiyama, A. and M. Nakanishi, Navigating the DNA methylation landscape of cancer. *Trends Genet*, 2021. 37(11): p. 1012–27.
28. Tran, H.T., et al., DNA methylation changes following 5-azacitidine treatment in patients with myelodysplastic syndrome. *J Korean Med Sci*, 2011. 26(2): p. 207–13.
29. Ye, C., et al., Epigenetic therapy: research progress of decitabine in the treatment of solid tumors. *Biochim Biophys Acta Rev Cancer*, 2024. 1879(2): p. 189066.
30. Peserico, A. and C. Simone, Physical and functional HAT/HDAC interplay regulates protein acetylation balance. *J Biomed Biotechnol*, 2011. 2011: p. 371832.
31. Smith, B.C. and J.M. Denu, Chemical mechanisms of histone lysine and arginine modifications. *Biochim Biophys Acta*, 2009. 1789(1): p. 45–57.
32. Seto, E. and M. Yoshida, Erasers of histone acetylation: the histone deacetylase enzymes. *Cold Spring Harb Perspect Biol*, 2014. 6(4): p. a018713.
33. Yang, X.J. and E. Seto, HATs and HDACs: from structure, function and regulation to novel strategies for therapy and prevention. *Oncogene*, 2007. 26(37): p. 5310–8.
34. Sapountzi, V. and J. Cote, MYST-family histone acetyltransferases: beyond chromatin. *Cell Mol Life Sci*, 2011. 68(7): p. 1147–56.
35. Gallinari, P., et al., HDACs, histone deacetylation and gene transcription: from molecular biology to cancer therapeutics. *Cell Res*, 2007. 17(3): p. 195–211.

36. Saha, A., et al., Class I histone deacetylases in retinal progenitors and differentiating ganglion cells. *Gene Expr Patterns*, 2018. 30: p. 37–48.
37. Marks, P.A., T. Miller, and V.M. Richon, Histone deacetylases. *Curr Opin Pharmacol*, 2003. 3(4): p. 344–51.
38. Selvi, B.R., et al., Tuning acetylation levels with HAT activators: therapeutic strategy in neurodegenerative diseases. *Biochim Biophys Acta*, 2010. 1799(10–12): p. 840–53.
39. Schneider, A., et al., Acetyltransferases (HATs) as targets for neurological therapeutics. *Neurotherapeutics*, 2013. 10(4): p. 568–88.
40. Greer, E.L. and Y. Shi, Histone methylation: a dynamic mark in health, disease and inheritance. *Nat Rev Genet*, 2012. 13(5): p. 343–57.
41. Husmann, D. and O. Gozani, Histone lysine methyltransferases in biology and disease. *Nat Struct Mol Biol*, 2019. 26(10): p. 880–9.
42. Cheng, J., et al., A role for H3K4 monomethylation in gene repression and partitioning of chromatin readers. *Mol Cell*, 2014. 53(6): p. 979–92.
43. Pan, M.R., et al., Orchestration of H3K27 methylation: mechanisms and therapeutic implication. *Cell Mol Life Sci*, 2018. 75(2): p. 209–23.
44. Rakow, S., et al., Assaying epigenome functions of PRMTs and their substrates. *Methods*, 2020. 175: p. 53–65.
45. Maron, M.I., et al., Independent transcriptomic and proteomic regulation by type I and II protein arginine methyltransferases. *iScience*, 2021. 24(9): p. 102971.
46. D'Oto, A., et al., Histone demethylases and their roles in cancer epigenetics. *J Med Oncol Ther*, 2016. 1(2): p. 34–40.
47. Fang, Y., G. Liao, and B. Yu, LSD1/KDM1A inhibitors in clinical trials: advances and prospects. *J Hematol Oncol*, 2019. 12(1): p. 129.
48. Yang, Y., M. Zhang, and Y. Wang, The roles of histone modifications in tumorigenesis and associated inhibitors in cancer therapy. *J Natl Cancer Cent*, 2022. 2(4): p. 277–90.
49. Yoo, K.H. and L. Hennighausen, EZH2 methyltransferase and H3K27 methylation in breast cancer. *Int J Biol Sci*, 2012. 8(1): p. 59–65.
50. Yi, J., et al., Histone demethylase UTX/KDM6A enhances tumor immune cell recruitment, promotes differentiation and suppresses medulloblastoma. *Cancer Let*, 2021. 499: p. 188–200.
51. Qi, W., et al., Selective inhibition of EZH2 by a small molecule inhibitor blocks tumor cells proliferation. *Proc Natl Acad Sci U S A*, 2012. 109(52): p. 21360–5.
52. Eich, M.-L., et al., EZH2-targeted therapies in cancer: hype or a reality. *Cancer Res*, 2020. 80(24): p. 5449–58.
53. Strahl, B.D. and C.D. Allis, The language of covalent histone modifications. *Nature*, 2000. 403(6765): p. 41–5.
54. Rossetto, D., N. Avvakumov, and J. Côté, Histone phosphorylation: a chromatin modification involved in diverse nuclear events. *Epigenetics*, 2012. 7(10): p. 1098–108.
55. Banerjee, T. and D. Chakravarti, A peek into the complex realm of histone phosphorylation. *Mol Cell Biol*, 2011. 31(24): p. 4858–73.
56. Zhao, H., et al., Detection of histone H2AX phosphorylation on Ser-139 as an indicator of DNA damage. *Current Protocols Cytometry*, 2019. 89(1): p. e55.

57. Wei, Y., et al., Phosphorylation of histone H3 is required for proper chromosome condensation and segregation. *Cell*, 1999. 97(1): p. 99–109.
58. Komar, D. and P. Juszczynski, Rebelled epigenome: histone H3S10 phosphorylation and H3S10 kinases in cancer biology and therapy. *Clin Epigenetics*, 2020. 12(1): p. 147.
59. Saha, A., et al., Site-specific phosphorylation of histone H1.4 is associated with transcription activation. *Int J Mol Sci*, 2020. **21**(22): p. 8861.
60. Liao, R. and C.A. Mizzen, Interphase H1 phosphorylation: regulation and functions in chromatin. *Biochim Biophys Acta*, 2016. 1859(3): p. 476–85.
61. Contreras, A., et al., The dynamic mobility of histone H1 is regulated by cyclin/CDK phosphorylation. *Mol Cell Biol*, 2003. 23(23): p. 8626–36.
62. Podhorecka, M., A. Skladanowski, and P. Bozko, H2AX phosphorylation: its role in DNA damage response and cancer therapy. *J Nucleic Acids*, 2010. **2010**: p. 920161.
63. Carmena, M., S. Ruchaud, and W.C. Earnshaw, Making the Auroras glow: regulation of Aurora A and B kinase function by interacting proteins. *Curr Opin Cell Biol*, 2009. 21(6): p. 796–805.
64. Zheng, Y., et al., Histone H1 phosphorylation is associated with transcription by RNA polymerases I and II. *J Cell Biol*, 2010. 189(3): p. 407–15.
65. O'Brien, S.K., et al., P-TEFb kinase complex phosphorylates histone H1 to regulate expression of cellular and HIV-1 genes*. *J Biol Chem*, 2010. 285(39): p. 29713–20.
66. Haanen, T.J., 3rd, C.M. O'Connor, and G. Narla, Biased holoenzyme assembly of protein phosphatase 2A (PP2A): from cancer to small molecules. *J Biol Chem*, 2022. 298(12): p. 102656.
67. Gil, R.S. and P. Vagnarelli, Protein phosphatases in chromatin structure and function. *Biochim Biophys Acta Mol Cell Res*, 2019. 1866(1): p. 90–101.
68. Kinner, A., et al., Gamma-H2AX in recognition and signaling of DNA double-strand breaks in the context of chromatin. *Nucleic Acids Res*, 2008. 36(17): p. 5678–94.
69. Cheung, W.L., et al., Apoptotic phosphorylation of histone H2B is mediated by mammalian sterile twenty kinase. *Cell*, 2003. 113(4): p. 507–17.
70. Ahn, S.-H., et al., Sterile 20 kinase phosphorylates histone H2B at Serine 10 during hydrogen peroxide-induced apoptosis in S. cerevisiae. *Cell*, 2005. 120(1): p. 25–36.
71. Keppler, B.R. and T.K. Archer, Chromatin-modifying enzymes as therapeutic targets–Part 2. *Expert Opin Ther Targets*, 2008. 12(11): p. 1457–67.
72. Borah, N.A. and M.M. Reddy, Aurora kinase B inhibition: a potential therapeutic strategy for cancer. *Molecules*, 2021. **26**(7): 1981.
73. Liu, R., et al., Post-translational modifications of histones: mechanisms, biological functions, and therapeutic targets. *MedComm (2020)*, 2023. 4(3): p. e292.
74. Meas, R. and P. Mao, Histone ubiquitylation and its roles in transcription and DNA damage response. *DNA Repair (Amst)*, 2015. 36: p. 36–42.
75. Yang, Q., et al., E3 ubiquitin ligases: styles, structures and functions. *Mol Biomed*, 2021. 2(1): p. 23.
76. Jadhav, T. and M.W. Wooten, Defining an embedded code for protein ubiquitination. *J Proteomics Bioinform*, 2009. 2: p. 316.

77. Tamburri, S., et al., Histone H2AK119 mono-ubiquitination is essential for polycomb-mediated transcriptional repression. *Mol Cell*, 2020. 77(4): p. 840–856.e5.
78. Oss-Ronen, L., T. Sarusi, and I. Cohen, Histone mono-ubiquitination in transcriptional regulation and its mark on life: emerging roles in tissue development and disease. *Cells*, 2022. **11**(15): p. 2404.
79. Cao, J. and Q. Yan, Histone ubiquitination and deubiquitination in transcription, DNA damage response, and cancer. *Front Oncol*, 2012. 2: p. 26.
80. Wu, L., et al., H2B ubiquitylation promotes RNA Pol II processivity via PAF1 and pTEFb. Mol Cell, 2014. 54(6): p. 920–31.
81. Snyder, N.A. and G.M. Silva, Deubiquitinating enzymes (DUBs): regulation, homeostasis, and oxidative stress response. *J Biol Chem*, 2021. 297(3): p. 101077.
82. Zheng, L.-L., et al., Recent advances in the development of deubiquitinase inhibitors as antitumor agents. *Eur J Med Chem*, 2024. 266: p. 116161.
83. Kaikkonen, M.U., M.T. Lam, and C.K. Glass, Non-coding RNAs as regulators of gene expression and epigenetics. *Cardiovasc Res*, 2011. 90(3): p. 430–40.
84. Chattopadhaya, A., et al., Urinary miRNAs in bladder cancer. *Clin Chim Acta*, 2024: p. 120113.
85. Ying, S.Y., D.C. Chang, and S.L. Lin, The microRNA (miRNA): overview of the RNA genes that modulate gene function. *Mol Biotechnol*, 2008. 38(3): p. 257–68.
86. Bartel, D.P., MicroRNAs: genomics, biogenesis, mechanism, and function. *Cell*, 2004. 116(2): p. 281–297.
87. Otmani, K. and P. Lewalle, Tumor suppressor miRNA in cancer cells and the tumor microenvironment: mechanism of deregulation and clinical implications. *Front Oncol*, 2021. 11: p. 708765.
88. Gambari, R., et al., Targeting oncomiRNAs and mimicking tumor suppressor miRNAs: new trends in the development of miRNA therapeutic strategies in oncology (Review). *Int J Oncol*, 2016. 49(1): p. 5–32.
89. Ma, Y., et al., The roles of the Let-7 family of MicroRNAs in the regulation of cancer stemness. *Cells*, 2021. **10**(9): p. 2415.
90. Mattick, J.S., et al., Long non-coding RNAs: definitions, functions, challenges and recommendations. *Nat Rev Mol Cell Biol*, 2023. 24(6): p. 430–47.
91. Sharma, S., A.A. Houfani, and L.J. Foster, Pivotal functions and impact of long non-coding RNAs on cellular processes and genome integrity. *J Biomed Sci*, 2024. 31(1): p. 52.
92. Cerase, A., et al., Xist localization and function: new insights from multiple levels. *Genome Biol*, 2015. 16(1): p. 166.
93. Wu, L., et al., Binding interactions between long noncoding RNA HOTAIR and PRC2 proteins. *Biochemistry*, 2013. 52(52): p. 9519–27.
94. Vance, K.W. and C.P. Ponting, Transcriptional regulatory functions of nuclear long noncoding RNAs. *Trends Genet*, 2014. 30(8): p. 348–55.
95. Wang, K.C. and H.Y. Chang, Molecular mechanisms of long noncoding RNAs. Mol Cell, 2011. 43(6): p. 904–14.
96. Huang, W., et al., LncRNA-mediated DNA methylation: an emerging mechanism in cancer and beyond. *J Exp Clin Cancer Res*, 2022. 41(1): p. 100.
97. Zhang, X., et al., Mechanisms and functions of long non-coding RNAs at multiple regulatory levels. *Int J Mol Sci*, 2019. **20**(22): p. 5573.

98. Joh, R.I., et al., Regulation of histone methylation by noncoding RNAs. *Biochim Biophys Acta*, 2014. 1839(12): p. 1385–94.
99. Samanta, S., et al., Epigenetic dysfunctional diseases and therapy for infection and inflammation. *Biochim Biophys Acta Mol Basis Dis*, 2017. 1863(2): p. 518–28.
100. Kumar, S., et al., Non-coding RNAs as mediators of epigenetic changes in malignancies. *Cancers (Basel)*, 2020. **12**(12): p. 3657.
101. Singh, D.D., et al., Clinical significance of microRNAs, long non-coding RNAs, and circRNAs in cardiovascular diseases. *Cells*, 2023. **12**(12): p. 1629. doi:10.3390/cells12121629.
102. Yu, B. and S. Wang, Angio-LncRs: LncRNAs that regulate angiogenesis and vascular disease. *Theranostics*, 2018. 8(13): p. 3654–75.
103. Bertogliat, M.J., K.C. Morris-Blanco, and R. Vemuganti, Epigenetic mechanisms of neurodegenerative diseases and acute brain injury. *Neurochem Int*, 2020. 133: p. 104642.
104. Li, Y.-B., et al., MicroRNAs: pioneering regulators in Alzheimer's disease pathogenesis, diagnosis, and therapy. *Translational Psychiatry*, 2024. 14(1): p. 367.
105. Feng, L., et al., Plasma long non-coding RNA BACE1 as a novel biomarker for diagnosis of Alzheimer disease. *BMC Neurol*, 2018. 18(1): p. 4.
106. Szymczak, F., et al., Gene expression signatures of target tissues in type 1 diabetes, lupus erythematosus, multiple sclerosis, and rheumatoid arthritis. *Sci Adv*, 2021. 7(2): p. eabd7600. doi:10.1126/sciadv.abd7600
107. Zhang, P., et al., The lncRNA Neat1 promotes activation of inflammasomes in macrophages. *Nat Commun*, 2019. 10(1): p. 1495.
108. Wang, Z., The guideline of the design and validation of MiRNA mimics. *Methods Mol Biol*, 2011. 676: p. 211–23.
109. Krützfeldt, J., et al., Silencing of microRNAs in vivo with 'antagomirs'. *Nature*, 2005. 438(7068): p. 685–9.
110. Zhu, Y., et al., RNA-based therapeutics: an overview and prospectus. *Cell Death Disease*, 2022. 13(7): p. 644.
111. Baylin, S.B. and P.A. Jones, Epigenetic determinants of cancer. *Cold Spring Harb Perspect Biol*, 2016. **8**(9): p. a019505. doi:10.1101/cshperspect.a019505.
112. Roman-Gomez, J., et al., Promoter hypermethylation of cancer-related genes: a strong independent prognostic factor in acute lymphoblastic leukemia. *Blood*, 2004. 104(8): p. 2492–8.
113. Zhao, R., et al., Implications of genetic and epigenetic alterations of CDKN2A (p16(INK4a)) in cancer. *EBioMedicine*, 2016. 8: p. 30–9.
114. Krishnan, R., P.S. Patel, and R. Hakem, BRCA1 and metastasis: outcome of defective DNA repair. *Cancers (Basel)*, 2021. **14**(1): p. a019505. doi:10.1101/cshperspect.a019505
115. Xin, B., et al., Critical role of Myc activation in mouse hepatocarcinogenesis induced by the activation of AKT and RAS pathways. *Oncogene*, 2017. 36(36): p. 5087–97.
116. Dakal, T.C., et al., Oncogenes and tumor suppressor genes: functions and roles in cancers. *MedComm*, 2024. 5(6): p. e582.
117. Pappalardo, X.G. and V. Barra, Losing DNA methylation at repetitive elements and breaking bad. *Epigenetics Chromatin*, 2021. 14(1): p. 25.

118. Audia, J.E. and R.M. Campbell, Histone modifications and cancer. *Cold Spring Harb Perspect Biol*, 2016. 8(4): p. a019521.
119. Shen, X., et al., EZH1 mediates methylation on histone H3 lysine 27 and complements EZH2 in maintaining stem cell identity and executing pluripotency. *Mol Cell*, 2008. 32(4): p. 491–502.
120. Zhang, Z., et al., Recent progress in DNA methyltransferase inhibitors as anticancer agents. *Front Pharmacol*, 2022. 13: p. 1072651.
121. Tan, S.H., et al., Emerging pathways to neurodegeneration: dissecting the critical molecular mechanisms in Alzheimer's disease and Parkinson's disease. *Biomed Pharmacother*, 2019. 111: p. 765–77.
122. Karpova, N.N., Role of BDNF epigenetics in activity-dependent neuronal plasticity. *Neuropharmacology*, 2014. 76: p. 709–18.
123. Xiao, X., X. Liu, and B. Jiao, Epigenetics: recent advances and its role in the treatment of Alzheimer's disease. *Front Neurol*, 2020. 11: p. 538301.
124. Park, J., et al., The role of histone modifications: from neurodevelopment to neurodiseases. *Signal Transduct Target Ther*, 2022. 7(1): p. 217.
125. Shukla, S. and B.L. Tekwani, Histone deacetylase inhibitors in neurodegenerative diseases, neuroprotection and neuronal differentiation. *Front Pharmacol*, 2020. 11: p. 537.
126. Chuang, D.M., et al., Multiple roles of HDAC inhibition in neurodegenerative conditions. *Trends Neurosci*, 2009. 32(11): p. 591–601.
127. Li, S., Z. Lei, and T. Sun, The role of microRNAs in neurodegenerative diseases: a review. *Cell Biol Toxicol*, 2023. 39(1): p. 53–83.
128. Zhang, J., et al., Recent advances in Alzheimer's disease: mechanisms, clinical trials and new drug development strategies. *Signal Transduct Target Ther*, 2024. 9(1): p. 211.
129. Titze-de-Almeida, S.S., et al., The promise and challenges of developing miRNA-based therapeutics for Parkinson's disease. *Cells*, 2020. 9(4): p 211. doi:10.1038/s41392-024-01911-3
130. Placek, K., J.L. Schultze, and A.C. Aschenbrenner, Epigenetic reprogramming of immune cells in injury, repair, and resolution. *J Clin Invest*, 2019. 129(8): p. 2994–3005.
131. Hewagama, A. and B. Richardson, The genetics and epigenetics of autoimmune diseases. *J Autoimmun*, 2009. 33(1): p. 3–11.
132. Song, Y., J. Li, and Y. Wu, Evolving understanding of autoimmune mechanisms and new therapeutic strategies of autoimmune disorders. *Signal Transduct Target Ther*, 2024. 9(1): p. 263.
133. Araki, Y. and T. Mimura, The histone modification code in the pathogenesis of autoimmune diseases. *Mediators Inflamm*, 2017. 2017: p. 2608605.
134. Surace, A.E.A. and C.M. Hedrich, The role of epigenetics in autoimmune/inflammatory disease. *Front Immunol*, 2019. 10: p. 1525.
135. Topper, M.J., et al., The emerging role of epigenetic therapeutics in immuno-oncology. *Nat Rev Clin Oncol*, 2020. 17(2): p. 75–90.
136. Wu, Y.-L., et al., Epigenetic regulation in metabolic diseases: mechanisms and advances in clinical study. *Signal Transduct Target Ther*, 2023. 8(1): p. 98.
137. Parveen, N. and S. Dhawan, DNA methylation patterning and the regulation of beta cell homeostasis. *Front Endocrinol (Lausanne)*, 2021. 12: p. 651258.
138. Ling, C. and T. Rönn, Epigenetics in human obesity and type 2 diabetes. *Cell Metab*, 2019. 29(5): p. 1028–44.

139. Dong, H., et al., Metabolic memory: mechanisms and diseases. *Signal Transduct Target Ther*, 2024. 9(1): p. 38.
140. Nigi, L., et al., MicroRNAs as regulators of insulin signaling: research updates and potential therapeutic perspectives in type 2 diabetes. *Int J Mol Sci*, 2018. **19**(12): p. 3705. doi:10.3390/ijms19123705
141. Alexander, R., H. Lodish, and L. Sun, MicroRNAs in adipogenesis and as therapeutic targets for obesity. *Expert Opin Ther Targets*, 2011. 15(5): p. 623–36.
142. Ramzan, F., M.H. Vickers, and R.F. Mithen, Epigenetics, microRNA and metabolic syndrome: a comprehensive review. *Int J Mol Sci*, 2021. **22**(9):p. 5047. doi:10.3390/ijms22095047
143. Shi, Y., et al., Epigenetic regulation in cardiovascular disease: mechanisms and advances in clinical trials. *Signal Transduct Target Ther*, 2022. 7(1): p. 200.
144. Abi Khalil, C., The emerging role of epigenetics in cardiovascular disease. *Ther Adv Chronic Dis*, 2014. 5(4): p. 178–87.
145. Liu, C.F. and W.H.W. Tang, *Epigenetics in cardiac hypertrophy and heart failure. JACC Basic Transl Sci*, 2019. 4(8): p. 976–93.
146. Felisbino, M.B. and T.A. McKinsey, Epigenetics in cardiac fibrosis: emphasis on inflammation and fibroblast activation. *JACC Basic Transl Sci*, 2018. 3(5): p. 704–15.
147. Sanchez-Fernandez, E., et al., Histone deacetylase inhibitors as a promising treatment against myocardial infarction: a systematic review. *J Clin Med*, 2024. **13**(24): p. 7797. doi:10.3390/jcm13247797
148. Vora, L.K., et al., Artificial intelligence in pharmaceutical technology and drug delivery design. *Pharmaceutics*, 2023. **15**(7): p. 1916. doi:10.3390/pharmaceutics15071916
149. Phimmachanh, M., et al., Histone deacetylases and histone deacetylase inhibitors in neuroblastoma. *Front Cell Dev Biol*, 2020. 8: p. 578770.
150. Bernhart, E., et al., Histone deacetylase inhibitors vorinostat and panobinostat induce G1 cell cycle arrest and apoptosis in multidrug resistant sarcoma cell lines. *Oncotarget*, 2017. 8(44): p. 77254–67.
151. Zhang, L.-Y., et al., Role of histone deacetylases and their inhibitors in neurological diseases. *Pharmacol Res*, 2024. 208: p. 107410.
152. Shi, M.-Q., et al., Advances in targeting histone deacetylase for treatment of solid tumors. *J Hematol Oncol*, 2024. 17(1): p. 37.
153. Deng, S., et al., HDAC3 inhibition upregulates PD-L1 expression in B-cell lymphomas and augments the efficacy of anti-PD-L1 therapy. *Mol Cancer Ther*, 2019. 18(5): p. 900–8.
154. Fadul, S.M., A. Arshad, and R. Mehmood, CRISPR-based epigenome editing: mechanisms and applications. *Epigenomics*, 2023. 15(21): p. 1137–55.
155. Akinci, E., et al., Using CRISPR to understand and manipulate gene regulation. *Development*, 2021. **148**(9): p. dev182667. doi:10.1242/dev.182667
156. Xie, N., et al., Novel epigenetic techniques provided by the CRISPR/Cas9 system. *Stem Cells Int*, 2018. 2018: p. 7834175.
157. Tejedor, J.R., et al., CRISPR/dCAS9-mediated DNA demethylation screen identifies functional epigenetic determinants of colorectal cancer. *Clin Epigenet*, 2023. 15(1): p. 133.

158. Li, T., et al., CRISPR/Cas9 therapeutics: progress and prospects. *Signal Transduct Target Ther*, 2023. 8(1): p. 36.
159. Cazaly, E., et al., Making sense of the epigenome using data integration approaches. *Front Pharmacol*, 2019. 10: p. 126.
160. Duan, L., et al., Nanoparticle delivery of CRISPR/Cas9 for genome editing. *Front Genet*, 2021. **12**: p. 673286. doi:10.3389/fgene.2021.673286
161. Mohammadian Farsani, A., et al., Lipid nanoparticles: the game-changer in CRISPR-Cas9 genome editing. *Heliyon*, 2024. 10(2): p. e24606.
162. Heyn, H., J. Méndez-González, and M. Esteller, Epigenetic profiling joins personalized cancer medicine. *Expert Rev Mol Diagn*, 2013. 13(5): p. 473–9.
163. Soto, J., et al., The impact of next-generation sequencing on the DNA methylation–based translational cancer research. *Translat Res*, 2016. 169: p. 1–18.e1.
164. Feehley, T., et al., Drugging the epigenome in the age of precision medicine. *Clin Epigenet*, 2023. 15(1): p. 6.
165. Palanca-Ballester, C., et al., Cancer epigenetic biomarkers in liquid biopsy for high incidence malignancies. *Cancers (Basel)*, 2021. **13**(12): p. 3016.
166. Arslan, E., J. Schulz, and K. Rai, Machine learning in epigenomics: insights into cancer biology and medicine. *Biochim Biophys Acta Rev Cancer*, 2021. 1876(2): p. 188588.
167. Borsoi, F.T., et al., Dietary polyphenols and their relationship to the modulation of non-communicable chronic diseases and epigenetic mechanisms: a mini-review. *Food Chem Mol Sci*, 2023. 6: p. 100155.
168. Busch, C., et al., Epigenetic activities of flavonoids in the prevention and treatment of cancer. *Clin Epigenet*, 2015. 7(1): p. 64.
169. Salehi, B., et al., Resveratrol: a double-edged sword in health benefits. *Biomedicines*, 2018. **6**(3): p. 91. doi:10.3390/biomedicines6030091
170. Hassan, F.U., et al., Curcumin as an alternative epigenetic modulator: mechanism of action and potential effects. *Front Genet*, 2019. 10: p. 514.
171. Kok, D.E.G., et al., The effects of long-term daily folic acid and vitamin B12 supplementation on genome-wide DNA methylation in elderly subjects. *Clin Epigenet*, 2015. 7(1): p. 121.
172. Woo, V. and T. Alenghat, Epigenetic regulation by gut microbiota. *Gut Microbes*, 2022. 14(1): p. 2022407.
173. Paul, B., et al., Influences of diet and the gut microbiome on epigenetic modulation in cancer and other diseases. *Clin Epigenet*, 2015. 7: p. 112.
174. Lorenzo, P.M., et al., Epigenetic effects of healthy foods and lifestyle habits from the Southern European Atlantic diet pattern: a narrative review. *Adv Nutr*, 2022. 13(5): p. 1725–47.
175. Santaló, J. and M. Berdasco, Ethical implications of epigenetics in the era of personalized medicine. *Clin Epigenet*, 2022. 14(1): p. 44.
176. Anderson, S.J., et al., Off-target drug effects resulting in altered gene expression events with epigenetic and "quasi-epigenetic" origins. *Pharmacol Res*, 2016. 107: p. 229–33.
177. Scherer, F., et al., High-throughput sequencing for noninvasive disease detection in hematologic malignancies. *Blood*, 2017. 130(4): p. 440–52.

178. Churko, J.M., et al., Overview of high throughput sequencing technologies to elucidate molecular pathways in cardiovascular diseases. *Circulation Res*, 2013. 112(12): p. 1613–23.
179. Song, L., et al., The performance of the SEPT9 gene methylation assay and a comparison with other CRC screening tests: a meta-analysis. *Sci Rep*, 2017. 7(1): p. 3032.
180. Butler, M., et al., MGMT status as a clinical biomarker in glioblastoma. *Trends Cancer*, 2020. 6(5): p. 380–91.
181. Mori, H., et al., Blood MAPT expression and methylation status in Alzheimer's disease. *PCN Rep*, 2022. 1(4): p. e65.
182. Xu, H., S. Li, and Y.S. Liu, Roles and mechanisms of DNA methylation in vascular aging and related diseases. *Front Cell Dev Biol*, 2021. 9: p. 699374.
183. Ballestar, E., A.H. Sawalha, and Q. Lu, Clinical value of DNA methylation markers in autoimmune rheumatic diseases. *Nat Rev Rheumatol*, 2020. 16(9): p. 514–24.
184. Sahoo, K. and V. Sundararajan, Methods in DNA methylation array dataset analysis: a review. *Comput Struct Biotechnol J*, 2024. 23: p. 2304–25.
185. Bhanu, N.V., S. Sidoli, and B.A. Garcia, Histone modification profiling reveals differential signatures associated with human embryonic stem cell self-renewal and differentiation. *Proteomics*, 2016. 16(3): p. 448–58.
186. Chervona, Y. and M. Costa, Histone modifications and cancer: biomarkers of prognosis? *Am J Cancer Res*, 2012. 2(5): p. 589–97.
187. Nativio, R., et al., An integrated multi-omics approach identifies epigenetic alterations associated with Alzheimer's disease. *Nat Genet*, 2020. 52(10): p. 1024–35.
188. Goud Alladi, C., et al., DNA methylation as a biomarker of treatment response variability in serious mental illnesses: a systematic review focused on bipolar disorder, schizophrenia, and major depressive disorder. *Int J Mol Sci*, 2018. **19**(10): p. 3026. doi:10.3390/ijms19103026
189. Wawrzczak-Bargieła, A., W. Bilecki, and M. Maćkowiak, Epigenetic targets in schizophrenia development and therapy. *Brain Sci*, 2023. 13(3): p. 426.
190. Soldi, M., et al., Chromatin proteomics reveals novel combinatorial histone modification signatures that mark distinct subpopulations of macrophage enhancers. *Nucleic Acids Res*, 2017. 45(21): p. 12195–213.
191. Hino, S., T. Sato, and M. Nakao, Chromatin immunoprecipitation sequencing (ChIP-seq) for detecting histone modifications and modifiers. *Methods Mol Biol*, 2023. 2577: p. 55–64.
192. Qian, H., et al., Decoding the promise and challenges of miRNA-based cancer therapies: an essential update on miR-21, miR-34, and miR-155. *Int J Med Sci*, 2024. 21(14): p. 2781–98.
193. Bostjancic, E., et al., MicroRNAs miR-1, miR-133a, miR-133b and miR-208 are dysregulated in human myocardial infarction. *Cardiology*, 2010. 115(3): p. 163–9.
194. Selvam, S. and V. Ayyavoo, Biomarkers in neurodegenerative diseases: a broad overview. *Explor Neuroprotect Ther*, 2024. 4(2): p. 119–47.
195. Hou, J., et al., Functions and mechanisms of lncRNA MALAT1 in cancer chemotherapy resistance. *Biomarker Res*, 2023. 11(1): p. 23.

196. Yang, Y., et al., Silencing of LncRNA-HOTAIR decreases drug resistance of non-small cell lung cancer cells by inactivating autophagy via suppressing the phosphorylation of ULK1. *Biochem Biophys Res Commun*, 2018. 497(4): p. 1003–10.
197. Zhang, Z., T. Yang, and J. Xiao, Circular RNAs: promising biomarkers for human diseases. *EBioMedicine*, 2018. 34: p. 267–74.
198. Bamodu, O.A., C.-C. Chung, and T.R. Pisanic, Harnessing liquid biopsies: exosomes and ctDNA as minimally invasive biomarkers for precision cancer medicine. *J Liquid Biopsy*, 2023. 2: p. 100126.
199. Li, L. and Y. Sun, Circulating tumor DNA methylation detection as biomarker and its application in tumor liquid biopsy: advances and challenges. *MedComm*, 2024. 5(11): p. e766.
200. Brito-Rocha, T., et al., Shifting the cancer screening paradigm: the rising potential of blood-based multi-cancer early detection tests. *Cells*, 2023. **12**(6): p. 935. doi:10.3390/cells12060935
201. Lee, J., et al., Multimodal circulating tumor DNA (ctDNA) colorectal neoplasia detection assay for asymptomatic and early-stage colorectal cancer (CRC). *J Clin Oncol*, 2021. 39(15_suppl): p. 3536.
202. Spector, B.L., et al., The methylome and cell-free DNA: current applications in medicine and pediatric disease. *Pediatric Res*, 2023. 94(1): p. 89–95.
203. Pollard, C.A., et al., Considering biomarkers of neurodegeneration in Alzheimer's disease: the potential of circulating cell-free DNA in precision neurology. *J Pers Med*, 2024. **14**(11): p. 1104. doi:10.3390/jpm14111104
204. Moser, T., et al., Bridging biological cfDNA features and machine learning approaches. *Trends Genet*, 2023. 39(4): p. 285–307.
205. Tivey, A., et al., Mining nucleic acid "omics" to boost liquid biopsy in cancer. *Cell Rep Med*, 2024. 5(9): p. 101736.
206. Kronfol, M.M., et al., The role of epigenomics in personalized medicine. *Expert Rev Precis Med Drug Dev*, 2017. 2(1): p. 33–45.
207. Bhat, A.A., et al., Integration of CRISPR/Cas9 with artificial intelligence for improved cancer therapeutics. *J Transl Med*, 2022. 20(1): p. 534.
208. Carraro, C., et al., Chromatin accessibility profiling of targeted cell populations with laser capture microdissection coupled to ATAC-seq. *Cell Reports Methods*, 2023. **3**(10): p. 100598. doi:10.1016/j.crmeth.2023.1005
209. Xie, Y., et al., *Spatial Chromatin Accessibility Sequencing Resolves High-Order Spatial Interactions of Epigenomic Markers*. 2024, eLife Sciences Publications, Ltd.
210. Rauschert, S., et al., Machine learning and clinical epigenetics: a review of challenges for diagnosis and classification. *Clin Epigenet*, 2020. 12(1): p. 51.
211. De Riso, G. and S. Cocozza, Artificial intelligence for epigenetics: towards personalized medicine. *Curr Med Chem*, 2021. 28(32): p. 6654–74.
212. Karahmet Sher, E., et al., Nanotechnology in medicine revolutionizing drug delivery for cancer and viral infection treatments. *Int J Pharm*, 2024. 660: p. 124345.
213. Basu, A. and V.K. Tiwari, Epigenetic reprogramming of cell identity: lessons from development for regenerative medicine. *Clin Epigenet*, 2021. 13(1): p. 144.

214. Papp, B. and K. Plath, Reprogramming to pluripotency: stepwise resetting of the epigenetic landscape. *Cell Res*, 2011. 21(3): p. 486–501.
215. Hwang, J.Y., K.A. Aromolaran, and R.S. Zukin, The emerging field of epigenetics in neurodegeneration and neuroprotection. *Nat Rev Neurosci*, 2017. 18(6): p. 347–61.
216. Kumar, V., et al., Understanding the role of histone deacetylase and their inhibitors in neurodegenerative disorders: current targets and future perspective. *Curr Neuropharmacol*, 2022. 20(1): p. 158–78.
217. Maity, S., et al., Epigenetic mechanisms in memory and cognitive decline associated with aging and Alzheimer's disease. *Int J Mol Sci*, 2021. **22**(22): p. 12280. doi:10.3390/ijms222212280
218. Paine, P.T., A. Nguyen, and A. Ocampo, Partial cellular reprogramming: a deep dive into an emerging rejuvenation technology. *Aging Cell*, 2024. 23(2): p. e14039.
219. Rothstein, M.A., Y. Cai, and G.E. Marchant, The ghost in our genes: legal and ethical implications of epigenetics. *Health Matrix Clevel*, 2009. 19(1): p. 1–62.
220. Dupras, C., et al., Potential (mis)use of epigenetic age estimators by private companies and public agencies: human rights law should provide ethical guidance. *Environ Epigenet*, 2019. 5(3): p. dvz018.
221. Williams, J.S., R.J. Walker, and L.E. Egede, Achieving equity in an evolving healthcare system: opportunities and challenges. *Am J Med Sci*, 2016. 351(1): p. 33–43.
222. Vick, A.D. and H.H. Burris, Epigenetics and health disparities. *Curr Epidemiol Rep*, 2017. 4(1): p. 31–7.

Chapter 4

Exploring the potential of nanomaterials and innovations in biomedical nanotechnology

A future perspective for public health

Nazek AL-GALLAS

Department of Biology, University of Hafr Al Batin, Hafr Al Batin, Kingdom of Saudi Arabia

4.1 INTRODUCTION

Today, nanotechnology is a life-changing field of biomedical science providing novel modalities for the prevention, diagnosis, and treatment of diseases with great potential to transform public health in the future [1]. At the nanoscale, which could be as little as 1 nm or up to 100 nm, manipulating materials gives rise to novel properties in such nanoparticles, nanoscale devices, and biomaterials. These innovations on the nanoscale extend native interactions with biological systems to allow unprecedented precision, targeting, and release of therapeutic agents in a timely fashion [2].

Nanotechnology was recently introduced as a new and rapidly developing field with great promise in biomedical applications, helping solve some major public health problems [1]. Nanotechnology ranges from early diagnostic tools that enhance disease diagnosis to targeted delivery systems for drugs that decrease side effects and increase their efficiency, thereby aiding more personalized, efficient, and accessible healthcare solutions [3]. The burden of chronic diseases, infections, and health inequalities is creating unprecedented demands for more sophisticated public health systems globally [4]. Even though research devoted to biomedical nanotechnology is rapidly progressing, this chapter reviews recent developments on how such advances may affect public health. This chapter will explore topics including the following key concepts:

- **Recent Advancements in Nanobiotechnology for Biomedical Applications:** Recent advancements in nanobiotechnology have revolutionized biomedical applications, enabling targeted drug delivery, precise diagnostics, and tissue regeneration. Innovations like nanoparticle drug carriers, highly sensitive nanosensors, and regenerative nanofibers are enhancing treatment efficacy and enabling more personalized, minimally invasive healthcare solutions [5].

DOI: 10.1201/9781003733010-4

- **Novel Nanomaterials and Their Uses in Biomedical Fields:** Novel nanomaterials, including carbon-based, metallic, polymeric, and lipid-based nanoparticles, are transforming biomedical fields. Their unique properties enable targeted drug delivery, advanced diagnostics, and improved tissue engineering, making treatments more precise and effective while paving the way for new approaches in personalized medicine [1].
- **Nanotechnology-Driven Diagnostics and Therapeutic Strategies:** Nanotechnology-driven diagnostics and therapeutic strategies offer unprecedented precision in disease detection and treatment. By using nanoparticles for targeted drug delivery and advanced imaging, these innovations improve early diagnosis, minimize side effects, and enable highly personalized, effective medical interventions [6].
- **Clinical Case Studies Highlighting Nanobiotechnology Applications:** Clinical case studies on nanobiotechnology demonstrate its real-world impact, showcasing advancements in targeted cancer therapies, improved drug delivery systems, and enhanced diagnostic accuracy. These applications highlight the potential of nanotechnology to improve patient outcomes, reduce side effects, and offer more personalized treatment approaches in clinical settings [7].
- **Future Directions and Challenges in Advancing the Field:** Future directions in nanobiotechnology focus on enhancing biocompatibility, targeting precision, and scalable production while addressing challenges like toxicity, regulatory approval, and ethical considerations. Overcoming these hurdles is essential to unlocking nanotechnology's full potential in providing safer, more effective, and widely accessible biomedical solutions [1].

4.2 RECENT ADVANCEMENTS IN NANOBIOTECHNOLOGY FOR BIOMEDICAL APPLICATIONS

Nanobiotechnology, the intersection of nanotechnology and biological sciences, has rapidly evolved to address complex challenges in modern medicine. This interdisciplinary field leverages nanoscale materials and techniques to enhance diagnostic accuracy, improve therapeutic efficacy, and enable advanced tissue engineering, ultimately transforming biomedical applications. With unique properties, such as high surface area-to-volume ratio, tunable optical characteristics, and enhanced reactivity, nanomaterials interact at the molecular and cellular levels, allowing for precise manipulation within biological systems [1,5].

In recent years, significant advancements in nanobiotechnology have led to breakthrough solutions for targeted drug delivery, noninvasive diagnostics, and regenerative medicine. Innovations such as nanoparticles for controlled drug release, nanoscale imaging agents, and scaffold-based nanomaterials for tissue engineering are revolutionizing personalized medicine by making treatments more efficient and reducing side effects. For example, nanoparticles designed to selectively target cancer cells minimize collateral damage to healthy tissues, while nanostructured biosensors enable early detection of diseases by identifying trace biomolecules in bodily fluids [8,9,10].

Furthermore, the development of nanovaccines, gene-editing tools, and hybrid nanomaterials is opening new avenues for treating infectious diseases, genetic disorders, and chronic illnesses [11,12]. This section explores the latest advancements in nanobiotechnology across various biomedical applications, highlighting innovations in nanomaterials, diagnostics, drug delivery systems, and regenerative therapies. The ongoing research and technological breakthroughs in nanobiotechnology offer transformative potential to improve patient outcomes, personalize treatments, and address pressing global health challenges, marking an exciting frontier in the future of healthcare [5].

4.2.1 Nanoscale drug delivery systems

Nanoscale drug delivery systems represent a revolutionary advancement in the targeted and controlled administration of therapeutic agents. These systems utilize nanoparticles, which can be engineered to improve drug solubility, stability, and bioavailability, as well as to enhance precision in targeting specific cells or tissues. By delivering drugs directly to disease sites, nanoscale drug delivery reduces systemic exposure, minimizes side effects, and improves therapeutic efficacy, which is particularly valuable for treating complex conditions such as cancer, neurological disorders, and infectious diseases [14,15].

4.2.1.1 Lipid-based nanoparticles

Lipid-based nanoparticles (LNPs), including liposomes and solid lipid nanoparticles, have gained prominence due to their biocompatibility and efficiency in encapsulating hydrophobic and hydrophilic drugs. Recent advancements in lipid-based nanoparticles have significantly expanded their applications in biomedicine, particularly for drug delivery and mRNA therapeutics.

mRNA Therapeutics and Vaccines: Research at the University of Pennsylvania introduced a novel method of optimizing ionizable lipids used in lipid nanoparticles for mRNA delivery. These lipids improve the stability and delivery efficiency of mRNA therapies, such as COVID-19 vaccines and gene-editing applications. The team utilized an iterative chemical evolution process to design biodegradable lipids with enhanced performance, reducing development timelines for new lipid formulations [16].

Cancer Therapy: LBNPs are becoming critical in targeted cancer therapies. Their biocompatibility and ability to encapsulate drugs like doxorubicin have proven effective in improving drug uptake while reducing systemic toxicity. Advances include leveraging the enhanced permeability and retention (EPR) effect, which allows nanoparticles to accumulate selectively in tumor tissues. These properties have positioned LBNPs as effective carriers for both hydrophilic and hydrophobic drugs, as well as nucleic acids [17].

General Drug Delivery: The adaptability of lipid nanoparticles has also been demonstrated in delivering siRNA and DNA-based therapies, such as the FDA-approved patisiran for hereditary amyloidosis. Ionizable lipids within these systems enhance cargo stability and controlled release, making LBNPs versatile platforms for various therapeutic applications [17].

4.2.1.2 Polymer-based nanoparticles

Polymers like polylactic acid (PLA) and polyethylene glycol (PEG) have shown significant advancements in creating biodegradable, biocompatible nanoparticles for drug delivery. These polymer-based nanoparticles are now tailored to degrade within specific timeframes, allowing for controlled, sustained release of therapeutic agents. Recent innovations include stimuli-responsive polymer nanoparticles that release drugs only when encountering certain pH or temperature conditions, enhancing the precision of treatments, particularly for localized diseases [18,19,20].

Applications in Cancer Therapy: Polymer-based nanoparticles are being explored extensively for cancer treatment due to their ability to deliver chemotherapeutics selectively, reduce side effects, and enhance efficacy. Both synthetic polymers like PLGA (poly(lactic-co-glycolic acid)) and natural polymers such as chitosan and gelatin are widely used. These systems allow controlled and targeted drug release, minimizing systemic toxicity. They are particularly effective for encapsulating hydrophobic drugs like paclitaxel and doxorubicin [19].

Biodegradable Polymers: The use of biodegradable polymers such as PLA and PCL (polycaprolactone) is growing due to their biocompatibility and approval by regulatory agencies like the FDA. These materials degrade into nontoxic byproducts, making them suitable for long-term applications in the body [19,20].

Emerging Research: Novel delivery platforms using polymeric nanoparticles are being developed for various diseases, including ocular conditions, nutraceutical delivery, and infectious diseases. The adaptability of polymeric materials for specific drugs and conditions makes them a versatile choice [18].

4.2.1.3 Nanomicelles for targeted delivery

Nanomicelles, often used for solubilizing hydrophobic drugs, have advanced with functionalization techniques that enhance their targeting capabilities. By attaching specific ligands to the micelle surfaces, researchers have enabled more precise delivery to cancerous or diseased cells. These ligand-functionalized nanomicelles have shown promise in delivering drugs directly to tumor cells, increasing therapeutic efficacy while minimizing harm to healthy tissues [21].

Advances in Nanomicelle Targeting Mechanisms: Nanomicelles, self-assembled structures formed from amphiphilic molecules, have been extensively studied for their ability to deliver therapeutic drugs to specific sites. They utilize stimuli-responsive systems (e.g., pH, temperature, and light) to ensure precise drug release. A recent review highlighted their potential for overcoming challenges in pharmacokinetics and biodistribution while emphasizing low toxicity and enhanced stability compared to conventional micelles [22].

Aptamer-Based Nanomicelles: One study demonstrated the use of aptamer-modified nanomicelles (e.g., C18-sgc8) for targeted drug delivery. These micelles showed selective binding to cancer cells overexpressing PTK7, a cell surface receptor, and exhibited efficient drug internalization. This approach emphasizes the utility of DNA-based micellar structures for precise targeting and enhanced therapeutic efficacy [23].

Polymer-Based Nanomicelles in Cancer Therapy: Researchers have explored polymeric nanomicelles for delivering anticancer agents like docetaxel, utilizing specific ligands such as trastuzumab for targeting HER2-positive cells. These systems showed pH-responsive drug release and superior anticancer activity while minimizing off-target effects [22].

4.2.2 Nanobiotechnology in diagnostics

Recent advancements in nanobiotechnology have also led to significant breakthroughs in diagnostic methodologies. Nanoparticles and nanodevices are instrumental in improving the sensitivity, specificity, and rapidity of diagnostic tools [24].

4.2.2.1 *Quantum dots for imaging and detection*

Quantum dots (QDs) are semiconductor nanocrystals renowned for their unique optical properties, including size-dependent fluorescence, exceptional photostability, and high quantum yield. These attributes have significantly advanced biological imaging and detection techniques. Recent Developments in Quantum Dot Imaging and Detection:

Biological Imaging Applications: QDs have been extensively utilized in biological imaging due to their ability to provide high-resolution, stable, and multiplexed fluorescence signals. Their exceptional photostability surpasses that of traditional fluorescent dyes, enabling prolonged imaging sessions without significant signal degradation. This advantage is particularly beneficial in tracking dynamic cellular processes over extended periods [25].

Noninvasive Detection of Oral Cancer: In the realm of noninvasive cancer detection, QDs have shown promise in imaging oral squamous cell carcinoma. Their application in this context offers potential therapeutic and diagnostic benefits, highlighting their versatility in medical imaging [26].

Advancements in Semiconductor Quantum Dots: Recent research has focused on the development of core–shell semiconductor nanocrystals, which enhance the optical properties of QDs. These advancements aim to improve the efficiency and specificity of QDs in various imaging applications, paving the way for more precise diagnostic tools [27].

Eco-Friendly Quantum Dots for Bioimaging: The shift toward environmentally friendly QDs has been a significant focus in bioimaging. Recent reviews summarize the progress of eco-friendly QDs, including their preparation, optical properties, and surface modifications, underscoring their potential in biomedical applications [28].

Single Quantum Dot Detection via Microtoroid Optical Resonators: Innovative detection methods have been explored, such as the use of microtoroid optical resonators to detect single 5 nm diameter QDs. This technique achieves a signal-to-noise ratio exceeding 10, demonstrating the potential for highly sensitive detection systems [29].

These developments underscore the transformative impact of QDs on imaging and detection technologies, offering enhanced sensitivity, specificity, and versatility across various applications.

4.2.2.2 *Gold nanoparticles in biosensors*

Gold nanoparticles (AuNPs) have remarkable optical and electronic properties, which have proven useful in biosensor applications. AuNP-based biosensors are now widely used for detecting pathogens, monitoring biomarkers, and assessing disease progression. Recent advances include functionalizing gold nanoparticles with specific antibodies or DNA sequences and improving their sensitivity and selectivity in detecting target molecules even in trace amounts. This technology is particularly valuable in early disease diagnosis, such as detecting cancer biomarkers or viral pathogens in blood samples. Also, AuNPs have been developed to detect specific bacterial species using localized surface plasmon resonance (LSPR) shifts. These biosensors offer high sensitivity and specificity based on nanoparticle shape and functionalization [30].

4.2.2.3 Magnetic nanoparticles for MRI enhancement

Magnetic nanoparticles (MNPs), such as iron oxide particles, are increasingly used to enhance magnetic resonance imaging (MRI) for disease diagnosis. By coupling MNPs with targeting molecules, researchers have improved the specificity of MRI for detecting tumors, plaques, and other disease sites. Innovations in coating and functionalizing MNPs have also enhanced their stability and reduced toxicity, allowing for longer circulation times and clearer imaging [31,32].

4.2.3 Nanotechnology in regenerative medicine

Nanobiotechnology's contributions to regenerative medicine have opened new avenues for tissue engineering and cell-based therapies, particularly through the development of nanomaterials that support cell growth and differentiation [33].

4.2.3.1 Nanofibers in tissue engineering

Nanofibers, which mimic the extracellular matrix (ECM), provide structural support and biological cues to cells in tissue engineering applications. Recent advancements in electrospinning techniques allow for the precise control of nanofiber properties, such as fiber diameter and porosity, which influence cell attachment, proliferation, and differentiation. These engineered nanofibers are now used in skin grafts, wound healing, and even more complex tissue structures, such as nerve regeneration scaffolds [34].

4.2.3.2 Carbon nanotubes for bone regeneration

Carbon nanotubes (CNTs) are highly versatile nanomaterials with excellent mechanical strength and electrical conductivity. Recent studies have incorporated CNTs into bone regeneration scaffolds, enhancing their strength and bioactivity. CNT-based scaffolds promote osteoblast attachment and differentiation, making them suitable for repairing bone defects and fractures. Additionally, CNTs can be functionalized with growth factors to stimulate bone growth further, providing a powerful tool for regenerative medicine [35,36].

4.2.3.3 *Hydrogel nanocomposites for soft tissue engineering*

Hydrogel nanocomposites, which combine hydrophilic polymers with nanoparticles, have shown promise in soft tissue engineering applications. These nanocomposites offer tunable mechanical and biological properties, providing an ideal environment for cell growth in tissues like cartilage, muscle, and liver. Recent advancements include developing injectable nanocomposite hydrogels that can solidify in situ, providing a minimally invasive approach to tissue repair and regeneration [37,38].

4.2.4 Immunotherapy and nanovaccines

The application of nanobiotechnology in immunotherapy and vaccine development has shown significant potential, particularly with the success of mRNA vaccines. Nanoparticles offer a versatile platform for delivering antigens, adjuvants, and genetic material to stimulate the immune system effectively [39].

4.2.4.1 *Lipid nanoparticles in mRNA vaccines*

The success of lipid nanoparticles in mRNA COVID-19 vaccines has catalyzed research into other nanovaccines. LNPs protect mRNA from degradation, enhance cell uptake, and improve immune responses. Current research aims to optimize LNPs for other infectious diseases and cancer vaccines, focusing on enhancing stability, targeting, and reducing potential side effects [40,41].

4.2.4.2 *Drugs for targeted immunotherapy*

Dendrimers, which are branched tree-like nanoparticles, offer a highly customizable structure for delivering therapeutic agents. Recent developments have focused on functionalizing dendrimers with immunomodulatory agents to target specific immune cells, providing a more directed and effective immune response against tumors. By delivering immune checkpoint inhibitors or cytokines directly to immune cells, dendrimer-based therapies hold promise for treating cancers and autoimmune diseases [42,43].

4.2.4.3 *Challenges facing immunotherapy and nanovaccines*

Immunotherapy and nanovaccines face challenges summarized in Table 4.1. These difficulties highlight the need for innovative approaches and ongoing research to optimize these promising biomedical technologies [44].

Table 4.1 Challenges facing immunotherapy and nanovaccines

Challenge	*Description*
Delivery system complexity	Precise targeting of immune cells without off-target effects is hindered by biological barriers and complexity.
Manufacturing and scalability	Producing nanovaccines with consistent quality and efficacy on a large scale is technically challenging.
Immune tolerance and safety	Avoiding unintended immune responses or excessive inflammation requires careful nanoparticle design.
Cost and accessibility	High development costs limit the widespread availability and affordability of these advanced therapies.

4.2.5 Gene therapy and nanocarriers

Gene therapy, a cutting-edge approach to treating genetic disorders, cancers, and other complex diseases, involves delivering specific genetic material to modify or correct faulty genes at the cellular level. This technique holds significant promise for addressing the root causes of diseases, potentially providing long-term and even curative treatments [45,46]. However, one of the main challenges in gene therapy is effectively delivering the genetic material to target cells without degradation or immune rejection, while minimizing off-target effects [47]. Nanocarriers, engineered nanoparticles designed for safe and targeted delivery of genes, have emerged as a crucial tool in overcoming these obstacles and enhancing the precision and efficacy of gene therapies [48].

Nanocarriers offer several advantages in gene delivery due to their customizable surfaces, biocompatibility, and ability to protect genetic material, such as DNA, RNA, or CRISPR components, from degradation in the bloodstream. They can be engineered to navigate biological barriers, selectively bind to specific cell types, and control the release of genetic material within target cells. Among the most widely studied nanocarriers are lipid-based nanoparticles, virus-like particles, polymeric nanoparticles, and dendrimers, each offering unique benefits in terms of stability, payload capacity, and delivery efficiency [49].

Recent advancements in nanocarrier technology have significantly improved gene therapy's potential, enabling targeted delivery to difficult-to-reach tissues, enhancing therapeutic outcomes, and reducing adverse effects. This chapter explores the role of nanocarriers in gene therapy, detailing the various types of nanocarriers, their mechanisms of action, and recent breakthroughs that are bringing gene therapy closer to widespread clinical use. As research progresses, nanocarriers are poised to become an essential component in making gene therapy a viable and accessible option for treating a broad range of genetic and acquired diseases [50].

4.2.5.1 *CRISPR delivery via nanoparticles*

The delivery of CRISPR–Cas9 components for gene editing has been significantly enhanced with nanoparticles. Lipid and polymer-based nanoparticles can encapsulate CRISPR components, protecting them from degradation and enhancing cell uptake. Recent advances include the development of targeting ligands on nanoparticle surfaces to direct CRISPR delivery to specific cell types, increasing the precision and reducing off-target effects in gene therapy applications [51].

4.2.5.2 *Virus-like nanoparticles for gene therapy*

Virus-like nanoparticles (VLPs) mimic the structure of viruses, making them highly effective in gene delivery. Recent advancements focus on engineering VLPs that lack viral genetic material, making them safer for therapeutic applications. VLPs are now being explored for delivering genes that address a range of genetic disorders, including cystic fibrosis, muscular dystrophy, and certain cancers, offering a safer and more controlled alternative to viral vectors [52].

Recent developments in nanobiotechnology have completely changed the biomedical industry by offering creative answers to persistent problems in immunotherapy, gene therapy, regenerative medicine, drug delivery, and diagnostics. Unprecedented control and specificity are provided by nanoparticles, nanofibers, and other nanoscale materials, allowing for more individualized, efficient treatments with fewer adverse effects. To fully realize the potential of nanobiotechnology in medicine, issues including toxicity, biocompatibility, and large-scale manufacturing must be resolved as research advances. Combining these developments has enormous potential for enhancing patient outcomes, developing tailored medicine, and creatively and precisely treating complicated medical problems [53].

4.3 NOVEL NANOMATERIALS AND THEIR USES IN BIOMEDICAL FIELDS

Recent advancements in nanotechnology have led to the development of a diverse range of nanomaterials, each with unique physical, chemical, and biological properties that make them suitable for biomedical applications. This section explores the structure, functionality, and biomedical uses of novel nanomaterials, with a focus on their roles in diagnostics, drug delivery, tissue engineering, regenerative medicine, and therapeutic innovations [54].

Nanomaterials are structures with at least one dimension measuring between 1 and 100 nm. At this scale, materials exhibit distinctive characteristics such as enhanced reactivity, increased surface area-to-volume ratio, and unique optical and magnetic properties. These attributes make nanomaterials particularly effective in interacting with biological systems at the cellular or molecular level. Biomedical applications of nanomaterials have grown considerably, with innovative applications that address some of the biggest challenges in modern medicine, including targeted drug delivery, noninvasive diagnostics, and improved regenerative therapies [1,13,26].

4.3.1 Types of novel nanomaterials and their biomedical applications

Nanotechnology has introduced a new era in biomedical science by enabling the development of nanomaterials with unique properties that operate at scales comparable to biological molecules. These nanomaterials, which include carbon-based structures, metal and metal oxide nanoparticles, polymers, lipid-based particles, and hybrid composites, exhibit highly specialized properties—such as enhanced surface reactivity, tunable optical characteristics, and exceptional mechanical strength—that make them invaluable for a variety of biomedical applications. Their versatility allows them to be designed for specific tasks, from delivering drugs precisely to targeted cells to enabling real-time diagnostic imaging [55].

Each type of nanomaterial offers distinct advantages in biomedical fields. For example, carbon-based nanomaterials like graphene provide robust scaffolding for tissue engineering, while metal nanoparticles such as gold and silver are highly effective in diagnostic and therapeutic applications due to their unique optical and antimicrobial properties. Polymeric nanoparticles enable controlled drug release, lipid-based nanostructures facilitate gene delivery, and hybrid nanomaterials combine functionalities to achieve multifaceted therapeutic effects [56].

The application of these novel nanomaterials has led to groundbreaking advances in drug delivery systems, diagnostic platforms, regenerative medicine, and cancer therapies. This section provides an in-depth overview of the most promising types of novel nanomaterials, examining their structure, functionality, and potential impact on biomedical fields. As the development of nanomaterials continues, they hold transformative potential for improving patient outcomes, advancing personalized medicine, and addressing critical challenges in global healthcare [56].

4.3.1.1 Carbon-based nanomaterials

Carbon-based nanomaterials, including carbon nanotubes (CNTs), graphene, and fullerenes, have unique mechanical strength, thermal conductivity, and electrical properties. These characteristics make them particularly suited for a range of biomedical applications [57].

- **Carbon Nanotubes (CNTs)**: CNTs are cylindrical nanostructures with a unique surface that can be functionalized to carry drugs, proteins, or nucleic acids. Their high surface area allows for substantial drug loading, while their mechanical stability supports applications in bone and cartilage regeneration. CNTs also serve as imaging agents due to their fluorescence properties and are increasingly used in photothermal therapy, where they convert light into heat to target cancer cells [58].
- **Graphene and Graphene Oxide**: Graphene, a single layer of carbon atoms arranged in a hexagonal lattice, is highly conductive and can be easily functionalized, making it suitable for biosensors, drug delivery, and tissue engineering. Graphene oxide (GO), a derivative with oxygen-containing functional groups, is hydrophilic and has demonstrated effectiveness as a carrier for chemotherapy drugs and as a scaffold material in bone regeneration [57,58].

4.3.1.2 Metal and metal oxide nanoparticles

Metal and metal oxide nanoparticles, such as gold, silver, iron oxide, and zinc oxide, are widely used in biomedical research due to their unique optical, magnetic, and antimicrobial properties [59].

- **Gold Nanoparticles (AuNPs)**: Known for their biocompatibility and tunable surface, AuNPs have become essential in diagnostic applications, particularly in biosensors and imaging, as well as in drug delivery. In photothermal therapy, gold nanoparticles convert light into localized heat, which can selectively destroy tumor cells. Recent advances include conjugating AuNPs with antibodies or ligands to target specific cells or tissues.
- **Silver Nanoparticles (AgNPs)**: With strong antimicrobial properties, silver nanoparticles are frequently used in wound dressings and coatings for medical devices to prevent infections. Silver nanoparticles also exhibit cytotoxic effects on cancer cells and are under investigation for potential anticancer therapies [59,60].
- **Iron Oxide Nanoparticles (IONPs)**: Iron oxide nanoparticles, especially superparamagnetic iron oxide nanoparticles (SPIONs), are highly valuable in imaging due to their magnetic properties, which enhance contrast in MRI. Additionally, SPIONs can be directed by magnetic fields to deliver drugs precisely, making them promising candidates for targeted drug delivery and hyperthermia treatment for cancer [59,60].

4.3.1.3 Polymeric nanoparticles

Polymeric nanoparticles are made from biodegradable and biocompatible polymers, such as PLA, polyglycolic acid (PGA), and poly (lactic-co-glycolic acid) (PLGA). These nanoparticles allow for controlled and sustained drug release and are commonly used in drug delivery systems [61].

- **Polymeric Micelles**: Composed of amphiphilic block copolymers, polymeric micelles form core–shell structures that encapsulate hydrophobic drugs, enhancing solubility and stability. Micelles can be functionalized for targeted delivery to specific cell types, particularly in cancer treatment [61].
- **Dendrimers**: These branched, tree-like polymeric nanoparticles have a high surface functionality, allowing for the attachment of multiple therapeutic agents, imaging compounds, or targeting ligands. Dendrimers are highly customizable and are used in drug delivery, gene therapy, and diagnostics [61].

4.3.1.4 Lipid-based nanoparticles

Lipid-based nanoparticles, including liposomes and solid lipid nanoparticles (SLNs), are among the most established nanomaterials in clinical use due to their biocompatibility and ability to encapsulate both hydrophilic and hydrophobic drugs [62].

- **Liposomes**: Liposomes are spherical vesicles with a phospholipid bilayer, widely used in drug delivery. They are particularly valuable for encapsulating chemotherapeutics, as they can reduce toxicity and increase drug accumulation in target tissues. Liposomal formulations of anticancer drugs (e.g., doxorubicin) have been approved for clinical use [62].
- **Solid Lipid Nanoparticles (SLNs)**: SLNs offer controlled drug release and high stability, making them suitable for long-term drug delivery. SLNs are commonly used in cosmetics and are now under research for applications in vaccine delivery and gene therapy [63].

4.3.1.5 Quantum dots (QDs) in biomedical applications

QDs are semiconductor nanoparticles with diameters typically ranging from 2 to 10 nm. Known for their exceptional optical properties, QDs exhibit bright fluorescence, high stability, and tunable emission wavelengths, which make them invaluable for biomedical imaging, diagnostics, and targeted therapeutic applications. Unlike traditional fluorescent dyes, QDs can emit light in a variety of colors depending on their size—smaller QDs emit higher-energy colors (e.g., blue), while larger QDs emit lower-energy colors (e.g., red). This feature allows QDs to be used in multiplexed imaging, where multiple biological targets are simultaneously observed in different colors. QDs have distinct advantages over conventional fluorescent markers [64]:

- **High Quantum Yield**: QDs emit brighter fluorescence than organic dyes due to their high quantum yield, enabling clear visualization even at low concentrations.
- **Photostability**: QDs are less susceptible to photobleaching, allowing for prolonged observation of biological processes, which is essential for real-time imaging and long-term studies.
- **Tunable Emission Wavelengths**: By adjusting the size and composition of QDs, researchers can precisely control their emission wavelength, enabling multicolor imaging in a single assay.

4.3.1.5.1 QDs applications in bioimaging

QDs have revolutionized bioimaging by providing highly sensitive, stable, and customizable fluorescent markers [65]:

- **Cellular and Molecular Imaging**: QDs can be conjugated with specific biomolecules, such as antibodies or peptides, to selectively target and label cells, proteins, or nucleic acids. This allows researchers to study cellular pathways, track cell migration, and monitor molecular interactions in real-time.
- **Tumor Imaging**: QDs can be modified with tumor-targeting ligands (e.g., antibodies against tumor-specific markers), enabling precise localization of tumor cells during imaging. This is particularly beneficial for detecting early-stage cancers, where traditional imaging may not provide sufficient resolution.
- ***In Vivo* Imaging**: Due to their stability and tunable properties, QDs are effective *in vivo* imaging applications, such as tracking cell distribution, identifying metastatic cancer sites, and mapping lymph nodes. Near-infrared (NIR) QDs are especially useful for *in vivo* imaging, as NIR light penetrates deeper into tissues, allowing visualization of internal organs and structures.

4.3.1.5.2 Diagnostic applications of quantum dots

QDs are being integrated into diagnostic platforms to improve sensitivity and specificity in detecting biomarkers associated with various diseases [66]:

- **Multiplexed Assays**: By using QDs of different sizes and colors, multiple biomarkers can be detected simultaneously in a single sample. This is highly valuable in diagnostic assays for complex diseases, such as cancer or autoimmune disorders, where several biomarkers must be assessed to confirm a diagnosis.
- **Point-of-Care Diagnostics**: QDs have been incorporated into portable diagnostic devices for rapid detection of pathogens, including viruses and bacteria, at low concentrations. These point-of-care diagnostics

are particularly useful in resource-limited settings or for rapid disease screening in the field.

- **FRET-Based Biosensors**: In fluorescence resonance energy transfer (FRET) biosensors, QDs serve as energy donors to detect biomolecular interactions. When a target molecule is present, it alters the energy transfer between the QD and a paired molecule, creating a detectable signal change. QD-based FRET biosensors have shown high sensitivity for detecting specific nucleic acids, proteins, and enzymes.

4.3.1.5.3 Quantum dots in targeted therapy

In addition to imaging and diagnostics, QDs are being explored in therapeutic applications, especially in targeted drug delivery and photodynamic therapy [67]:

- **Drug Delivery**: QDs can be engineered to carry therapeutic molecules, such as small-molecule drugs, peptides, or nucleic acids. By conjugating QDs with targeting ligands, they can deliver drugs specifically to diseased cells, minimizing off-target effects and enhancing treatment efficacy.
- **Photodynamic Therapy (PDT)**: In PDT, QDs act as photosensitizers that, upon exposure to light, produce reactive oxygen species (ROS) to damage and kill cancer cells. QDs are advantageous in PDT due to their tunable fluorescence and stability, which enable precise control of ROS production in targeted areas, reducing harm to surrounding healthy tissue.

4.3.1.5.4 Challenges and future directions

While QDs have demonstrated remarkable potential, certain challenges must be addressed for their widespread biomedical application [68]:

- **Toxicity**: Many QDs are composed of heavy metals, such as cadmium, which pose toxicity risks to cells and tissues. To address this, researchers are developing cadmium-free QDs, using elements like silicon and indium, or creating biocompatible coatings to reduce toxicity.
- **Biodegradability**: QDs are not naturally biodegradable, which can lead to long-term accumulation in the body. Innovations in QD composition and surface modification are being pursued to make them safer for prolonged *in vivo* use.
- **Regulatory Approval**: Due to potential toxicity and biocompatibility issues, QDs face regulatory hurdles before clinical adoption. Ongoing studies are evaluating the safety and efficacy of QDs in animal models, and future research aims to establish standardized protocols for clinical-grade QDs.

We can say that QDs represent a significant advancement in the field of nanobiotechnology, with applications that span bioimaging, diagnostics, and targeted therapy. Their unique optical properties—such as tunable fluorescence, photostability, and high quantum yield—make them invaluable for multiplexed imaging and highly sensitive diagnostics. As research continues to address the challenges of toxicity and biodegradability, QDs are likely to play an increasingly central role in personalized medicine, allowing for more accurate diagnosis, targeted therapies, and real-time monitoring of disease progression [68].

4.3.1.6 Hybrid nanomaterials

Hybrid nanomaterials combine two or more types of nanomaterials to leverage the strengths of each component, resulting in multifunctional systems for biomedical applications [69].

A. **Gold–Silica Nanoshells**: These core–shell nanostructures combine the optical properties of gold with the stability of silica, allowing for use in photothermal therapy and imaging. Nanoshells can be functionalized for targeted cancer therapies, where they accumulate in tumor tissues and generate heat upon near-infrared irradiation [70].
B. **Magnetic–Lipid Hybrid Nanoparticles**: By integrating magnetic and lipid components, these nanoparticles can be directed to specific sites using external magnetic fields and deliver drugs or genes to targeted cells. Magnetic–lipid nanoparticles are under investigation for applications in targeted cancer therapy and localized drug delivery [71].

4.3.2 Applications of nanomaterials in biomedical fields

Nanomaterials offer unparalleled advantages in biomedical applications due to their ability to interact at the cellular and molecular levels with high precision. Key applications include drug delivery, diagnostics, regenerative medicine, antimicrobial treatments, and cancer therapies. The unique properties of nanomaterials, such as high surface-to-volume ratio, tunable size, and customizable surfaces, make them highly effective for targeted biomedical interventions [55].

4.3.2.1 Drug delivery systems

Nanomaterials have revolutionized drug delivery by providing targeted and controlled release of therapeutic agents. They enhance drug stability, bioavailability, and effectiveness while reducing side effects, as nanomaterials can be engineered to specifically target diseased cells [72].

- **Polymeric Nanoparticles**: These include biodegradable polymers like poly(lactic-co-glycolic acid) (PLGA) and polyethylene glycol (PEG). Polymeric nanoparticles encapsulate drugs, allowing controlled release and reducing systemic toxicity. For example, PEGylated nanoparticles have been used to improve the pharmacokinetics of anticancer drugs like doxorubicin [73].
- **Lipid-Based Nanoparticles**: Lipid nanoparticles, such as liposomes, have been widely used for delivering both hydrophilic and hydrophobic drugs. Liposomal formulations of chemotherapeutic drugs reduce off-target effects and increase drug accumulation at tumor sites. Lipid nanoparticles were also key in mRNA vaccine delivery, as seen in COVID-19 vaccines [62].
- **Nanomicelles**: These are surfactant-based structures that can encapsulate hydrophobic drugs, increasing their solubility. Nanomicelles with targeting ligands are used in cancer treatment, allowing for greater drug accumulation within tumors and reduced impact on healthy tissues [22].

4.3.2.2 Diagnostic applications

Nanomaterials enhance the sensitivity, specificity, and speed of diagnostic tests, enabling early detection and improved monitoring of diseases.

- **Quantum Dots (QDs)**: QDs are semiconductor nanoparticles with bright, stable fluorescence, allowing for highly sensitive imaging and multiplexed biomarker detection. They are used to label cancer cells and track cellular interactions in real-time, aiding in the study of disease progression [64].
- **Gold Nanoparticles (AuNPs)**: Due to their optical properties, AuNPs are widely used in biosensors and imaging. They are used in colorimetric assays, where binding to a target biomolecule changes the color of a solution, making them valuable for rapid and portable diagnostics in detecting pathogens like HIV or malaria [74].
- **Magnetic Nanoparticles (MNPs)**: Magnetic nanoparticles, particularly iron oxide nanoparticles, enhance MRI by increasing contrast in images. They are often functionalized with antibodies or ligands to target specific tissues, allowing for more precise imaging in detecting tumors or vascular diseases [75].
- **Nanosensors**: Nanosensors are used for detecting small quantities of biomolecules, such as proteins, DNA, or metabolites. They offer quick, highly sensitive diagnostics, which are valuable in detecting infectious diseases, cancer biomarkers, and even monitoring glucose levels in diabetic patients [76].

4.3.2.3 Regenerative medicine and tissue engineering

Nanomaterials support cell growth, differentiation, and tissue repair, making them highly valuable in regenerative medicine and tissue engineering [77]:

- **Nanofibers**: Electrospun nanofibers mimic the ECM, providing structural support for cell attachment and proliferation. They are used in wound dressings and scaffolds for skin, bone, and nerve regeneration. Nanofibers loaded with growth factors can enhance tissue repair and healing [78].
- **Hydrogel Nanocomposites**: Hydrogels incorporated with nanoparticles are highly porous, allowing for nutrient exchange and cell migration, crucial for soft tissue engineering. They have applications in cartilage and muscle regeneration, and recent advances include injectable nanocomposite hydrogels for minimally invasive tissue repair [79].
- **Carbon Nanotubes (CNTs)**: CNTs exhibit high mechanical strength and electrical conductivity, supporting bone regeneration and nerve repair. CNTs have been incorporated into bone scaffolds to enhance osteoblast differentiation, and their conductivity promotes neural cell regeneration, potentially useful in nerve tissue engineering [80].
- **Graphene and Graphene Oxide**: Graphene-based materials are biocompatible and conductive, making them ideal for tissue engineering applications, especially for neural tissue engineering where electrical conductivity is essential. They also support stem cell differentiation and have been explored for bone and cartilage tissue regeneration [81].

4.3.2.4 Antimicrobial applications

Nanomaterials with inherent antimicrobial properties are being applied in wound dressings, coatings for medical devices, and treatments for infections [82].

- **Silver Nanoparticles (AgNPs)**: Silver nanoparticles exhibit strong antimicrobial effects against a wide range of pathogens, including bacteria, viruses, and fungi. AgNPs disrupt microbial cell membranes and inhibit enzymatic activity. They are used in wound dressings, catheters, and implant coatings to reduce infection risk [83].
- **Zinc Oxide (ZnO) and Copper Oxide (CuO) Nanoparticles**: These nanoparticles have demonstrated antimicrobial properties, which are especially useful in healthcare settings. ZnO and CuO nanoparticles are incorporated into surface coatings to reduce microbial colonization on medical devices and hospital surfaces [84].
- **Chitosan Nanoparticles**: Derived from chitin, chitosan has natural antimicrobial and wound-healing properties. Chitosan nanoparticles

are used in wound dressings, drug delivery for treating skin infections, and coating surgical tools, reducing infection risk and promoting faster healing [84].

4.3.2.5 Cancer therapy

Nanomaterials provide multiple pathways for targeting and treating cancer:

- **Gold Nanoparticles in Photothermal Therapy**: Gold nanoparticles can absorb light and convert it into heat, allowing for targeted destruction of cancer cells in photothermal therapy (PTT). Functionalized gold nanoparticles accumulate in tumor cells and, upon laser exposure, generate localized heat that kills the cancerous cells while sparing surrounding healthy tissue [85].
- **Magnetic Nanoparticles in Hyperthermia Treatment**: Magnetic nanoparticles can generate heat when exposed to an alternating magnetic field. By targeting tumors with magnetic nanoparticles, hyperthermia can destroy cancer cells or enhance the effectiveness of chemotherapy and radiation therapy [86].
- **Nanocarriers for Chemotherapy Drugs**: Nanocarriers, such as liposomes, polymeric nanoparticles, and micelles, deliver chemotherapeutic agents directly to tumors, reducing systemic toxicity. For instance, liposomal formulations of doxorubicin and paclitaxel allow for higher drug concentrations at tumor sites, enhancing efficacy and minimizing side effects [87].
- **Photodynamic Therapy (PDT) with Quantum Dots**: QDs can be used as photosensitizers in PDT, where they produce ROS upon light activation. ROS induce oxidative stress in targeted cancer cells, leading to cell death. QD-based PDT offers precise control, reducing collateral damage to healthy tissues [88].

4.3.2.6 Gene therapy

Nanomaterials have indeed emerged as a transformative tool in the field of gene delivery, offering several advantages over conventional methods. Here are some key points highlighting how nanomaterials enhance gene delivery and their therapeutic potential [56]:

4.3.2.6.1 Protection of genetic material [89]

- **Encapsulation**: Nanomaterials, such as liposomes, dendrimers, and polymeric nanoparticles, can encapsulate nucleic acids (like DNA and RNA), shielding them from enzymatic degradation and environmental stresses. This increases the stability of genetic material during storage and transport.

- **Controlled Release**: Nanocarriers can be engineered to release their payload in a controlled manner, ensuring that genetic material reaches target cells intact.

4.3.2.6.2 Enhanced cellular uptake [90]

- **Surface Modification**: The surface properties of nanomaterials can be tailored by adding targeting moieties (like ligands, antibodies, or peptides) that facilitate selective binding to specific cell types, improving uptake efficiency.
- **Endocytosis Facilitation**: Certain nanomaterials can exploit cellular uptake mechanisms, such as endocytosis, enhancing the internalization of genetic material into cells.

4.3.2.6.3 Improved targeting and distribution [90]

- **Passive and Active Targeting**: Nanoparticles can utilize the EPR effect to accumulate tumor tissues. Additionally, by functionalizing their surface, nanoparticles can actively target specific cell types, minimizing off-target effects and enhancing therapeutic efficacy.
- **Size and Shape Optimization**: The physicochemical properties (size, shape, charge) of nanomaterials can be optimized to navigate biological barriers, influencing their biodistribution and cellular uptake.

4.3.2.6.4 Therapeutic applications [91]

- **Gene Therapy**: Nanomaterials facilitate the delivery of corrective genes for genetic disorders, providing new avenues for treating diseases such as cystic fibrosis, hemophilia, or certain cancers.
- **CRISPR–Cas9 Systems**: Nanoparticles are being developed to deliver CRISPR components efficiently into cells for gene-editing applications.
- **Vaccination**: Nanomaterials can enhance mRNA vaccine delivery, improving immunity against infectious diseases by improving cellular responses.

4.3.2.6.5 Biocompatibility and safety [92]

- **Material Selection**: Biodegradable and biocompatible nanomaterials are preferred to minimize toxicity and promote safe degradation after delivering their payload.
- **Reduced Immunogenicity**: By carefully designing nanocarriers, it's possible to reduce the immune response against them, enhancing their effectiveness in delivering therapeutic genes.

In summary, nanomaterials represent a promising approach to gene delivery, addressing some of the key limitations of traditional methods. Their ability to protect genetic material, enhance cellular uptake, and provide targeted therapies paves the way for innovative treatments in various fields, from oncology to genetic disorders. Continued research and development in this area hold significant potential for advancing therapeutic options in modern medicine.

4.3.2.7 Immunotherapy and vaccine delivery

Nanomaterials play a pivotal role in developing advanced vaccines and immunotherapies by enhancing immune responses and improving antigen delivery.

- **Lipid Nanoparticles in mRNA Vaccines**: Lipid nanoparticles protect mRNA from degradation and promote cellular uptake, as demonstrated in COVID-19 vaccines. This success has catalyzed the development of nanovaccines targeting other diseases, including cancer, where mRNA delivers instructions to immune cells to recognize and destroy cancer cells [93].
- **Gold Nanoparticles in Vaccine Adjuvants**: Gold nanoparticles are used as adjuvants to boost immune responses in vaccines. Functionalized gold nanoparticles enhance antigen presentation, promoting stronger immune activation and improving vaccine efficacy against diseases like influenza and hepatitis [94].
- **Virus-Like Particles (VLPs)**: VLPs mimic the structure of viruses without containing viral genetic material, making them safe for delivering antigens. VLPs elicit strong immune responses and are used in vaccines against diseases like HPV, with research ongoing for applications in cancer immunotherapy [95].

The application of nanomaterials in biomedical fields is transforming modern medicine, enabling precise diagnostics, targeted therapies, and effective disease prevention. With continued innovation, nanomaterials hold the potential to revolutionize healthcare by improving patient outcomes, reducing side effects, and paving the way for personalized medicine. As research progresses, overcoming challenges such as toxicity, scalability, and regulatory approval will be essential to fully realize the impact of nanomaterials on global health [96].

4.3.3 Challenges and future directions for applications of nanomaterials in biomedical fields

The integration of nanomaterials into biomedical fields has led to significant advancements in diagnostics, drug delivery, tissue engineering, and regenerative medicine. However, several challenges continue to hinder the widespread adoption of nanomaterials in clinical applications, including concerns related to safety, scalability, regulatory approval, and biocompatibility. Addressing these obstacles is essential to fully realize the potential of nanomaterials in improving patient outcomes and advancing precision medicine. This section explores the key challenges in nanomedicine and outlines future directions to overcome them [96].

4.3.3.1 Challenges

The application of nanomaterials in biomedical fields brings immense potential, but it also presents several significant challenges that hinder their widespread adoption in clinical practice. One of the foremost concerns is toxicity and biocompatibility. Due to their nanoscale size, many nanomaterials interact readily with biological systems, raising concerns about cytotoxicity, immune reactions, and potential long-term health effects. For instance, heavy metal nanoparticles, such as cadmium-based QDs, can release toxic ions over time, which may lead to oxidative stress, inflammation, and DNA damage in cells. Moreover, nonbiodegradable nanomaterials like certain carbon nanotubes or metal-based nanoparticles can accumulate in organs like the liver, spleen, and kidneys, where they may persist indefinitely, posing additional risks [97,98,99,100].

Another challenge lies in the clearance and distribution of nanomaterials within the body. While small enough to interact with cellular and molecular structures, nanoparticles often accumulate in nontargeted organs due to the body's natural filtration systems. This off-target accumulation, particularly in organs such as the liver and spleen, may not only reduce therapeutic efficacy but also lead to unintended side effects and potential toxicity. Moreover, some nanomaterials lack efficient pathways for excretion and are not easily metabolized, increasing the risk of prolonged retention and associated toxicological issues [101].

Standardizing the production and quality of nanomaterials presents additional obstacles. The synthesis of nanomaterials often results in batch variability, where small differences in size, shape, or surface properties can significantly affect performance, safety, and reliability. Ensuring batch consistency is crucial for clinical applications, as even minor variations can influence the therapeutic outcome. Moreover, the complex manufacturing processes required for nanoscale precision make large-scale production costly and technically demanding, limiting accessibility and scalability [97,101].

Regulatory and ethical challenges also impede the clinical translation of nanomaterials. Regulatory bodies currently lack specific frameworks tailored to nanomaterials, making it difficult to establish standards for evaluating their safety, efficacy, and long-term effects. This regulatory gap can lead to delays in approval and uncertainty regarding safe clinical use. Ethical considerations, including potential unknown long-term impacts and patient consent for treatments involving nanotechnology, further complicate the development and acceptance of nanomaterial-based therapies [96,102].

These challenges highlight the need for continued research, improved synthesis and quality control processes, and collaborative efforts to develop regulatory guidelines specific to nanomaterials. Addressing these issues will be crucial to fully realizing the potential of nanomaterials in biomedical fields, enabling their safe and effective integration into patient care.

4.3.3.2 Future directions for applications of nanomaterials in biomedical fields

The future of nanomaterials in biomedical applications holds transformative potential, with advancements poised to address current challenges and enable innovative solutions in healthcare. A primary focus in future research is the development of biodegradable and self-clearing nanomaterials, which would allow for the safe metabolism and excretion of nanoparticles after they have served their purpose. Materials such as PLGA (poly(lactic-co-glycolic acid)) and PEG (polyethylene glycol), as well as naturally derived nanomaterials, offer promising solutions by reducing long-term toxicity concerns associated with nonbiodegradable particles. Additionally, researchers are exploring nanomaterials that respond to specific physiological triggers, such as pH changes or temperature shifts, to enable controlled disassembly and clearance after targeted delivery [103].

Improving targeting precision is another critical direction for nanomaterial research, as enhanced specificity could minimize side effects and increase therapeutic efficacy. The functionalization of nanoparticles with ligands—antibodies, peptides, or other molecules that selectively bind to target cells—has shown promise in fields like cancer therapy, where precision is crucial. Future developments will likely include "smart" nanomaterials that respond to environmental stimuli for on-demand drug release, further refining targeted treatment options in oncology and other localized therapies [104].

Nanomaterials are also expected to play a significant role in the advancement of personalized medicine. Given their customizable nature, nanomaterials can be tailored to suit individual patient profiles, enabling more effective treatment regimens. In cancer treatment, for instance, nanoparticle carriers could be personalized based on tumor biomarkers, enhancing specificity and therapeutic success rates. Similarly, nanomaterial-based diagnostics, such as nanosensors designed to detect unique biomarkers, could facilitate early disease diagnosis and monitoring, allowing for treatment plans tailored to the patient's specific health needs [105].

Expanding the range of biocompatible materials will remain a focus for researchers, as alternatives to traditional heavy-metal-based nanomaterials are essential for reducing toxicity. Heavy metal-free options, including carbon dots, silicon nanoparticles, and biomimetic nanostructures, offer safer profiles while retaining the desirable optical and structural properties required for medical applications. The use of naturally derived materials or biomimetic designs that replicate biological structures is likely to increase, as these materials often show superior compatibility with human biology and present fewer long-term health risks [101].

Furthermore, integrating nanomaterials with other cutting-edge technologies, such as CRISPR gene editing and artificial intelligence (AI), could significantly broaden their impact on biomedical fields. AI, for example, can optimize nanoparticle design, predict interactions within the body, and accelerate the discovery of new nanomaterials. When paired with CRISPR, nanomaterials can serve as precise delivery mechanisms for gene-editing tools, opening new possibilities for treating genetic disorders and diseases with underlying genetic components [106].

To achieve these advancements, there will be a need for regulatory frameworks specifically tailored to the unique properties of nanomaterials. Ensuring safety and standardization is essential for widespread clinical adoption, and coordinated efforts between researchers, industry stakeholders, and regulatory bodies will be crucial in establishing guidelines for the clinical use of nanomaterials. As these future directions unfold, nanomaterials are set to become an integral part of healthcare, making treatments more effective, personalized, and safe, while also opening new avenues in precision medicine and diagnostics.

4.4 NANOTECHNOLOGY-DRIVEN DIAGNOSTICS AND THERAPEUTIC STRATEGIES

Nanotechnology has revolutionized diagnostics and therapeutic strategies in medicine, providing tools with unprecedented precision, sensitivity, and effectiveness. By operating at the nanoscale, nanotechnology allows for direct interaction with biological molecules, cells, and tissues, enabling earlier detection of diseases, targeted delivery of therapeutics, and reduced side effects. These advancements have contributed to the development of more personalized, minimally invasive approaches, significantly improving outcomes in treating diseases such as cancer, cardiovascular disorders, and infectious diseases [107].

4.4.1 Nanotechnology in diagnostics

Nanotechnology has enhanced diagnostic tools by making them more sensitive, faster, and capable of detecting biomarkers at lower concentrations than traditional methods. By leveraging the unique properties of nanoparticles such as high surface area, optical fluorescence, and magnetic responsiveness, nanotechnology has enabled early and precise disease detection, even at the molecular level [107].

4.4.1.1 Nanosensors and biomarker detection

Nanosensors are designed to detect trace amounts of specific biomolecules, making them invaluable in early diagnosis. For example, nanoparticles functionalized with antibodies or ligands can bind specifically to cancer biomarkers, generating signals detectable by imaging devices. These nanosensors have been applied in cancer diagnosis, where the detection of tumor markers in blood or tissues can reveal early-stage cancer before symptoms emerge. Similarly, nanosensors are also being developed for monitoring glucose levels, proteins associated with neurodegenerative diseases, and infectious disease markers [108].

4.4.1.2 Quantum dots in imaging

QDs are semiconductor nanoparticles that emit bright, stable fluorescence and can be tuned to specific wavelengths based on their size. This property makes them ideal for high-resolution imaging in diagnostics, as they enable real-time tracking of cellular processes and can illuminate specific cell types or molecules in complex biological environments. QDs are particularly useful in multiplexed imaging, where they label multiple biomarkers simultaneously with distinct colors, improving the accuracy of diagnosis in conditions like cancer and cardiovascular disease [25].

4.4.1.3 Magnetic nanoparticles in MRI enhancement

Magnetic nanoparticles (MNPs), especially superparamagnetic iron oxide nanoparticles (SPIONs), are used to enhance the contrast in MRI. When functionalized with targeting ligands, these nanoparticles accumulate in diseased tissues, such as tumors or plaques in blood vessels, improving the specificity of imaging. This enhancement enables the detailed visualization of abnormal tissues, facilitating the early detection of cancers, atherosclerosis, and other pathologies that might not be visible with traditional MRI [109].

4.4.2 Nanotechnology in therapeutic strategies

Nanotechnology has also transformed therapeutic strategies by improving drug delivery, enhancing therapeutic efficacy, and enabling new treatment modalities that were previously challenging or impossible with traditional approaches [107].

4.4.2.1 Targeted drug delivery systems

Nanotechnology allows for the design of drug carriers that deliver therapeutic agents directly to diseased cells while minimizing exposure to healthy tissues, thereby reducing side effects. Nanoparticles can be functionalized with ligands, such as antibodies or peptides, that bind specifically to receptors on target cells. For example, liposomes, polymeric nanoparticles, and dendrimers are widely used in cancer therapy to deliver chemotherapeutic drugs specifically to tumor cells. Targeted drug delivery reduces the toxicity of chemotherapy and enhances its effectiveness by increasing drug concentration within tumors, making it a promising strategy in oncology and other fields [110].

4.4.2.2 Nanoparticle-based gene therapy

Nanoparticles have emerged as effective carriers for gene therapy, where genetic material is delivered to correct or modify the expression of faulty genes. Lipid-based nanoparticles (LNPs) have been pivotal in delivering mRNA in vaccines, such as COVID-19 vaccines, and hold potential for treating genetic diseases. These nanoparticles protect nucleic acids from degradation and promote cellular uptake, allowing for controlled gene expression in target cells. Moreover, nanoparticles like dendrimers and VLPs are being developed to deliver CRISPR–Cas9 components for gene editing, opening doors for precision treatment of genetic disorders [92].

4.4.2.3 Photothermal and photodynamic therapy

Nanotechnology has introduced innovative therapeutic techniques such as photothermal therapy (PTT) and photodynamic therapy (PDT), which are particularly promising in cancer treatment. In PTT, nanoparticles such as gold or carbon-based particles absorb light energy and convert it into heat, selectively destroying cancer cells without affecting surrounding tissues. Photodynamic therapy (PDT) involves nanoparticles that generate ROS upon exposure to light, inducing oxidative stress that kills targeted cells. These therapies offer a minimally invasive approach to treating localized tumors, providing greater precision and fewer side effects compared to traditional treatments [111].

4.4.2.4 Immunotherapy and nanovaccines

Nanotechnology is advancing immunotherapy and vaccine development by enhancing the delivery and efficacy of immunostimulatory agents. Nanoparticles can be engineered to deliver antigens directly to immune cells, stimulating robust immune responses. For example, lipid nanoparticles used in mRNA vaccines protect the genetic material and improve cellular uptake, as seen in COVID-19 vaccines. Additionally, gold and polymeric nanoparticles are being studied as vaccine adjuvants to enhance immune responses in cancer immunotherapy and infectious disease vaccines [112].

4.4.3 Future potential and impact on personalized medicine

Nanotechnology-driven diagnostics and therapeutic strategies have opened new avenues for personalized medicine. By tailoring treatments based on individual biomarker profiles, nanotechnology enables precise, patient-specific approaches that improve therapeutic outcomes and minimize adverse effects. For example, nanosensors that detect specific genetic or protein biomarkers can guide personalized treatment plans, while nanoparticle-based drug delivery can target unique cellular markers associated with a patient's tumor. These advances bring precision medicine closer to reality, allowing treatments to be tailored to individual patient needs and characteristics, ultimately improving patient outcomes and quality of life [113].

Nanotechnology is driving significant advancements in diagnostics and therapeutic strategies, offering more precise, sensitive, and personalized approaches to healthcare. Through innovative tools like nanosensors, targeted drug delivery systems, and novel therapeutic techniques, nanotechnology is reshaping the diagnosis and treatment of a range of diseases. As research progresses and these technologies are refined, nanotechnology is set to play an increasingly central role in modern medicine, with the potential to transform patient care and address some of the most challenging issues in healthcare [3,114].

4.5 CLINICAL CASE STUDIES HIGHLIGHTING NANOBIOTECHNOLOGY APPLICATIONS

Clinical case studies highlight the transformative impact of nanobiotechnology in medicine, showcasing its applications in cancer treatment, infectious disease management, and regenerative medicine. For instance, in cancer therapy, nanoparticle-based drug delivery systems have been employed to target tumor cells specifically, reducing side effects commonly associated with chemotherapy. Liposomal doxorubicin, a nanotechnology-enhanced formulation, has shown success in clinical trials by improving drug concentration within tumors while sparing healthy tissues [115].

4.5.1 Nanoparticle-based cancer therapies

One of the most widely studied applications of nanobiotechnology in clinical medicine is in cancer therapy, where nanoparticle-based drug delivery systems are used to target tumor cells while sparing healthy tissue.

A prominent example is the clinical use of liposomal doxorubicin (marketed as Doxil), a chemotherapy drug encapsulated in lipid nanoparticles. This formulation allows doxorubicin to concentrate specifically within tumor sites, leveraging the EPR effect of tumors. Clinical trials have demonstrated that liposomal doxorubicin reduces cardiotoxicity, a major side effect of traditional doxorubicin, while improving therapeutic outcomes for patients with breast cancer, ovarian cancer, and Kaposi's sarcoma [116].

Another case study involves gold nanoparticles (AuNPs) in photothermal therapy, where gold particles are introduced into the bloodstream and accumulate in tumor tissues. Upon exposure to near-infrared light, AuNPs generate heat, selectively killing cancer cells while minimizing damage to surrounding healthy tissues. Clinical trials in head and neck cancers have shown promising results, with AuNP-based photothermal therapy offering a minimally invasive option with fewer side effects than conventional radiation therapy [117].

4.5.2 Rapid diagnostics for infectious diseases

The COVID-19 pandemic underscored the need for rapid and accurate diagnostics, a demand that nanobiotechnology has been uniquely equipped to meet. Nanotechnology-based diagnostic tools, such as gold nanoparticle lateral flow assays, have been successfully deployed for COVID-19 detection in clinical settings. These tests function similarly to pregnancy tests but incorporate gold nanoparticles coated with antibodies that bind specifically to SARS-CoV-2 antigens. In addition to providing results in minutes, these assays are cost-effective and easy to use, making them particularly valuable in resource-limited settings. This diagnostic technology is also being adapted for other infectious diseases, including tuberculosis and HIV, demonstrating the versatility of nanobiotechnology in global public health [118].

Another example involves quantum dot-based fluorescence assays, which enhance the sensitivity of diagnostics by binding to disease biomarkers with high specificity. QDs are semiconductor nanoparticles that emit bright fluorescence under ultraviolet light, making it easier to detect low-abundance biomarkers. These assays have shown potential in detecting early-stage infections, allowing clinicians to make timely treatment decisions and improve patient outcomes [119].

4.5.3 Regenerative medicine and tissue engineering

Nanobiotechnology also plays a critical role in regenerative medicine, where nanomaterials are used to develop scaffolds that support tissue growth and repair. For example, nanofiber scaffolds composed of electrospun polymers mimic the ECM and have been used to aid in skin regeneration for burn victims.

Clinical studies on nanofiber-based dressings have shown that these materials not only support cell attachment and proliferation but also facilitate wound healing by providing structural integrity and promoting vascularization. Such dressings have been tested in clinical settings for patients with severe burns or chronic wounds, demonstrating faster healing times and reduced infection rates compared to conventional dressings [120].

In bone regeneration, nanocomposite materials containing hydroxyapatite nanoparticles have been used in clinical trials for repairing bone defects. Hydroxyapatite, a naturally occurring mineral in bones, enhances biocompatibility and supports osteoblast activity, encouraging bone formation and integration with existing bone tissue. Case studies have shown that patients receiving these nanocomposite implants experience quicker recovery times and improved bone density in affected areas, offering a promising solution for complex fractures and osteoporotic bone defects [121].

4.5.4 Nanovaccines and immunotherapy

Nanobiotechnology has also led to the development of nanovaccines, which use nanoparticles as carriers for antigens or genetic material to stimulate targeted immune responses. The success of lipid nanoparticle (LNP)-based COVID-19 mRNA vaccines is a key example, demonstrating the efficacy of LNPs in protecting mRNA from degradation, enhancing cellular uptake, and delivering genetic instructions to immune cells. These vaccines have proven highly effective in generating immunity against SARS-CoV-2 and are now a foundation for other nanovaccine developments aimed at diseases such as influenza, Zika virus, and certain cancers. Nanovaccines show particular promise in cancer immunotherapy, where they can be engineered to deliver tumor antigens directly to immune cells, promoting a stronger, more targeted response to tumor cells [112,122].

4.5.5 Challenges and considerations in clinical implementation

While these clinical case studies highlight the remarkable potential of nanobiotechnology, certain challenges remain in translating laboratory successes into routine clinical practice. Concerns related to the long-term biocompatibility, toxicity, and clearance of nanomaterials need to be rigorously addressed. Regulatory hurdles also present challenges, as there are currently limited frameworks for assessing the safety and efficacy of nanotechnology-based therapies and diagnostics. Finally, the high cost of production and the complexity of scaling up nanomaterial manufacturing can limit accessibility [112,115,118,119]. Addressing these challenges will be essential to making nanobiotechnology more widely available and clinically feasible. Continued research, interdisciplinary collaboration, and the development of standardized safety protocols will be key to realizing the full potential of nanobiotechnology in patient care.

4.6 FUTURE DIRECTIONS AND CHALLENGES IN ADVANCING BIOMEDICAL NANOTECHNOLOGY

Biomedical nanotechnology is poised to revolutionize the healthcare industry by enabling more precise diagnostics, targeted drug delivery, and personalized treatment options. Here's an overview of key future directions and challenges that lie ahead in this field [123].

4.6.1 Future directions in advancing biomedical nanotechnology

4.6.1.1 Targeted drug delivery [124]

- **Targeted Therapy**: Researchers are focusing on engineering nanoparticles to deliver drugs directly to diseased cells, such as cancer cells, which could reduce side effects and enhance therapeutic efficacy.
- **Controlled Drug Release**: There is ongoing development of nanoparticles for controlled, sustained drug release, which promises more consistent therapeutic effects and less dosing frequency.
- **Gene Therapy**: Nanotechnology may facilitate the delivery of genetic materials, such as RNA or CRISPR components, for treating genetic disorders or boosting immune responses in diseases like cancer.

4.6.1.2 Nanodiagnostics [125]

- **Point-of-Care Diagnostics**: Portable diagnostic devices employing nanoparticles may allow for rapid disease detection at the point of care, significantly improving early diagnosis.
- **Biosensors and Imaging**: Advances in sensitive biosensors and enhanced imaging techniques using nanomaterials like gold or QDs could lead to earlier and more accurate disease detection.

4.6.1.3 Tissue engineering and regenerative medicine [126]

- **Scaffolds and 3D Bioprinting**: The development of nanomaterials for tissue scaffolding and custom tissue implants is key to advancing regenerative medicine and organ transplantation.
- **Stem Cell Therapy**: Enhancing stem cell therapies with nanomaterials could improve cell adhesion, growth, and differentiation for effectively regenerating damaged tissues.

4.6.1.4 *Nanorobots for precision medicine [127]*

- **Surgical and Diagnostic Nanobots:** The potential for nanorobots in performing minimally invasive procedures or providing real-time diagnostic feedback represents an exciting frontier for precision medicine.

4.6.2 Challenges in advancing biomedical nanotechnology

4.6.2.1 *Safety and toxicity [128]*

- Understanding and ensuring biocompatibility are critical as nanoparticles can be unpredictably reactive with biological systems. There's a need for comprehensive research into their long-term effects and safe degradation or elimination from the body.

4.6.2.2 *Manufacturing and scalability [129]*

- The high cost and complexity of producing nanoparticles at scale pose a significant barrier. Standardizing production processes is essential for commercial viability and consistent efficacy.

4.6.2.3 *Regulatory and ethical issues [130]*

- Current regulatory frameworks are adapting slowly to accommodate nanotechnology-based therapies, which necessitate extensive testing and documentation. Ethical issues also arise, particularly with applications in gene editing and personalized medicine.

4.6.2.4 *Public perception [131]*

- Establishing trust in nanomedicine requires addressing public concerns about safety, privacy, and ethical implications. Transparent communication and education are key to gaining acceptance.

Addressing these challenges through interdisciplinary research, robust testing, and international collaboration will be crucial for unlocking the full potential of biomedical nanotechnology and transforming healthcare.

4.7 CONCLUSION

In conclusion, nanotechnology represents a transformative frontier in the biomedical sciences, offering groundbreaking innovations for disease prevention, diagnosis, and treatment. By manipulating materials at the nanoscale ranging from 1 to 100 nm, scientists can harness unique properties that significantly enhance interactions with biological systems. These advancements pave the way for unprecedented precision in targeting therapeutic agents, improving the efficacy of treatments, and addressing complex public health challenges [7].

The progress made in nanobiotechnology has introduced remarkable advancements, such as targeted drug delivery systems, highly sensitive diagnostics, and regenerative materials that are revolutionizing healthcare. Novel nanomaterials and innovative applications in diagnostics and therapeutics exemplify how nanotechnology is reshaping the landscape of personalized medicine, making healthcare solutions more efficient and accessible [132].

Moreover, clinical case studies illustrate the real-world impact of nanobiotechnology, confirming its potential to enhance patient outcomes, alleviate side effects, and provide tailored treatment options. However, the journey ahead is not without challenges. Future advancements must tackle critical issues surrounding biocompatibility, scalability, and safety while navigating complex regulatory landscapes and ethical considerations [115,116].

The promise of nanotechnology in biomedical applications is immense, and by addressing these challenges, the field can continue to innovate and improve public health outcomes. As research progresses, interdisciplinary collaboration and robust regulatory frameworks will be vital in integrating nanotechnology into routine clinical practice, ultimately revolutionizing healthcare for future generations. Through concerted efforts, nanotechnology holds the key to unlocking tailored healthcare solutions that meet the evolving demands of an increasingly complex healthcare landscape [133].

REFERENCES

1. Weerarathna IN, Luharia A, Tivaskar S, Nankong FA, Raymond D (2023) Emerging applications of biomedical science in pandemic prevention and control: a review. *Cureus*. 15: e44075. https://doi.org/10.7759/cureus.44075
2. Scott EA, Allen SD (2020) 17- Immunostimulatory materials, Editor(s): Kinam Park, In *Woodhead Publishing Series in Biomaterials, Biomaterials for Cancer Therapeutics* (Second Edition), Woodhead Publishing, Pages 471–497, ISBN 9780081029831, https://doi.org/10.1016/B978-0-08-102983-1.00017-X
3. Zhuo Y, Zhao YG, Zhang Y (2024) Enhancing drug solubility, bioavailability, and targeted therapeutic applications through magnetic nanoparticles. *Molecules*. 29: 4854. https://doi.org/10.3390/molecules29204854
4. Baker RE, Mahmud AS, Miller IF, Rajeev M, Rasambainarivo F, Rice BL, Takahashi S, Tatem AJ, Wagner CF, Wang L-F, Wesolowski A, Metcalf CJE (2022) Infectious disease in an era of global change. *Nat Rev Microbiol*. 20: 193–205. https://doi.org/10.1038/s41579-021-00639-z
5. Malik S, Muhammad K, Waheed Y (2023) Emerging applications of nanotechnology in healthcare and medicine. *Molecules*. 28: 6624. https://doi.org/10.3390/molecules28186624

6. Hamdan S, Pastar I, Drakulich S, Dikici E, Tomic-Canic M, Deo S, Daunert S (2017) Nanotechnology-driven therapeutic interventions in wound healing: potential uses and applications. *ACS Cent Sci.* 163–175. https://doi.org/10.1021/acscentsci.6b00371
7. Dutt Y, Pandey RP, Dutt M, Gupta A, Vibhuti A, Vidic J, Raj VS, Chang CM, Priyadarshini A (2023) Therapeutic applications of nanobiotechnology. *J Nanobiotechnol.* 21:148. https://doi.org/10.1186/s12951-023-01909-z
8. Babanejad N, Nabid MR, Farhadian A, Dorkoosh F, Zarrintaj P, Saeb MR, Mozafari M (2019) Sustained delivery of olanzapine from sunflower oil-based polyol-urethane nanoparticles synthesised through a cyclic carbonate ring-opening reaction. *IET Nanobiotechnol.* 13: 703–711. https://doi.org/10.1049/iet-nbt.2018.5440
9. Karnwal A, Sharma V, Kumar G, Jassim AY, Dohroo A, Sivanesan I (2024) Transforming medicine with nanobiotechnology: nanocarriers and their biomedical applications. *Pharmaceutics.* 16: 1114. https://doi.org/10.3390/pharmaceutics16091114
10. Nosrati H, Aramideh Khouy R, Nosrati A, Khodaei M, Banitalebi-Dehkordi M, Ashrafi-Dehkordi K, Sanami S, Alizadeh Z (2021) Nanocomposite scaffolds for accelerating chronic wound healing by enhancing angiogenesis. *J* Nanobiotechnol. 19: 1. https://doi.org/10.1186/s12951-020-00755-7
11. Bhardwaj P, Bhatia E, Sharma S, Ahamad N, Banerjee R (2020) Advancements in prophylactic and therapeutic nanovaccines. *Acta Biomater.* 108: 1–21. https://doi.org/10.1016/j.actbio.2020.03.020
12. Torres-Sangiao E, Holban AM, Gestal MC (2016) Advanced nanobiomaterials: vaccines, diagnosis and treatment of infectious diseases. *Molecules.* 21: 867. https://doi.org/10.3390/molecules21070867
13. Soni A, Bhandari MP, Tripathi GK, Bundela P, Khiriya PK, Khare PS, Kashyap MK, Dey A, Vellingiri B, Sundaramurthy S, Suresh A, Pérez de la Lastra JM (2023) Nano-biotechnology in tumour and cancerous disease: a perspective review. *J Cell Mol Med.* 27: 737–762. https://doi.org/10.1111/jcmm.17677
14. Patra JK, Das G, Fraceto LF, Campos EVR, Rodriguez-Torres MDP, Acosta-Torres LS, Diaz-Torres LA, Grillo R, Swamy MK, Sharma S, Habtemariam S, Shin HS (2018) Nano based drug delivery systems: recent developments and future prospects. *J Nanobiotechnol.* 16: 71. https://doi.org/10.1186/s12951-018-0392-8
15. Prakash S (2023) Nano-based drug delivery system for therapeutics: a comprehensive review. *Biomed Phys Eng Express.* 9. https://doi.org/10.1088/2057-1976/acedb2
16. Kim J, Jozic A, Lin Y, Eygeris Y, Bloom E, Tan X, Acosta C, MacDonald KD, Welsher KD, Sahay G. (2022) Engineering lipid nanoparticles for enhanced intracellular delivery of mRNA through inhalation. *ACS Nano.* 16(9): 14792–14806. https://doi.org/10.1021/acsnano.2c05647
17. Waheed I, Ali A, Tabassum H, Khatoon N, Lai WF, Zhou X (2024) Lipid-based nanoparticles as drug delivery carriers for cancer therapy. *Front Oncol.* 14: 1296091. https://doi.org/10.3389/fonc.2024.1296091
18. Begines B, Ortiz T, Pérez-Aranda M, Martínez G, Merinero M, Argüelles-Arias F, Alcudia A (2020) Polymeric nanoparticles for drug delivery: recent developments and future prospects. *Nanomaterials (Basel).* 10: 1403. https://doi.org/10.3390/nano10071403

19. Gagliardi A, Giuliano E, Venkateswararao E, Fresta M, Bulotta S, Awasthi V, Cosco D (2021) Biodegradable polymeric nanoparticles for drug delivery to solid tumors. *Front Pharmacol.* 12: 601626. https://doi.org/10.3389/fphar.2021.601626
20. Ding L, Agrawal P, Singh SK, Chhonker YS, Sun J, Murry DJ (2024) Polymer-based drug delivery systems for cancer therapeutics. *Polymers (Basel).* 16: 843. https://doi.org/10.3390/polym16060843
21. Du X, Yada E, Terai Y, Takahashi T, Nakanishi H, Tanaka H, Akita H, Itaka K (2023) Comprehensive evaluation of lipid nanoparticles and polyplex nanomicelles for muscle-targeted mRNA delivery. *Pharmaceutics.* 15: 2291. https://doi.org/10.3390/pharmaceutics15092291
22. Tawfik SM, Azizov S, Elmasry MR, Sharipov M, Lee YI (2020) Recent advances in nanomicelles delivery systems. *Nanomaterials (Basel).* 11: 70. https://doi.org/10.3390/nano11010070
23. Chen G, Mao D, Wang X, Chen J, Gu C, Huang S, Yang Y, Zhang F, Tan W (2023) Aptamer-based self-assembled nanomicelle enables efficient and targeted drug delivery. *J Nanobiotechnol.* 21: 415. https://doi.org/10.1186/s12951-023-02164-y
24. Jain KK (2007) Applications of nanobiotechnology in clinical diagnostics. *Clin Chem.* 53: 2002–2009. https://doi.org/10.1373/clinchem.2007.090795
25. Huang S, Huang G (2024) The utilization of quantum dot labeling as a burgeoning technique in the field of biological imaging. *RSC Adv.* 14(29): 20884–20897. https://doi.org/10.1039/d4ra04402a
26. Shetty B, Chauhan RS, Tirupathi S, Krishnapriya N, Patil L, Rathi N (2024) Quantum dots in noninvasive imaging of oral squamous cell carcinomas: a scoping literature review. *J Cancer Res Ther.* 20(3): 745–749. https://doi.org/10.4103/jcrt.jcrt_2203_21
27. Mondal J, Lamba R, Yukta Y, Yadav R, Kumar R, Pani B, Singh B (2024) Advancements in semiconductor quantum dots: expanding frontiers in optoelectronics, analytical sensing, biomedicine, and catalysis. *J Mater Chem C.* 12(39). https://doi.org/10.1039/D4TC01396D
28. Malavika JP, Shobana C, Sundarraj S, Ganeshbabu M, Kumar P, Selvan RK (2022). Green synthesis of multifunctional carbon quantum dots: an approach in cancer theranostics. *Biomater Adv.* 136: 212756. https://doi.org/10.1016/j.bioadv.2022.212756
29. Hao S, Suebka S, Su J (2024) Single 5-nm quantum dot detection via microtoroid optical resonator photothermal microscopy. *Light Sci Appl.* 13: 195. https://doi.org/10.1038/s41377-024-01536-9
30. Wenck C, Leopoldt D, Habib M, Hegermann J, Stiesch M, Doll-Nikutta K, Heisterkamp A, Torres-Mapa ML (2024) Colorimetric detection of oral bacteria using functionalized gold nanoparticles as a plasmonic biosensor array. *Nanoscale Adv.* 6: 1447–1459. https://doi.org/10.1039/d3na00477e
31. Dadfar SM, Roemhild K, Drude NI, von Stillfried S, Knüchel R, Kiessling F, Lammers T (2019) Iron oxide nanoparticles: diagnostic, therapeutic and theranostic applications. *Adv Drug Deliv Rev.* 138: 302–325. https://doi.org/10.1016/j.addr.2019.01.005
32. Luo GF, Zhang XZ (2024) Magnetic nanoparticles for use in bioimaging. *Biomater Sci.* https://doi.org/10.1039/d4bm01145g

33. Alzate-Correa D, Lawrence WR, Salazar-Puerta A, Higuita-Castro N, Gallego-Perez D (2022) Nanotechnology-driven cell-based therapies in regenerative medicine. *AAPS J*. 24: 43. https://doi.org/10.1208/s12248-022-00692-3
34. Ding H, Cheng Y, Niu X, Hu Y (2021) Application of electrospun nanofibers in bone, cartilage and osteochondral tissue engineering. *J Biomater Sci Polym Ed*. 32: 536–561. https://doi.org/10.1080/09205063.2020.1849922
35. Liu H, Chen J, Qiao S, Zhang W (2021) Carbon-based nanomaterials for bone and cartilage regeneration: a review. *ACS Biomater Sci Eng*. 7: 4718–4735. https://doi.org/10.1021/acsbiomaterials.1c00759
36. Newman P, Minett A, Ellis-Behnke R, Zreiqat H (2013) Carbon nanotubes: their potential and pitfalls for bone tissue regeneration and engineering. *Nanomedicine*. 9: 1139–58. https://doi.org/10.1016/j.nano.2013.06.001
37. Liu D, Li L, Shi BL, Shi B, Li MD, Qiu Y, Zhao D, Shen QD, Zhu ZZ (2022) Ultrasound-triggered piezocatalytic composite hydrogels for promoting bacterial-infected wound healing. *Bioact Mater*. 24: 96–111. https://doi.org/10.1016/j.bioactmat.2022.11.023
38. Henn D, Chen K, Fischer K, Rauh A, Barrera JA, Kim YJ, Martin RA, Hannig M, Niedoba P, Reddy SK, Mao HQ, Kneser U, Gurtner GC, Sacks JM, Schmidt VJ (2020) Tissue engineering of axially vascularized soft-tissue flaps with a poly-(ε-caprolactone) nanofiber-hydrogel composite. *Adv Wound Care (New Rochelle)*. 9: 365–377. https://doi.org/10.1089/wound.2019.0975
39. Das A, Ali N (2021) Nanovaccine: an emerging strategy. *Expert Rev Vaccines*. 20: 1273–1290. https://doi.org/10.1080/14760584.2021.1984890
40. Hald Albertsen C, Kulkarni JA, Witzigmann D, Lind M, Petersson K, Simonsen JB (2022) The role of lipid components in lipid nanoparticles for vaccines and gene therapy. *Adv Drug Deliv Rev*. 188: 114416. https://doi.org/10.1016/j.addr.2022.114416
41. Wilson B, Geetha KM (2022) Lipid nanoparticles in the development of mRNA vaccines for COVID-19. *J Drug Deliv Sci Technol*. 74: 103553. https://doi.org/10.1016/j.jddst.2022.103553
42. Ye F, Dewanjee S, Li Y, Jha NK, Chen ZS, Kumar A, Vishakha, Behl T, Jha SK, Tang H (2023) Advancements in clinical aspects of targeted therapy and immunotherapy in breast cancer. *Mol Cancer*. 22: 105. https://doi.org/10.1186/s12943-023-01805-y
43. Song C, Shi J, Zhang P, Zhang Y, Xu J, Zhao L, Zhang R, Wang H, Chen H (2022) Immunotherapy for Alzheimer's disease: targeting β-amyloid and beyond. *Transl Neurodegener*. 11: 18. https://doi.org/10.1186/s40035-022-00292-3
44. Hassan HAFM, Diebold SS, Smyth LA, Walters AA, Lombardi G, Al-Jamal KT (2019) Application of carbon nanotubes in cancer vaccines: achievements, challenges and chances. *J Control Release*. 297: 79–90. https://doi.org/10.1016/j.jconrel.2019.01.017
45. Khaparde A, Mathias GP, Poornachandra B, Thirumalesh MB, Shetty R, Ghosh A (2024) Gene therapy for retinal diseases: from genetics to treatment. *Indian J Ophthalmol*. 72: 1091–1101. https://doi.org/10.4103/IJO.IJO_2902_23
46. Mertz TM, Sharma S, Chabes A, Shcherbakova PV (2015) Colon cancer-associated mutator DNA polymerase δ variant causes expansion of dNTP pools increasing its own infidelity. *Proc Natl Acad Sci U S A*. 112: E2467–E2476. https://doi.org/10.1073/pnas.1422934112

47. Jiang L, Wang D, He Y, Shu Y (2023) Advances in gene therapy hold promise for treating hereditary hearing loss. *Mol Ther.* 31: 934–950. https://doi.org/10.1016/j.ymthe.2023.02.001
48. Pandey Y, Ambwani S (2023) Nano metal based herbal theranostics for cancer management: coalescing nature's boon with nanotechnological advancement. *Curr Pharm Biotechnol.* 23: 30–46. https://doi.org/10.2174/1389201022666210122141724
49. Ha D, Yang N, Nadithe V (2016) Exosomes as therapeutic drug carriers and delivery vehicles across biological membranes: current perspectives and future challenges. *Acta Pharm Sin B.* 6: 287–296. https://doi.org/10.1016/j.apsb.2016.02.001
50. Chaudhari R, Patel V, Kumar A (2024) Cutting-edge approaches for targeted drug delivery in breast cancer: beyond conventional therapies. *Nanoscale Adv.* 6: 2270–2286. https://doi.org/10.1039/d4na00086b
51. Cheng Q, Wei T, Farbiak L, Johnson LT, Dilliard SA, Siegwart DJ (2020) Selective organ targeting (SORT) nanoparticles for tissue-specific mRNA delivery and CRISPR-Cas gene editing. *Nat Nanotechnol.* 15: 313–320. https://doi.org/10.1038/s41565-020-0669-6
52. Tan A, Rajadas J, Seifalian AM (2013) Exosomes as nano-theranostic delivery platforms for gene therapy. *Adv Drug Deliv Rev.* 65: 357–367. https://doi.org/10.1016/j.addr.2012.06.014
53. Pérez-Herrero E, Fernández-Medarde A (2015) Advanced targeted therapies in cancer: drug nanocarriers, the future of chemotherapy. *Eur J Pharm Biopharm.* 93: 52–79. https://doi.org/10.1016/j.ejpb.2015.03.018
54. Sahu MK, Yadav R, Tiwari SP (2023) Recent advances in nanotechnology. *Int J Nanomater Nanotechnol Nanomed.* 9(2): 015–023. https://doi.org/10.17352/2455-3492.000053
55. Bhagat DS, Gurnule WB, Bumbrah GS, Koinkar P, Chawla PA (2023) Recent advances in biomedical applications of biogenic nanomaterials. *Curr Pharm Biotechnol.* 24: 86–100. https://doi.org/10.2174/1389201023666220513101628
56. Wu G, Hui X, Hu L, Bai Y, Rahaman A, Yang XF, Chen C (2022) Recent advancement of bioinspired nanomaterials and their applications: a review. *Front Bioeng Biotechnol.* 10: 952523. https://doi.org/10.3389/fbioe.2022.952523
57. Maiti D, Tong X, Mou X, Yang K (2019) Carbon-based nanomaterials for biomedical applications: a recent study. *Front Pharmacol.* 9: 1401. https://doi.org/10.3389/fphar.2018.01401
58. Cui F, Li T, Wang D, Yi S, Li J, Li X (2022) Recent advances in carbon-based nanomaterials for combating bacterial biofilm-associated infections. *J Hazard Mater.* 431: 128597. https://doi.org/10.1016/j.jhazmat.2022.128597
59. Alavi M, Kamarasu P, McClements DJ, Moore MD (2022) Metal and metal oxide-based antiviral nanoparticles: properties, mechanisms of action, and applications. *Adv Colloid Interface Sci.* 306: 102726. https://doi.org/10.1016/j.cis.2022.102726
60. Mukherjee D, Sil M, Goswami A, Lahiri D, Nag M (2023) Effectiveness of metal and metal oxide nanoparticles against bacterial biofilms: perspectives and limitations. *J Basic Microbiol.* 63: 971–985. https://doi.org/10.1002/jobm.202300013

61. Neto WS, Jensen AT, Ferreira GR, Valadares LF, Gambetta R, Gonçalves SB, Machado FA (2015) Survey on synthesis processes of structured materials for biomedical applications: iron-based magnetic nanoparticles, polymeric materials and polymerization processes. *Curr Pharm.* 21: 5336–5358. https://doi.org/10.2174/1381612821666150917093031
62. Xu Y, Fourniols T, Labrak Y, Préat V, Beloqui A, des Rieux A (2022) Surface modification of lipid-based nanoparticles. *ACS Nano.* 16: 7168–7196. https://doi.org/10.1021/acsnano.2c02347
63. Al-Sayadi GMH, Verma A, Choudhary Y, Sandal P, Patel P, Singh D, Gupta GD, Kurmi BD (2023) Solid lipid nanoparticles (SLNs): advancements in modification strategies toward drug delivery vehicle. *Pharm Nanotechnol.* 11: 138–154. https://doi.org/10.2174/2211738511666221026163303
64. Le N, Kim K (2023) Current advances in the biomedical applications of quantum dots: promises and challenges. *Int J Mol Sci.* 24: 12682. https://doi.org/10.3390/ijms241612682
65. Yukawa H, Sato K, Baba Y (2023) Theranostics applications of quantum dots in regenerative medicine, cancer medicine, and infectious diseases. *Adv Drug Deliv Rev.* 200: 114863. https://doi.org/10.1016/j.addr.2023.114863
66. Kamila S, McEwan C, Costley D, Atchison J, Sheng Y, Hamilton GR, Fowley C, Callan JF (2016) Diagnostic and therapeutic applications of quantum dots in nanomedicine. *Top Curr Chem.* 370: 203–224. https://doi.org/10.1007/978-3-319-22942-3_7
67. Babu LT, Paira P (2017) Current application of quantum dots (QD) in cancer therapy: a review. *Mini Rev Med Chem.* 17: 1406–1415. https://doi.org/10.2174/1389557517666170315125504
68. Cui L, Li CC, Tang B, Zhang CY (2018) Advances in the integration of quantum dots with various nanomaterials for biomedical and environmental applications. *Analyst.* 143: 2469–2478. https://doi.org/10.1039/c8an00222c
69. Wang L, Arrabito G (2015) Hybrid, multiplexed, functional DNA nanotechnology for bioanalysis. *Analyst.* 140: 5821–5848. https://doi.org/10.1039/c5an00861a
70. Kadria-Vili Y, Neumann O, Zhao Y, Nordlander P, Martinez GV, Bankson JA, Halas NJ (2022) Gd_2O_3-mesoporous silica/gold nanoshells: a potential dual T_1/T_2 contrast agent for MRI-guided localized near-IR photothermal therapy. *Proc Natl Acad Sci U S A.* 119: e2123527119. https://doi.org/10.1073/pnas.2123527119
71. García-Hevia L, Casafont Í, Oliveira J, Terán N, Fanarraga ML, Gallo J, Bañobre-López M (2021) Magnetic lipid nanovehicles synergize the controlled thermal release of chemotherapeutics with magnetic ablation while enabling non-invasive monitoring by MRI for melanoma theranostics. *Bioact Mater.* 8: 153–164. https://doi.org/10.1016/j.bioactmat.2021.06.009
72. Beach MA, Nayanathara U, Gao Y, Zhang C, Xiong Y, Wang Y, Such GK (2024) Polymeric nanoparticles for drug delivery. *Chem Rev.* 124: 5505–5616. https://doi.org/10.1021/acs.chemrev.3c00705
73. Ahlawat J, Henriquez G, Narayan M (2018) Enhancing the delivery of chemotherapeutics: role of biodegradable polymeric nanoparticles. *Molecules.* 23: 2157. https://doi.org/10.3390/molecules23092157

74. Niżnik Ł, Noga M, Kobylarz D, Frydrych A, Krośniak A, Kapka-Skrzypczak L, Jurowski K (2024) Gold nanoparticles (AuNPs)-toxicity, safety and green synthesis: a critical review. *Int J Mol Sci* 25: 4057. https://doi.org/10.3390/ijms25074057
75. Chmykhalo V, Belanova A, Belousova M, Butova V, Makarenko Y, Khrenkova V, Soldatov A, Zolotukhin P (2021) Microbial-based magnetic nanoparticles production: a mini-review. *Integr Biol (Camb).* 13: 98–107. https://doi.org/10.1093/intbio/zyab005
76. Rong G, Tuttle EE, Neal Reilly A, Clark HA (2019) Recent developments in nanosensors for imaging applications in biological systems. *Annu Rev Anal Chem (Palo Alto Calif).* 12: 109–128. https://doi.org/10.1146/annurev-anchem-061417-125747
77. Berthiaume F, Maguire TJ, Yarmush ML (2011) Tissue engineering and regenerative medicine: history, progress, and challenges. *Annu Rev Chem Biomol Eng.* 2: 403–430. https://doi.org/10.1146/annurev-chembioeng-061010-114257
78. Unal S, Arslan S, Yilmaz BK, Oktar FN, Ficai D, Ficai A, Gunduz O (2020) Polycaprolactone/gelatin/hyaluronic acid electrospun scaffolds to mimic glioblastoma extracellular matrix. *Materials (Basel).* 13: 2661. https://doi.org/10.3390/ma13112661
79. Xiang Y, Peng Y, Li S, Luo L, Zan X, Kumeria T, Ye Q (2024) Local spinal cord injury treatment using a dental pulp stem cell encapsulated H_2S releasing multifunctional injectable hydrogel. *Adv Healthc Mater.* 13: e2302286. https://doi.org/10.1002/adhm.202302286
80. Pei B, Wang W, Dunne N, Li X (2019) Applications of carbon nanotubes in bone tissue regeneration and engineering: superiority, concerns, current advancements, and prospects. *Nanomaterials (Basel).* 9: 1501. https://doi.org/10.3390/nano9101501
81. Gostaviceanu A, Gavrilaş S, Copolovici L, Copolovici DM (2024) Graphene-oxide peptide-containing materials for biomedical applications. *Int J Mol Sci.*25: 10174. https://doi.org/10.3390/ijms251810174
82. Dediu V, Ghitman J, Gradisteanu Pircalabioru G, Chan KH, Iliescu FS, Iliescu C (2023) Trends in photothermal nanostructures for antimicrobial applications. *Int J Mol Sci.* 24: 9375. https://doi.org/10.3390/ijms24119375
83. Raza S, Ansari A, Siddiqui NN, Ibrahim F, Abro MI, Aman A (2021) Biosynthesis of silver nanoparticles for the fabrication of non cytotoxic and antibacterial metallic polymer based nanocomposite system. *Sci Rep.* 11: 10500. https://doi.org/10.1038/s41598-021-90016-w.
84. Aouadi A, Hamada Saud D, Rebiai A, Achouri A, Benabdesselam S, Mohamed Abd El-Mordy F, Pohl P, Ahmad SF, Attia SM, Abulkhair HS, Ararem A, Messaoudi M (2024) Introducing the antibacterial and photocatalytic degradation potentials of biosynthesized chitosan, chitosan-ZnO, and chitosan-ZnO/PVP nanoparticles. *Sci Rep.* 14: 14753. https://doi.org/10.1038/s41598-024-65579-z
85. Hlapisi N, Songca SP, Ajibade PA (2024) Capped plasmonic gold and silver nanoparticles with porphyrins for potential use as anticancer agents: a review. *Pharmaceutics.* 6: 1268. https://doi.org/10.3390/pharmaceutics16101268
86. Yusefi M, Shameli K, Jahangirian H, Teow SY, Afsah-Hejri L, Mohamad Sukri SNA, Kuča K (2023) How magnetic composites are effective anticancer therapeutics? A comprehensive review of the literature. *Int J Nanomed.* 18: 3535–3575. https://doi.org/10.2147/IJN.S375964

87. Qu F, Geng R, Liu Y, Zhu J (2022) Advanced nanocarrier- and microneedle-based transdermal drug delivery strategies for skin diseases treatment. *Theranostics*. 12: 3372–3406. https://doi.org/10.7150/thno.69999
88. Uprety B, Abrahamse H (2022) Semiconductor quantum dots for photodynamic therapy: recent advances. *Front Chem*. 10: 946574. https://doi.org/10.3389/fchem.2022.946574
89. Caulfield T, Gold ER, Cho MK (2000) Patenting human genetic material: refocusing the debate. *Nat Rev Genet*. 1: 227–231. https://doi.org/10.1038/35042087
90. Wang JH, Gessler DJ, Zhan W, Gallagher TL, Gao G (2024) Adeno-associated virus as a delivery vector for gene therapy of human diseases. *Signal Transduct Target Ther*. 9: 78. https://doi.org/10.1038/s41392-024-01780-w
91. Parambi DGT, Alharbi KS, Kumar R, Harilal S, Batiha GE, Cruz-Martins N, Magdy O, Musa A, Panda DS, Mathew B (2022) Gene therapy approach with an emphasis on growth factors: theoretical and clinical outcomes in neurodegenerative diseases. *Mol Neurobiol*. 59: 191–233. https://doi.org/10.1007/s12035-021-02555-y
92. Hosseinkhani H, Domb AJ, Sharifzadeh G, Nahum V (2023) Gene therapy for regenerative medicine. *Pharmaceutics*.15: 856. https://doi.org/10.3390/pharmaceutics15030856
93. Parvin N, Joo SW, Mandal TK (2024) Enhancing vaccine efficacy and stability: a review of the utilization of nanoparticles in mRNA vaccines. *Biomolecules*. 14: 1036. https://doi.org/10.3390/biom14081036
94. Zhao T, Cai Y, Jiang Y, He X, Wei Y, Yu Y, Tian X (2023) Vaccine adjuvants: mechanisms and platforms. *Signal Transduct Target Ther*. 8: 283. https://doi.org/10.1038/s41392-023-01557-7
95. Tariq H, Batool S, Asif S, Ali M, Abbasi BH (2022) Virus-like particles: revolutionary platforms for developing vaccines against emerging infectious diseases. *Front Microbiol*. 12: 790121. https://doi.org/10.3389/fmicb.2021.790121
96. Harun-Ur-Rashid M, Jahan I, Foyez T, Imran AB (2023) Bio-inspired nanomaterials for micro/nanodevices: a new era in biomedical applications. *Micromachines (Basel)*. 14: 1786. https://doi.org/10.3390/mi14091786
97. Ma W, Zhan Y, Zhang Y, Mao C, Xie X, Lin Y (2021) The biological applications of DNA nanomaterials: current challenges and future directions. *Signal Transduct Target Ther*. 6: 351. https://doi.org/10.1038/s41392-021-00727-9
98. Gupta P, Chattopadhaya A, Gautam V (2024) Myco-nanotechnological approach to synthesize gold nanoparticles using a fungal endophyte, Penicillium oxalicum, and unravelling its antibacterial activity and anti-breast cancer role via metabolic reprogramming. *Biomed Mater*, 19(6): 065030.
99. Gupta P Singh S., Rai N., Verma A., Tiwari. H., Kamble SC., Gautam HK., Gautam V. (2024). Unveiling the cytotoxic and anti-proliferative potential of green-synthesized silver nanoparticles mediated by Colletotrichum gloeosporioides. *RSC Adv*. 14(6): 4074–4088.
100. Gupta P, et al. (2022). Green-based approach to synthesize silver nanoparticles using the fungal endophyte Penicillium oxalicum and their antimicrobial, antioxidant, and in vitro anticancer potential. *ACS Omega*. 7(50): 46653–46673.

101. Barhoum A, García-Betancourt ML, Jeevanandam J, Hussien EA, Mekkawy SA, Mostafa M, Omran MM, S Abdalla M, Bechelany M (2022) Review on natural, incidental, bioinspired, and engineered nanomaterials: history, definitions, classifications, synthesis, properties, market, toxicities, risks, and regulations. *Nanomater (Basel).* 12: 177. https://doi.org/10.3390/nano12020177
102. Ma X, Tian Y, Yang R, Wang H, Allahou LW, Chang J, Williams G, Knowles JC, Poma A (2024) Nanotechnology in healthcare, and its safety and environmental risks. *J Nanobiotechnol.* 22: 715. https://doi.org/10.1186/s12951-024-02901-x
103. Singh S, Tiwari H, Verma A, Gupta P, Chattopadhaya A, Singh A, Singh S, Kumar B, Mandal A, Kumar R, Yadav AK, Gautam HK, Gautam V (2024). Sustainable synthesis of novel green-based nanoparticles for therapeutic interventions and environmental remediation. *ACS Synthet Biol.* 13(7): 1994–2007.
104. Gautam V., Kumar R., Das Manandhar K., Kamble SC (2024). Navigating the Nanoscale Frontier: An In-Depth Introduction to the World of Nanomedicine, in *Nanomedicine: Innovations, Applications, and Breakthroughs in the Quest for Health and Medicine's Future*, V. Gautam, et al., Editors. 1st ed. 2024, Springer Nature Switzerland: Cham. p. 531–559. https://doi.org/10.1007/978-3-031-72467-1
105. Tripathi A, Pandey VK, Sharma G, Sharma AR, Taufeeq A, Jha AK, Kim JC (2024) Genomic insights into dementia: precision medicine and the impact of gene-environment interaction. *Aging Dis.* 15: 2113–2135. https://doi.org/10.14336/AD.2024.0322
106. Feng S, Xie X, Liu J, Li A, Wang Q, Guo D, Li S, Li Y, Wang Z, Guo T, Zhou J, Tang DYY, Show PL (2023) A potential paradigm in CRISPR/Cas systems delivery: at the crossroad of microalgal gene editing and algal-mediated nanoparticles. *J Nanobiotechnol.* 21: 370. https://doi.org/10.1186/s12951-023-02139-z
107. Liu Y, Lin Z, Wang Y, Chen L, Wang Y, Luo C (2024) Nanotechnology in inflammation: cutting-edge advances in diagnostics, therapeutics and theranostics. *Theranostics.* 14: 2490–2525. https://doi.org/10.7150/thno.91394
108. Darwish MA, Abd-Elaziem W, Elsheikh A, Zayed AA (2024) Advancements in nanomaterials for nanosensors: a comprehensive review. *Nanoscale Adv.* 6: 4015–4046. https://doi.org/10.1039/d4na00214h
109. Bonlawar J, Setia A, Challa RR, Vallamkonda B, Mehata AK, Vaishali, Viswanadh MK, Muthu MS (2024) Targeted nanotheransotics: integration of preclinical MRI and CT in the molecular imaging and therapy of advanced diseases. *Nanotheranostics.* 8: 401–426. https://doi.org/10.7150/ntno.95791
110. Li S, Chen L, Fu Y (2023) Nanotechnology-based ocular drug delivery sysems: recent advances and future prospects. *J Nanobiotechnol.* 21: 232. https://doi.org/10.1186/s12951-023-01992-2
111. Aebisher D, Rogóż K, Myśliwiec A, Dynarowicz K, Wiench R, Cieślar G, Kawczyk-Krupka A, Bartusik-Aebisher D (2024) The use of photodynamic therapy in medical practice. *Front Oncol.* 14: 1373263. https://doi.org/10.3389/fonc.2024.1373263
112. Tang Y, Li L (2024) The application of nanovaccines in autoimmune diseases. *Int J Nanomed.* 19: 367–388. https://doi.org/10.2147/IJN.S440612

113. Ho D, Quake SR, McCabe ERB, Chng WJ, Chow EK, Ding X, Gelb BD, Ginsburg GS, Hassenstab J, Ho CM, Mobley WC, Nolan GP, Rosen ST, Tan P, Yen Y, Zarrinpar A (2020) Enabling technologies for personalized and precision medicine. *Trends Biotechnol.* 38: 497–518. https://doi.org/10.1016/j.tibtech.2019.12.021
114. Petrovic S, Bita B, Barbinta-Patrascu ME (2024) Nanoformulations in pharmaceutical and biomedical applications: green perspectives. *Int J Mol Sci.* 25: 5842. https://doi.org/10.3390/ijms25115842
115. Joseph TM, Kar Mahapatra D, Esmaeili A, Piszczyk Ł, Hasanin MS, Kattali M, Haponiuk J, Thomas S (2023) Nanoparticles: taking a unique position in medicine. *Nanomater (Basel).* 13: 574. https://doi.org/10.3390/nano13030574
116. Debnath SK, Debnath M, Ghosh A, Srivastava R, Omri A (2024) Targeting tumor hypoxia with nanoparticle-based therapies: challenges, opportunities, and clinical implications. *Pharmaceuticals (*Basel*).*17:1389. doi: 10.3390/ph17101389.
117. Ielo I, Rando G, Giacobello F, Sfameni S, Castellano A, Galletta M, Drommi D, Rosace G, Plutino MR (2021) Synthesis, chemical-physical characterization, and biomedical applications of functional gold nanoparticles: a review. *Molecules.* 26: 5823. https://doi.org/10.3390/molecules26195823
118. Mahmoudi A, Tavakoly Sany SB, Ahari Salmasi M, Bakhshi A, Bustan A, Heydari S, Rezayi M, Gheybi F (2023) Application of nanotechnology in air purifiers as a viable approach to protect against Corona virus. *IET Nanobiotechnol.* 17: 289–301. https://doi.org/10.1049/nbt2.12132
119. Ahmad A, Imran M, Ahsan H (2023) Biomarkers as biomedical bioindicators: approaches and techniques for the detection, analysis, and validation of novel biomarkers of diseases. *Pharmaceutics.* 15: 1630. https://doi.org/10.3390/pharmaceutics15061630
120. Qiao K, Xu L, Tang J, Wang Q, Lim KS, Hooper G, Woodfield TBF, Liu G, Tian K, Zhang W, Cui X (2022) The advances in nanomedicine for bone and cartilage repair. *J Nanobiotechnol.* 20: 141. https://doi.org/10.1186/s12951-022-01342-8
121. Maleki-Ghaleh H, Siadati MH, Fallah A, Koc B, Kavanlouei M, Khademi-Azandehi P, Moradpur-Tari E, Omidi Y, Barar J, Beygi-Khosrowshahi Y, Kumar AP, Adibkia K (2021) Antibacterial and cellular behaviors of novel zinc-doped hydroxyapatite/graphene nanocomposite for bone tissue engineering. *Int J Mol Sci.* 22: 9564. https://doi.org/10.3390/ijms22179564
122. Fang X, Lan H, Jin K, Gong D, Qian J (2022) Nanovaccines for cancer prevention and immunotherapy: an update review. *Cancers (Basel).* 14: 3842. https://doi.org/10.3390/cancers14163842
123. Mohd Maidin NN, Buyong MR, A Rahim R, Mohamed MA (2021) Dielectrophoresis applications in biomedical field and future perspectives in biomedical technology. *Electrophoresis.* 42: 2033–2059. https://doi.org/10.1002/elps.202100043
124. Hu Q, Li H, Wang L, Gu H, Fan C (2019) DNA nanotechnology-enabled drug delivery systems. *Chem Rev.* 119: 6459–6506. https://doi.org/10.1021/acs.chemrev.7b00663

125. Gorjikhah F, Davaran S, Salehi R, Bakhtiari M, Hasanzadeh A, Panahi Y, Emamverdy M, Akbarzadeh A (2016) Improving “lab-on-a-chip” techniques using biomedical nanotechnology: a review. *Artif Cells Nanomed Biotechnol.* 44: 1609–1614. https://doi.org/10.3109/21691401.2015.1129619.
126. Li L, Zhang X, Zhou J, Zhang L, Xue J, Tao W (2022) Non-invasive thermal therapy for tissue engineering and regenerative medicine. *Small.* 18: e2107705. https://doi.org/10.1002/smll.202107705
127. Li M, Xi N, Wang Y, Liu L (2021) Progress in nanorobotics for advancing biomedicine. *IEEE Trans Biomed Eng.* 68: 130–147. https://doi.org/10.1109/TBME.2020.2990380. Epub 2020 Dec 21.
128. Najahi-Missaoui W, Arnold RD, Cummings BS (2020) Safe nanoparticles: are we there yet? *Int J Mol Sci.* 22: 385. https://doi.org/10.3390/ijms22010385
129. Rostami M, Beheshtizadeh N, Ranjbar FE, Najafi N, Ahmadi A, Ahmadi P, Rostamabadi H, Pazhouhnia Z, Assadpour E, Mirzanajafi-Zanjani M, Kisomi MF, Kharazmi MS, Jafari SM (2023) Recent advances in electrospun protein fibers/nanofibers for the food and biomedical applications. *Adv Colloid Interface Sci.* 311: 102827. https://doi.org/10.1016/j.cis.2022.102827
130. Das RK, Drolet BC (2022) Lessons from Theranos: restructuring biomedical innovation. *J Med Syst.* 46: 25. https://doi.org/10.1007/s10916-022-01813-3
131. Yetisen AK, Coskun AF, England G, Cho S, Butt H, Hurwitz J, Kolle M, Khademhosseini A, Hart AJ, Folch A, Yun SH (2016) Art on the nanoscale and beyond. *Adv Mater.* 28: 1724–1742. https://doi.org/10.1002/adma.201502382
132. Kim ES, Ahn EH, Chung E, Kim DH (2013) Recent advances in nanobiotechnology and high-throughput molecular techniques for systems biomedicine. *Mol Cells.* 36: 477–484. https://doi.org/10.1007/s10059-013-0304-6
133. Staggers N, McCasky T, Brazelton N, Kennedy R (2008) Nanotechnology: the coming revolution and its implications for consumers, clinicians, and informatics. *Nurs Outlook.* 56: 268–274. https://doi.org/10.1016/j.outlook.2008.06.004

Chapter 5

Fungal endophytes as a treasure trove of bioactive compounds

Implications for anticancer drug development

Neil Ferland, Rana Chakraborty, Katelyn Tran Nguyen, Dawn Littlefield, and Indrani Mukherjee

Department of Pediatrics, University of Miami, Miami, FL

5.1 INTRODUCTION

Fungal endophytes are plant symbiotes that live within plant tissue without harming their hosts [1] and have been detected intracellularly within plant tissue, particularly in tree bark [2]. This symbiotic relationship between the fungi and host plants provides the fungus with an ideal environment for growth and survival. In turn, the fungus secretes several bioactive compounds and secondary metabolites, which improve growth, development, and immunity of host plants [1]. Endophytes produce a wide variety of metabolites that can be isolated and utilized in medical treatments. These include terpenes, alkaloids, flavonoids, and steroids, which possess antitumor, antibacterial, and anti-inflammatory properties. It is these bioactive compounds that can be isolated and used in the development of cancer treatments [3–5].

The medicinal properties of plants have been known and utilized by humans for centuries, with the oldest written evidence dating back 5,000 years [6]. Fungal endophytes were first isolated in 1904, but their potential usage in medicine was largely unexplored. That was the case until the discovery and isolation of *Taxomyces andreanae*, a species of endophytic fungi found in the Pacific yew tree (*Taxus brevifolia*). This was a landmark discovery in the field of fungal research, as *T. andreanae* is a natural producer of the anticancer agent paclitaxel [7,8]. The discovery of this taxane in an endophytic fungus was groundbreaking. Paclitaxel was previously only known as a plant product, but the finding that it was produced instead by a fungus increased interest in this field and led to the isolation of new fungal metabolites [9,10].

As these metabolites are derived from natural sources, the collection and isolation of endophytic fungi come with significant challenges. First, the

DOI: 10.1201/9781003733010-5

production of metabolites from endophytic fungi relies on the symbiotic relationship of two distinct organisms and the time it takes for the fungus to produce its metabolites [11]. Furthermore, the volume of these bioactive metabolites naturally produced is very low, too low to be reliably used in drug production [12,13].

This chapter will discuss the wide variety of anticancer compounds produced by endophytic fungi and the biological mechanisms by which these compounds act in common anticancer therapeutics. Here, we will also examine the synthetic compounds produced in laboratories that mimic the action of naturally produced metabolites, as well as emerging techniques used to maximize metabolite production from endophytes. Together, these topics will bridge the naturally occurring medical compounds with innovative biomedical advances to maximize the potential of fungal endophytes in modern medicine.

5.2 SOURCES OF FUNGAL ENDOPHYTES

Many modern anticancer therapeutics are derived from metabolites produced by fungal endophytes [14]. These important compounds are produced by a variety of fungal species, which we highlight in this section.

5.2.1 Paclitaxel

A taxane that was identified from the inner bark of the Pacific yew tree (*T. brevifolia*) and from the fungus *T. andreanae*. Paclitaxel is an antitumor agent and is approved for the treatment of breast cancer, ovarian cancer, and Kaposi's sarcoma by the U.S. Food and Drug Administration (FDA) [14–16].

5.2.2 Podophyllotoxin

An aryl tetralin lignan derived from the gymnosperms and angiosperms of various plants. The plant species *Sinopodophyllum* serves as the primary natural source. Many fungal species produce podophyllotoxin, including *Phialocephala fortinii* and *Fusarium oxysporum* [14]. Podophyllotoxin derivatives are approved as treatments for cancers such as lung cancer, testicular cancer, and leukemia [17].

5.2.3 Camptothecin

An alkaloid first isolated from the bark of *Camptotheca acuminata*. Fungal species that produce this compound include *Entrophospora infrequens* and

Table 5.1 Important metabolites and representative fungal producers

Metabolite	*Fungal species*	*Target cancers*	*Common drug*
Paclitaxel	*Taxomyces andreanae*	Breast, ovary, Kaposi's sarcoma	Taxol
Podophyllotoxin	*Phialocephala fortinii, Fusarium oxysporum*	Solid tumors, leukemia	Etoposide, teniposide
Camptothecin	*Entrophospora infrequens, Fomitopsis sp*	Metastatic colorectal, pancreatic, cervical, ovarian	Topotecan, irinotecan
Vinca alkaloids	*Fusarium oxysporum, Fusarium solani*	Leukemias, lymphomas, multiple myeloma, small-cell lung cancer	Vincristine, vinblastine

Fomitopsis sp. [18,19]. The FDA has approved camptothecin-based drugs that target several cancers, including colorectal, pancreatic, cervical, and ovarian cancers [20].

5.2.4 Vinca alkaloids

A group of alkaloids, notably vinblastine and vincristine, are isolated from endophytic fungi found in the Madagascar periwinkle (*Catharanthus roseus*) [21,22]. These alkaloids are commonly produced by *Fusarium oxysporum* and *Fusarium solani* fungi [14]. Vinca alkaloid treatments are used for leukemia, lymphoma, multiple myeloma, and small-cell lung cancer [23] (Table 5.1).

5.3 SYNTHETIC PRODUCTION OF ENDOPHYTIC COMPOUNDS

Utilizing metabolites from endophytic fungi in their natural state presents many unique challenges. For example, camptothecin was initially difficult to use for its cytotoxic capabilities because it is water-insoluble and can cause severe gastrointestinal toxicity [24–26]. Additionally, metabolite yields from endophytic fungi are typically very low. For example, the typical yields of podophyllotoxin give microgram-level yields per liter of dried mycelium [27]. Additionally, podophyllotoxin must first be converted to etoposide, which is the functional cancer drug. This conversion is only about 50%–60% efficient, meaning kilogram-scale levels of podophyllotoxin are needed to make an effective cancer drug. These issues further demonstrate the need to create synthetic alternatives to these naturally occurring metabolites [28].

5.3.1 Paclitaxel

10-deacetyl-baccatin III is a paclitaxel precursor isolated from fungal species such as *Gliocladium* sp. inhabiting the European yew tree (*T. baccata*) [14,29]. This precursor can be chemically converted to baccatin III through an acetylation reaction [30–32]. Baccatin III can then be converted to paclitaxel through the coupling of a side chain (*N*-benzoyl-β-phenylisoserine) to baccatin III [33–35].

5.3.2 Podophyllotoxin

Etoposide and teniposide are semisynthetic derivatives of podophyllotoxin used as anticancer medication [14]. Podophyllotoxin is converted to these compounds in a multi-step process. It begins with demethylating a methoxy group and creating an isomer through stereochemistry conversion [36]. This isomer is then coupled with a glucose derivative with varying acetal groups (ethylidene for etoposide and thenylidene for teniposide) [37,38], resulting in the final derivative. Etoposide and teniposide are effective in different cancer treatments. Etoposide is used to target small-cell lung cancer, testicular cancer, and lymphomas, while teniposide is primarily used to treat leukemia and brain tumors [39–42].

5.3.3 Camptothecin

Topotecan and irinotecan are semisynthetic analogs of camptothecin [43,44]. To synthesize topotecan, camptothecin is first converted to 10-hydroxycamptothecin through oxidation and hydroxylation reactions [45,46]. A dimethylaminomethyl group is then added, yielding topotecan. The final step is converting this compound to a hydrochloride salt so it can be used clinically [47]. Irinotecan synthesis involves adding an ethyl group and an oxidation reaction to produce an active metabolite. Finally, a side chain is added, resulting in irinotecan [48–50].

5.3.4 Vinca alkaloids

Alkaloids such as vinblastine and vincristine are very scarce in natural environments. So, these molecules are semisynthetically synthesized using the precursors vindoline and catharanthine produced by endophytic fungi [51,52]. To synthesize the alkaloids, catharanthine and vindoline are coupled together to form anhydrovinblastine. An oxidation reaction converts this molecule to vinblastine. Vinblastine can then be converted to vincristine by additional oxidation reactions [53,54] (Table 5.2).

Table 5.2 Yields of secondary metabolites naturally from fungal endophytes

Fungal endophyte	*Secondary metabolite*	*Typical yield (ug/L)*	*Annual global demand*	*Drug produced*
Taxomyces andreanae	Paclitaxel	0.024–0.05 [14]	~400 kg/year [55].	Taxol
Phialocephala fortinii	Podophyllotoxin	0.5–189 [27]	Unclear or not made public, but a large market size exists for podophyllotoxin as a drug intermediate. While the exact global demand for this compound is not readily available, it is likely large quantities are required, like that of other metabolites [56].	Etoposide and teniposide
Aspergillus terreus	Camptothecin	90–200 [57]	~3,000 kg/year [58].	Topotecan and irinotecan
Fusarium oxysporum	Vinca alkaloids	60–80 [59]	~2,000 kg of dried leaves from Catharanthus roseus is required for 1 g vincristine, and ~500 kg is required for 1 g vinblastine [60].	Oncovin (vincristine), Velban (vinblastine)

5.4 FUNCTIONAL ROLES IN CANCER

To understand the significance of endophytic fungi in the development of anticancer drugs, it is necessary to understand the molecular mechanisms upon which these bioactive metabolites act.

5.4.1 Interactions with microtubules

Many metabolites produced by endophytic fungi target microtubules during the cell cycle. These bioactive compounds function to either stabilize microtubules or prevent their polymerization. This function can arrest the cell cycle, preventing cell division and leading to apoptosis.

5.4.1.1 Paclitaxel

This compound reduces the critical concentration of free tubulin required for microtubule polymerization, meaning that the minimum concentration of free tubulin needed for microtubule polymerization is reduced [61,62].

This makes microtubule polymerization easier and increases the percentage of tubulin subunits assembled in the cell [63]. Paclitaxel also binds and stabilizes microtubules, preventing their depolymerization by factors such as cold temperature and calcium ions [64]. This mechanism blocks cell cycle progression, prevents mitosis, and inhibits the growth of cancer cells [65].

5.4.1.2 Vinblastine and vincristine

These drugs are both terpenoid indole alkaloids derived from the same precursors [51,52]. Although they are chemically different and are used to target different cancers, their biological action is similar. These drugs bind the *Vinca* domain on β-tubulin, inhibiting microtubule polymerization [66,67]. They also suppress the dynamic instability at the plus ends of microtubules, resulting in microtubule dissolution, preventing the formation of spindle fibers [68,69] and arresting cells in the M phase of mitosis. Thus, vinblastine and vincristine inhibit the growth of cancer cells by preventing progression through mitosis through targeting microtubule polymerization [70,71] (Figure 5.1).

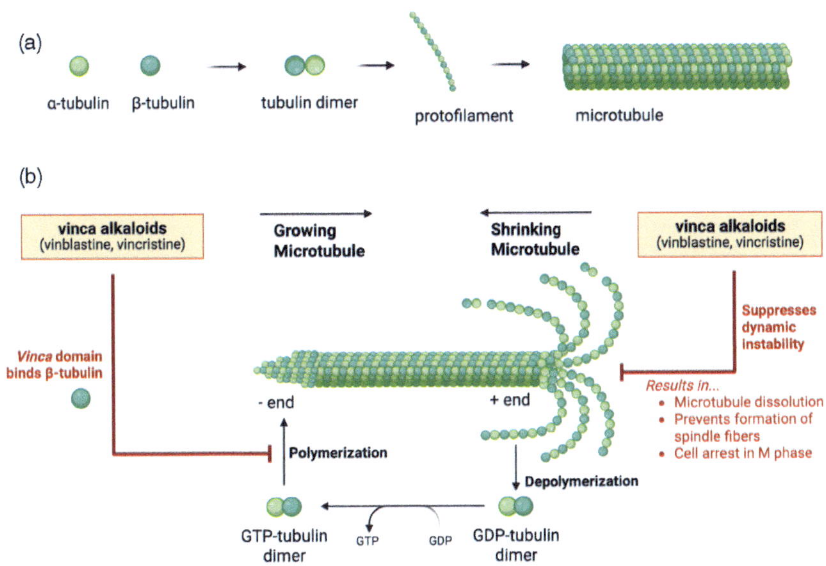

Figure 5.1 Molecular function of vinca alkaloids. Schematic of the components of microtubules. Microtubule polymerization is reduced by binding to the *Vinca* domain on the β-tubulin subunit. Plus, end instability results in the dissolution of spindle fibers, arresting cancer cells in M phase.

5.4.2 Targeting topoisomerase I and II

Topoisomerases are enzymes that relieve DNA supercoiling during replication and transcription by introducing single-stranded DNA breaks (topoisomerase I) or double-stranded DNA breaks (topoisomerase II) [72]. Fungal metabolites can target these enzymes to induce apoptosis in cancer cells.

5.4.2.1 Etoposide and teniposide

The derivatives of podophyllotoxin function differently than their parent compound. While podophyllotoxin is an inhibitor of microtubule polymerization (as it is not used in cancer medication, it is not discussed in this section) [73], its derivatives, etoposide and teniposide, inhibit topoisomerase II [74]. These drugs prevent topoisomerase II from ligating the double-stranded DNA breaks it induces to relieve supercoiling during cellular replication. This function is referred to as "poisoning" because it stabilizes the DNA breaks rather than inhibiting the function of the enzyme [75,76]. This poisoning results in persistent double-stranded breaks in DNA, arresting cells in the G2 or S phase of the cell cycle and triggering cellular senescence or apoptosis [74,77] (Figure 5.2).

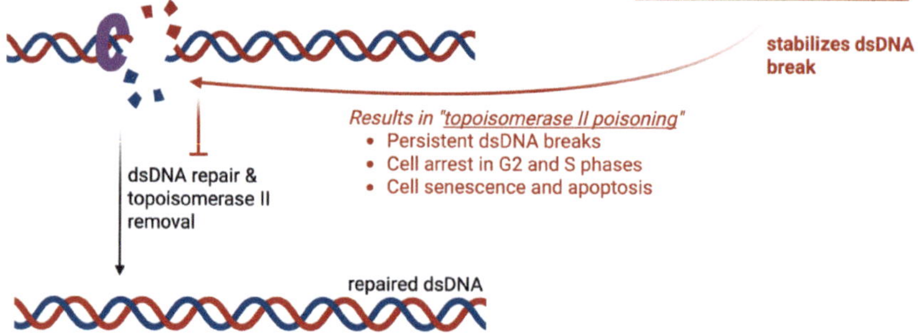

Figure 5.2 Molecular function of podophyllotoxin derivatives. (a) Double-stranded DNA breaks are stabilized, resulting in topoisomerase II poisoning. (b) Topoisomerase II is prevented from ligating the DNA breaks it induces, resulting in cell death.

5.4.2.2 *Camptothecin*

Camptothecin targets topoisomerase I [78,79]. The molecular function of camptothecin is like etoposide and teniposide, as it reversibly binds topoisomerase I cleavage complexes and stabilizes them, preventing topoisomerase I from ligating DNA breaks [80]. This leads to an accumulation of single-stranded DNA breaks, which are converted to double-stranded breaks during replication, triggering apoptosis and cell death [81].

5.5 EMERGING TECHNIQUES TO MAXIMIZE METABOLITE PRODUCTION FROM FUNGAL ENDOPHYTES

A major drawback to using natural compounds in therapeutics is how limited their production is. This is particularly true with fungal endophytes, as the natural production of metabolites from these sources is far smaller than what is required for drug development [27]. This problem has led to the creation of innovative new techniques to maximize the production of these metabolites.

5.5.1 Activating silent biosynthetic gene clusters

The cytotoxic compounds produced by endophytic fungi discussed in this chapter are referred to as "secondary metabolites." This is because the primary metabolites produced by endophytes, such as carbohydrates and proteins, are essential for the growth and survival of the organism [82–84]. Secondary metabolites, including those discussed in this chapter, are not necessary for survival and are produced in smaller amounts [85]. These secondary metabolites are coded by groups of genes termed biosynthetic gene clusters (BGCs) [86,87]. However, these clusters tend to be silenced by chromatin repression, and the natural signaling cascades that lead to their expression are still unknown [88]. Therefore, lab-based methods to express BGCs and increase metabolite production are a major focus for the future of cancer therapeutics based on fungal endophytes.

5.5.1.1 *Culture-based techniques*

Novel cell culture techniques are being developed to change environmental conditions and stimulate BGC activation.

5.5.1.1.1 One strain, many compounds (OSMAC)

One strain, many compounds (OSMAC) is an emerging technique and popular method to generate the production of a wide array of secondary metabolites from fungi [89,90]. This strategy involves changing the environmental parameters of the culture to induce a stress-like state and activate BGCs [91]. Common variables include:

- Osmolarity or trace elements (like metal ions)
- pH
- Temperature
- Oxygen
- Solid or liquid culture medium [91]

5.5.1.1.2 Co-culture

Co-culturing techniques are being utilized to stimulate BGCs in endophytes [92–94]. This technique involves culturing endophytic fungi with a differentmicroorganism, such as bacteria, resulting in crosstalk between the co-cultured organisms and inducing production of bioactive compounds from BGCs [88]. It has also been shown that co-cultivation of two different species of endophytic fungi can result in the production of previously unidentified metabolites [95]. This demonstrates the viability of this technique in stimulating the production of known compounds and also in generating previously unidentified secondary metabolites from fungi.

5.5.1.2 Epigenetic-based techniques

BGCs are often termed "silent" due to their inactivation by chromatin repression [96], a process that can be undone by manipulating the epigenome of target chromatin sequences [85]. The addition of small-molecule epigenetic modifiers can stimulate this effect. Common modifiers include:

Histone Deacetylase (HDAC) Inhibitors: HDACs remove acetyl groups from histones, converting the open and transcriptionally active euchromatin to condensed, genetically silent heterochromatin [88]. HDAC inhibitors such as suberoylanilide hydroxamic acid (SAHA) and valproic acid have been shown to increase the production of secondary metabolites. These inhibitors cause hyperacetylation of chromatin, decondensing it and making it transcriptionally active [88,97].

DNA Methyltransferase (DNMT) Inhibitors: DNMTs add a methyl group to 5′ carbon on cysteine residues in DNA, causing genetic silencing by preventing transcription machinery from accessing DNA. The use of DNMT inhibitors, such as 5-azacytidine and decitabine, results in hypomethylated gene clusters and can activate previously silenced BGCs [98,99].

5.6 CHALLENGES

Although endophytic fungi have been known to carry these antitumor compounds for decades, there remain many challenges in utilizing these metabolites in drug production. One of the most obvious is the limited natural production of these compounds. For example, paclitaxel comprises about 0.01%–0.03% of *T. brevifolia*, the Pacific yew tree, and overharvesting of these natural sources can lead to environmental destruction [100]. While there are emerging techniques that allow for the activation of silent BGCs to increase the production of bioactive compounds from these fungi, they are not without limitations. The specific production of target metabolites is difficult using methods such as OSMAC and co-culturing, as activating silent BGCs does not reliably produce a target metabolite. Additionally, these culture-based techniques are time-consuming and expensive, requiring laboratories to have the necessary resources to perform multiple experiments in an attempt to produce a specific metabolite [101]. Additionally, culturing these fungi presents further challenges, as endophytes are reliant on specific signals and factors derived from their host organism. Many of these biochemical signals and important host-derived factors remain poorly understood, creating additional challenges in culturing endophytic fungi [101].

5.7 CONCLUSION

Fungal endophytes truly represent a treasure trove of potent secondary metabolites and bioactive compounds that can revolutionize the future of therapeutics. The current knowledge of this field has already produced some of the most used cancer medications in the world. Drugs like paclitaxel (brand name Taxol), irinotecan (brand name Camptosar), and topotecan (brand name Hycamtin) are frequently used chemotherapy drugs and are all derived from endophytic fungi. While using the true, natural metabolites produced by these fungal species presents many challenges, innovations in the field have allowed for the production of synthetic analogs that function similarly and are modified to reduce toxicity.

Secondary metabolites from fungal endophytes are potent antitumor agents, shown to both inhibit and stabilize microtubule polymerization to arrest cancer cells in the M phase of the cell cycle and target both topoisomerase I and II, leading to an accumulation of double-stranded DNA breaks in the cancer cell genome, causing apoptosis. While these metabolites have different molecular targets, they all have the same end result—killing cancer cells. The future of this field relies on enhancing and optimizing emerging techniques to maximize the production of metabolites from cultured endophytes. While popular chemotherapeutic drugs are derived from synthetic analogs of secondary metabolites or produced from natural precursors to mimic the action of the fungal compound, these metabolites and precursors are still necessary to produce synthetic products. Because of this, it is necessary to enhance and optimize techniques like OSMAC to maximize efficiency and allow for the targeted production of specific metabolites in large quantities, allowing for increased therapeutic production and novel metabolite discovery.

REFERENCES

1. Aamir, M., Rai, K.K., Zehra, A., Kumar, S., Yadav, M., Shukla, V., Upadhyay, R.S.: Fungal endophytes: Classification, diversity, ecological role, and their relevance in sustainable agriculture. *Microbial Endophytes: Prospects for Sustainable Agriculture.* 291–323 (2020). https://doi.org/10.1016/B978-0-12-818734-0.00012-7
2. Wen, J., Okyere, S.K., Wang, S., Wang, J., Xie, L., Ran, Y., Hu, Y.: Endophytic fungi: An effective alternative source of plant-derived bioactive compounds for pharmacological studies. *Journal of Fungi.* 20;8(2):205 (2022). doi: 10.3390/jof8020205
3. Zheng, R., Li, S., Zhang, X., Zhao, C.: Biological activities of some new secondary metabolites isolated from endophytic fungi: A review study. *International Journal of Molecular Sciences.* 22(2):959 (2021). DOI: 10.3390/ijms22020959
4. Tsipinana, S., Husseiny, S., Alayande, K.A., Raslan, M., Amoo, S., Adeleke, R.: Contribution of endophytes towards improving plant bioactive metabolites: A rescue option against red-taping of medicinal plants. *Frontiers in Plant Science.* 14 (2023). https://doi.org/10.3389/fpls.2023.1248319
5. Kumari, P., Deepa, N., Trivedi, P.K., Singh, B.K., Srivastava, V., Singh, A.: Plants and endophytes interaction: A "secret wedlock" for sustainable biosynthesis of pharmaceutically important secondary metabolites. *Microbial Cell Factories.* 22(1):226 (2023). doi: 10.1186/s12934-023-02234-8.
6. Petrovska, B.B.: Historical review of medicinal plants' usage. *Pharmacognosy Reviews.* (11):1-5 (2012). doi: 10.4103/0973-7847.95849
7. Flores-Bustamante, Z.R., Rivera-Ordũa, F.N., Martínez-Cárdenas, A., Flores-Cotera, L.B.: Microbial paclitaxel: Advances and perspectives. *The Journal of Antibiotics.* 63(8):460-7 (2010). doi: 10.1038/ja.2010.83
8. Gärditz, K.F., Czesnick, H.: Paclitaxel - a Product of Fungal Secondary Metabolism or an Artefact? *Planta Medica.* 90(9):726–735 (2024). doi: 10.1055/a-2309-6298
9. Gupta, A., Meshram, V., Gupta, M., Goyal, S., Qureshi, K.A., Jaremko, M., Shukla, K.K.: Fungal endophytes: Microfactories of novel bioactive compounds with therapeutic interventions; a comprehensive review on the biotechnological developments in the field of fungal endophytic biology over the last decade. *Biomolecules.* (2023). 13(7): 1038. https://doi.org/10.3390/biom13071038
10. Singh, A., Singh, D.K., Kharwar, R.N., White, J.F., Gond, S.K.: Fungal endophytes as efficient sources of plant-derived bioactive compounds and their prospective applications in natural product drug discovery: Insights, avenues, and challenges. *Microorganisms.* 9(1):197 (2021). doi: 10.3390/microorganisms9010197
11. Adeleke, B.S., Babalola, O.O.: Pharmacological potential of fungal endophytes associated with medicinal plants: A review. *Journal of Fungi.* 7(2):147 (2021). doi: 10.3390/jof7020147
12. Chandra, H., Yadav, A., Prasad, R., Kalra, S.J.S., Singh, A., Bhardwaj, N., Gupta, K.K.: Fungal endophytes from medicinal plants acting as natural therapeutic reservoir. *The Microbe.* 3 (2024). https://doi.org/10.1016/j.microb.2024.100073

13. Tiwari, P., Bae, H.: Endophytic fungi: Key insights, emerging prospects, and challenges in natural product drug discovery. *Microorganisms.* 10(2):360 (2022). doi: 10.3390/microorganisms10020360
14. Uzma, F., Mohan, C.D., Hashem, A., Konappa, N.M., Rangappa, S., Kamath, P. V., Singh, B.P., Mudili, V., Gupta, V.K., Siddaiah, C.N., Chowdappa, S., Alqarawi, A.A., Abd-Allah, E.F.: Endophytic fungi-alternative sources of cytotoxic compounds: A review. *Frontiers in Pharmacology.* 9:309 (2018). doi: 10.3389/fphar.2018.00309
15. Weaver, B.A.: How Taxol/paclitaxel kills cancer cells. *Molecular Biology of the Cell.* 25(18):2677–2681 (2014). doi: 10.1091/mbc.E14-04-0916
16. Wiseman, L.R., Spencer, C.M., Conte, P.F.: Paclitaxel an update of its use in the treatment of metastatic breast cancer and ovarian and other gynaecological cancers. *Drugs & Aging.* 12(4):305-34 (1998). doi: 10.2165/00002512-199812040-00005
17. Fan, H.Y., Zhu, Z.L., Xian, H.C., Wang, H.F., Chen, B.J., Tang, Y.J., Tang, Y.L., Liang, X.H.: Insight into the molecular mechanism of podophyllotoxin derivatives as anticancer drugs. *Frontiers in Cell and Developmental Biology.* 9. (2021). doi: 10.3389/fcell.2021.709075
18. Kusari, S., Zühlke, S., Spiteller, M.: An endophytic fungus from Camptotheca acuminata that produces camptothecin and analogues. *J Nat Prod.* 72, 2–7 (2009). https://doi.org/10.1021/np800455b
19. Yang, M., Wang, Q., Liu, Y., Hao, X., Wang, C., Liang, Y., Chen, J., Xiao, Y., Kai, G.: Divergent camptothecin biosynthetic pathway in Ophiorrhiza pumila. *BMC Biol.* 19, (2021). https://doi.org/10.1186/s12915-021-01051-y
20. Venditto, V.J., Simanek, E.E.: Cancer therapies utilizing the camptothecins: A review of the in vivo literature. *Molecular Pharmacology.* 7(2):307–349 (2010). doi: 10.1021/mp900243b
21. Palem, P.P.C., Kuriakose, G.C., Jayabaskaran, C.: An endophytic fungus, talaromyces radicus, isolated from catharanthus roseus, produces vincristine and vinblastine, which induce apoptotic cell death. *PLoS One.* 10, (2015). https://doi.org/10.1371/journal.pone.0144476
22. Taub, J.W., Buck, S.A., Xavier, A.C., Edwards, H., Matherly, L.H., Ge, Y.: The evolution and history of Vinca alkaloids: From the Big Bang to the treatment of pediatric acute leukemia. *Pediatric Blood & Cancer.* 71(11) (2024). doi: 10.1002/pbc.31247
23. Rowinsky E.K.: The Vinca Alkaloids. In: Kufe DW, Pollock RE, and Weichselbaum RR (eds.) *Holland-Frie Cancer Medicine.* BC Decker (2003). Available from: https://www.ncbi.nlm.nih.gov/books/NBK12718/
24. Abigerges, D., Chabot, G.G., Armand, J.-P., H6rait, P., Gouyette, A., Gandia, D.: Phase I and pharmacologic studies of the camptothecin analog irinotecan administered every 3 weeks in cancer patients. *Journal of Clinical Oncology.* 13(1):210-21 (1995). doi: 10.1200/JCO.1995.13.1.210
25. Gong, J., Zhang, W., Balthasar, J.P.: Camptothein-based anti-cancer therapies and strategies to improve their therapeutic index. *Cancers.* 17(6):1032 (2025). doi: 10.3390/cancers17061032
26. Martino, E., Della Volpe, S., Terribile, E., Benetti, E., Sakaj, M., Centamore, A., Sala, A., Collina, S.: The long story of camptothecin: From traditional medicine to drugs. *Bioorg Med Chem Lett.* 27, 701–707 (2017). https://doi.org/10.1016/J.BMCL.2016.12.085

27. Eyberger, A.L., Dondapati, R., Porter, J.R.: Endophyte fungal isolates from Podophyllum peltatum produce podophyllotoxin. *J Nat Prod.* 69, 1121–1124 (2006).
28. Shah, Z., Farooq Gohar, U., Jamshed, I., Mushtaq, A., Mukhtar, H., Zia-Ui-Haq, M., Toma, S.I., Manea, R., Moga, M., Popovici, B.: Podophyllotoxin: history, recent advances and future prospects. *Biomolecules.* 11, 603 (2021). https://doi.org/10.3390/biom11040603
29. Sreekanth, D., Syed, A., Sarkar, S., Sarkar, D., Santhakumari, B., Ahmad, A., Khan, M.I.: Production, purification, and characterization of taxol and 10-DABIII from a new endophytic fungus Gliocladium sp. isolated from the Indian yew tree, Taxus baccata. *J Microbiol Biotechnol.* 19, 1342–1347 (2009). doi: 10.4014/jmb.0904.4041
30. Saičić, R.N., Matovi, R., Cekovic, Z.: Semisynthesis of Taxol® : An improved procedure for the isolation of 10-deacetylbaccatin III. *Journal of the Serbian Chemical Society.* 64(9):497-503 (1999). DOI:10.2298/JSC9909497S
31. Wang, H., Zhang, B.Y., Gong, T., Chen, T.J., Chen, J.J., Yang, J.L., Zhu, P.: Construction of acetyl-CoA and DBAT hybrid metabolic pathway for acetylation of 10-deacetylbaccatin III to baccatin III. *Acta Pharm Sin B.* 11, 3322–3334 (2021). https://doi.org/10.1016/J.APSB.2021.03.029
32. Patel, R.N., Banerjee, A., Nanduri, V.: Enzymatic acetylation of 10-deacetylbaccatin III to baccatin III by C-10 deacetylase from Nocardioides luteus SC 13913. *Enzyme Microb Technol.* 27, 371–375 (2000). https://doi.org/10.1016/S0141-0229(00)00235-0
33. US6,307,071B1. Gibson, F. S. Synthesis of paclitaxel from Baccatin III by protection of the 7 hydroxyl of Baccatin III using a strong base and an electrophile. (2000).
34. Zhang, S., Ye, T., Liu, Y., Hou, G., Wang, Q., Zhao, F., Li, F., Meng, Q.: Research advances in clinical applications, anticancer mechanism, total chemical synthesis semi-synthesis and biosynthesis of paclitaxel. *Molecules.* 28(22):7517 (2023). https://doi.org/10.3390/molecules28227517
35. Xie, X., Zhai, Y., Cheng, H., Wei, W.H., Ren, M.: From Taxus to paclitaxel: Opportunities and challenges for urban agriculture to promote human health. *Plant Physiol Biochem.* 220, 109502 (2025). https://doi.org/10.1016/J.PLAPHY.2025.109502
36. Liu, Y.Q., Tian, J., Qian, K., Zhao, X.B., Morris-Natschke, S.L., Yang, L., Nan, X., Tian, X., Lee, K.H.: Recent progress on C-4-modified podophyllotoxin analogs as potent antitumor agents. *Med Res Rev.* 35, 1–62 (2015). https://doi.org/10.1002/med.21319
37. Schultz, B.J., Kim, S.Y., Lau, W., Sattely, E.S.: Total biosynthesis for milligram-scale production of etoposide intermediates in a plant chassis. *J Am Chem Soc.* 141, 19231–19235 (2019). https://doi.org/10.1021/jacs.9b10717
38. Guo, Y., Chen, B., Guo, J., Jiang, P., Wang, J., Sun, W.: Design, synthesis and biological evaluation of novel podophyllotoxin derivatives as tubulin-targeting anticancer agents. *Pharm Biol.* 62, 233–249 (2024). https://doi.org/10.1080/13880209.2024.2318350
39. Clark, P.I., Slevin, M.L.: The clinical pharmacology of etoposide and teniposide. *Clin Pharmacokinet.* 12, 223–252 (2012).
40. Bjorkholm, M.: Etoposide and teniposide in the treatment of acute leukemia. *Medical Oncology.* 7(1):3-10 (1990). doi: 10.1007/BF03000484

41. Jang, J.Y., Kim, D., Im, E., Kim, N.D.: Etoposide as a key therapeutic agent in lung cancer: Mechanisms, efficacy, and emerging strategies. *International Journal of Molecular Sciences.* 26(2):796 (2025). doi: 10.3390/ijms26020796
42. Pagani, M.: The complex clinical picture of presumably allergic side effects to cytostatic drugs: Symptoms, pathomechanism, reexposure, and desensitization. *Med Clin North Am.* 94, 835–852 (2010). https://doi.org/10.1016/j.mcna.2010.03.002
43. Li, F., Jiang, T., Li, Q., Ling, X.: Camptothecin (CPT) and its derivatives are known to target topoisomerase I (Top1) as their mechanism of action: did we miss something in CPT analogue molecular targets for treating human disease such as cancer? *American Journal of Cancer Research.* 7(12):2350–2394 (2017).
44. Khaiwa, N., Maarouf, N.R., Darwish, M.H., Alhamad, D.W.M., Sebastian, A., Hamad, M., Omar, H.A., Orive, G., Al-Tel, T.H.: Camptothecin's journey from discovery to WHO Essential Medicine: Fifty years of promise. *Eur J Med Chem.* 223, 113639 (2021). https://doi.org/10.1016/J.EJMECH.2021.113639
45. Nguyen, T.A.M., Nguyen, T.D., Leung, Y.Y., McConnachie, M., Sannikov, O., Xia, Z., Dang, T.T.T.: Discovering and harnessing oxidative enzymes for chemoenzymatic synthesis and diversification of anticancer camptothecin analogues. *Commun Chem.* 4, (2021). https://doi.org/10.1038/s42004-021-00602-2
46. Tan, H., Wang, G., Li, J., Meng, G., Liu, Z., Dong, M., Li, Y., Ju, D., Zhang, Q.: Synthesis of novel 10-hydroxycamptothecin derivatives utilizing topotecan hydrochloride as ortho-quinonemethide precursor. *Bioorg Med Chem.* 23, 118–125 (2015). https://doi.org/10.1016/J.BMC.2014.11.020
47. Scinopharm Taiwan, Ltd. "Process for making topotecan." WIPO Patent WO2008127606A1, 23 Oct. 2008. Google Patents, patents.google.com/patent/WO2008127606A1/en.
48. Wang, Bo. "Method for producing irinotecan." China Patent CN102070643A, 25 May 2011. Google Patents, patents.google.com/patent/CN102070643A/en.
49. Kciuk, M., Marciniak, B., Kontek, R.: Irinotecan: Still an important player in cancer chemotherapy: A comprehensive overview. *International Journal of Molecular Sciences.* 21(14):4919 (2020). doi: 10.3390/ijms21144919
50. Si, J., Zhao, X., Gao, S., Huang, D., Sui, M.: Advances in delivery of Irinotecan (CPT-11) active metabolite 7-ethyl-10-hydroxycamptothecin. *Int J Pharm.* 568, 118499 (2019). https://doi.org/10.1016/J.IJPHARM.2019.118499
51. Sears, J.E., Boger, D.L.: Total synthesis of vinblastine, related natural products, and key analogues and development of inspired methodology suitable for the systematic study of their structure-function properties. *Acc Chem Res.* 48, 653–662 (2015). https://doi.org/10.1021/ar500400w
52. Shukla, R., Singh, A., Singh, K.K.: Vincristine-based nanoformulations: A preclinical and clinical studies overview. *Drug Delivery and Translational Research.* 14(1):1-16 (2024). doi: 10.1007/s13346-023-01389-6.
53. Ishikawa, H., Colby, D.A., Seto, S., Va, P., Tam, A., Kakei, H., Rayl, T.J., Hwang, I., Boger, D.L.: Total synthesis of vinblastine, vincristine, related natural products, and key structural analogues. *J Am Chem Soc.* 131, 4904–4916 (2009). https://doi.org/10.1021/ja809842b

54. Zhu, J., Wang, M., Wen, W., Yu, R.: Biosynthesis and regulation of terpenoid indole alkaloids in Catharanthus roseus. *Pharmacognosy Reviews*. 9(17): 24–28 (2015). doi: 10.4103/0973-7847.156323
55. Farr, Ken. Genus-level approach to Taxus species. NDF Workshop Case Studies: WG 1 – Trees, Case Study 6, 2008, www.conabio.gob.mx/institucion/cooperacion_internacional/TallerNDF/Links-Documentos/Casos%20de%20Estudio/Trees/WG1%20CS6.pdf.
56. Podophyllotoxin Market Size, Share, Growth, and Industry Analysis, By Type (Purity 95%, Purity 98%, Others, Podophyllotoxin), By Application (Chemical Reagents, Pharmaceutical Intermediates, Others), Regional Insights and Forecast to 2033, https://www.marketreportsworld.com/market-reports/podophyllotoxin-market-14717376
57. El-Sayed, A.S.A., ElSayed, A.I., Wadan, K.M., El-Saadany, S.S., Abd El-Hady, N.A.A.: Camptothecin bioprocessing from Aspergillus terreus, an endophyte of Catharanthus roseus: Antiproliferative activity, topoisomerase inhibition and cell cycle analysis. *Microb Cell Fact*. 23, (2024). https://doi.org/10.1186/s12934-023-02270-4
58. Camptothecin-Discovery, Clinical Perspectives and Biotechnology. *Nat Prod Chem Res*. 3, (2015). https://doi.org/10.4172/2329-6836.1000175
59. Kumar, A., Patil, D., Rajamohanan, P.R., Ahmad, A.: Isolation, purification and characterization of vinblastine and vincristine from endophytic fungus Fusarium oxysporum isolated from Catharanthus roseus. *PLoS One*. 8, (2013). https://doi.org/10.1371/journal.pone.0071805
60. Zhang, J., Hansen, L.G., Gudich, O., Viehrig, K., Lassen, L.M.M., Schrübbers, L., Adhikari, K.B., Rubaszka, P., Carrasquer-Alvarez, E., Chen, L., D'Ambrosio, V., Lehka, B., Haidar, A.K., Nallapareddy, S., Giannakou, K., Laloux, M., Arsovska, D., Jørgensen, M.A.K., Chan, L.J.G., Kristensen, M., Christensen, H.B., Sudarsan, S., Stander, E.A., Baidoo, E., Petzold, C.J., Wulff, T., O'Connor, S.E., Courdavault, V., Jensen, M.K., Keasling, J.D.: A microbial supply chain for production of the anti-cancer drug vinblastine. *Nature*. 609, 341–347 (2022). https://doi.org/10.1038/s41586-022-05157-3
61. Sharma, S., Lagisetti, C., Poliks, B., Coates, R.M., Kingston, D.G.I., Bane, S.: Dissecting paclitaxel-microtubule association: Quantitative assessment of the 2′-OH group. *Biochemistry*. 52, 2328–2336 (2013). https://doi.org/10.1021/bi400014t
62. Ezzo, M., Etienne-Manneville, S.: Microtubule-targeting agents: Advances in tubulin binding and small molecule therapy for gliomas and neurodegenerative diseases. *International Journal of Molecular Sciences*. 26(15):7652 (2025). doi: 10.3390/ijms26157652
63. Verma, S., Kumar, N., Verma, V.: Role of paclitaxel on critical nucleation concentration of tubulin and its effects thereof. *Biochem Biophys Res Commun*. 478, 1350–1354 (2016). https://doi.org/10.1016/J.BBRC.2016.08.127
64. Foland, T.B., Dentler, W.L., Suprenant, K.A., Gupta, M.L., Himes, R.H.: Paclitaxel-induced microtubule stabilization causes mitotic block and apoptotic-like cell death in a paclitaxel-sensitive strain of Saccharomyces cerevisiae. *Yeast*. 22, 971–978 (2005). https://doi.org/10.1002/yea.1284
65. Leung, J.C., Cassimeris, L.: Reorganization of paclitaxel-stabilized microtubule arrays at mitotic entry: roles of depolymerizing kinesins and severing proteins. *Cancer Biol Ther*. 20, 1337–1347 (2019). https://doi.org/10.1080/15384047.2019.1638678

66. Saju, A., Chen, P.P., Weng, T.H., Tsai, S.Y., Tanaka, A., Tseng, Y.T., Chang, C.C., Wang, C.H., Shimamoto, Y., Hsia, K.C.: HURP binding to the vinca domain of β-tubulin accounts for cancer drug resistance. *Nat Commun.* 15, (2024). https://doi.org/10.1038/s41467-024-53139-y
67. Gupta, S., Bhattacharyya, B.: Antimicrotubular drugs binding to vinca domain of tubulin. *Molecular and Cellular Biochemistry.* 253(1-2):41-7 (2003). doi: 10.1023/a:1026045100219
68. Dhamodharan, R., Jordan, M.A., Thrower, D., Wilson, L., Wadsworth, P.: Vinblastine suppresses dynamics of individual microtubules in living interphase cells. *Molecular and Cellular Biochemistry.* ;6(9):1215-29 (1995). doi: 10.1091/mbc.6.9.1215
69. Bates, D., Eastman, A.: Microtubule destabilising agents: Far more than just antimitotic anticancer drugs. *British Journal of Clinical Pharmacology.* 83(2):255-268 (2017). doi: 10.1111/bcp.13126
70. Jordan, M.A., Horwitz, S.B., Lobert, S., Correia, J.J.: Exploring the mechanisms of action of the novel microtubule inhibitor vinflunine. *Semin Oncol.* 35, (2008). https://doi.org/10.1053/j.seminoncol.2008.01.009
71. Chen, R. J., Arora, R. D., & Menezes, R. G. (2026). Vinca alkaloid toxicity. In StatPearls Editorial Board (Eds.), StatPearls. StatPearls Publishing. https://www.ncbi.nlm.nih.gov/books/NBK557842/
72. Tricoli, J.V., Sahai, B.M., Mccormick, P.J., Jarlinski, S.J., Bertram, J.S., Kowalski, D.: DNA Topoisomerase I and II activities during cell proliferation and the cell cycle in cultured mouse embryo fibroblast (C3H 10T1/2) cells. *Experimental Cell Research.* 158(1):1-14 (1985). doi: 10.1016/0014-4827(85)90426-4
73. Chen, J.Y., Tang, Y.A., Li, W.S., Chiou, Y.C., Shieh, J.M., Wang, Y.C.: A synthetic podophyllotoxin derivative exerts anti-cancer effects by inducing mitotic arrest and pro-apoptotic ER stress in lung cancer preclinical models. *PLoS One.* 8, (2013). https://doi.org/10.1371/journal.pone.0062082
74. Gordaliza, M., Castro, M.A., del Corral, J.M., Feliciano, A.S.: Antitumor properties of podophyllotoxin and related compounds. *Curr Pharm Des.* 6, 1811–1839 (2000).
75. Pommier, Y., Leo, E., Zhang, H., Marchand, C.: DNA topoisomerases and their poisoning by anticancer and antibacterial drugs. *Chemistry & Biology.* 17(5):421-33 (2010). doi: 10.1016/j.chembiol.2010.04.012
76. Vann, K.R., Oviatt, A.A., Osheroff, N.: Topoisomerase II poisons: Converting essential enzymes into molecular scissors. *Biochemistry.* 60(21):1630-1641 (2021). doi: 10.1021/acs.biochem.1c00240
77. Rappa, G., Lorico, A., Sartorelli, A.C.: Potentiation by novobiocin of the cytotoxic activity of etoposide (VP-16) and teniposide (VM-26). *Int J Cancer.* 51, 780–787 (1992). https://doi.org/10.1002/ijc.2910510519
78. Tesauro, C., Simonsen, A.K., Andersen, M.B., Petersen, K.W., Kristoffersen, E.L., Algreen, L., Hansen, N.Y., Andersen, A.B., Jakobsen, A.K., Stougaard, M., Gromov, P., Knudsen, B.R., Gromova, I.: Topoisomerase I activity and sensitivity to camptothecin in breast cancer-derived cells: A comparative study. *BMC Cancer.* 19, (2019). https://doi.org/10.1186/s12885-019-6371-0

79. Wu, N., Wu, X.W., Agama, K., Pommier, Y., Du, J., Li, D., Gu, L.Q., Huang, Z.S., An, L.K.: A novel DNA topoisomerase I inhibitor with different mechanism from camptothecin induces G2/M phase cell cycle arrest to K562 cells. *Biochemistry.* 49, 10131–10136 (2010). https://doi.org/10.1021/bi1009419
80. O'Leary, J., Muggia, F.M.: Camptothecins: A review of their development and schedules of administration. *Eur J Cancer.* 34, 1500–1508 (1998). https://doi.org/10.1016/S0959-8049(98)00229-9
81. Pommier, Y.: Topoisomerase I inhibitors: Camptothecins and beyond. *Nature Reviews Cancer.* 6(10):789-802 (2006). doi: 10.1038/nrc1977
82. Parihar, R., Singh, U., Das, A., Baishya, B., Singh, V., Ahirwar, S.C., Islahi, S., Sen, M., Mittal, V.: Identification of primary metabolites in fungal species of Trichophyton mentagrophyte and Trichophyton rubrum by NMR spectroscopy. *Mycoses.* 67, (2024). https://doi.org/10.1111/myc.13699
83. Wadhwa, K., Kapoor, N., Kaur, H., Abu-Seer, E.A., Tariq, M., Siddiqui, S., Yadav, V.K., Niazi, P., Kumar, P., Alghamdi, S.: A comprehensive review of the diversity of fungal secondary metabolites and their emerging applications in healthcare and environment. *Mycobiology.* 52(6):335-387 (2024). doi: 10.1080/12298093.2024.2416736
84. Pinar, O., Rodríguez-Couto, S.: Biologically active secondary metabolites from white-rot fungi. *Frontiers in Chemistry.* 12 (2024). doi: 10.3389/fchem.2024.1363354
85. Alam, B., Lǐ, J., Gě, Q., Khan, M.A., Gōng, J., Mehmood, S., Yuán, Y., Gǒng, W.: Endophytic fungi: From symbiosis to secondary metabolite communications or vice versa? *Frontiers in Plant Science.* 12 (2021). doi: 10.3389/fpls.2021.791033
86. Ben Slama, H., Bouket, A.C., Alenezi, F.N., Pourhassan, Z., Golińska, P., Oszako, T., Belbahri, L.: Potentials of endophytic fungi in the biosynthesis of versatile secondary metabolites and enzymes. *Forests.* 12(12):1784 (2021). https://doi.org/10.3390/f12121784
87. Hashem, A.H., Attia, M.S., Kandil, E.K., Fawzi, M.M., Abdelrahman, A.S., Khader, M.S., Khodaira, M.A., Emam, A.E., Goma, M.A., Abdelaziz, A.M.: Bioactive compounds and biomedical applications of endophytic fungi: A recent review. *Microbial Cell Factories.* 22(1):107 (2023). doi: 10.1186/s12934-023-02118-x
88. Pillay, L.C., Nekati, L., Makhwitine, P.J., Ndlovu, S.I.: Epigenetic activation of silent biosynthetic gene clusters in endophytic fungi using small molecular modifiers. *Frontiers in Microbiology.* 13 (2022). doi: 10.3389/fmicb.2022.815008
89. de Andrade, C.P., Lacerda, C.D., Valente, R.A., Rocha, L.S. de H., de Souza, A.T.F., Pereira, D.Í. de M., Barbosa, L.K., Fantin, C., Duvoisin Junior, S., Albuquerque, P.M.: An OSMAC strategy for the production of antimicrobial compounds by the amazonian fungi Talaromyces pinophilus CCM-UEA-F0414 and Penicillium paxilli CCM-UEA-F0591. *Antibiotics.* 14, (2025). https://doi.org/10.3390/antibiotics14080756
90. Zhang, Y., Feng, L., Hemu, X., Tan, N.H., Wang, Z.: OSMAC strategy: A promising way to explore microbial cyclic peptides. *Eur J Med Chem.* 268, 116175 (2024). https://doi.org/10.1016/J.EJMECH.2024.116175
91. Pinedo-Rivilla, C., Aleu, J., Durán-Patrón, R.: Cryptic metabolites from marine-derived microorganisms using OSMAC and epigenetic approaches. *Marine Drugs.* 20(2):84 (2022). doi: 10.3390/md20020084

92. Wu, J., Ye, J., Cen, J., Chen, Y., Xu, J.: Induction of three new secondary metabolites by the co-culture of endophytic fungi Phomopsis asparagi DHS-48 and Phomopsis sp. DHS-11 isolated from the Chinese Mangrove Plant Rhizophora mangle. *Mar Drugs*. 22, (2024). https://doi.org/10.3390/md22080332
93. Tironi, L.S., Carletto, L.B., Silva, E.O., Schripsema, J., Luiz, J.H.H.: Endophytic fungi co-culture: An alternative source of antimicrobial substances. *Microorganisms*. 12(12) (2024). doi: 10.3390/microorganisms12122413
94. Knowles, S.L., Raja, H.A., Roberts, C.D., Oberlies, N.H.: Fungal–fungal co-culture: A primer for generating chemical diversity. *Nat Prod Rep*. 39, 1557–1573 (2022). https://doi.org/10.1039/D1NP00070E
95. Marmann, A., Aly, A.H., Lin, W., Wang, B., Proksch, P.: Co-cultivation: A powerful emerging tool for enhancing the chemical diversity of microorganisms. *Marine Drugs*. 12(2):1043-65 (2014). doi: 10.3390/md12021043
96. Reen, F.J., Romano, S., Dobson, A.D.W., O'Gara, F.: The sound of silence: Activating silent biosynthetic gene clusters in marine microorganisms. *Marine Drugs*. 13(8):4754–4783 (2015). doi: 10.3390/md13084754
97. Beau, J., Mahid, N., Burda, W.N., Harrington, L., Shaw, L.N., Mutka, T., Kyle, D.E., Barisic, B., Van Olphen, A., Baker, B.J.: Epigenetic tailoring for the production of anti-infective cytosporones from the marine fungus Leucostoma persoonii. *Mar Drugs*. 10, 762–774 (2012). https://doi.org/10.3390/md10040762
98. Bond, D.R., Burnard, S.M., Uddipto, K., Hunt, K.V., Harvey, B.M., Steffens Reinhardt, L., Lawlor-O'Neill, C., Roper, E.A., Humphries, S., Murray, H.C., Mannan, A., Dun, M.D., de Bock, C.E., Bowden, N.A., Enjeti, A.K., Verrills, N.M., Riveros, C., Lê Cao, K.A., Lee, H.J.: Hypomethylating agents induce epigenetic and transcriptional heterogeneity with implications for acute myeloid leukemia cell self-renewal. *Leukemia*. 39, 2275–2280 (2025). https://doi.org/10.1038/s41375-025-02693-5
99. Abdelhamid, R.M., Soliman, E.R.S., Mohamed, E.T., Elsaba, Y.M.: Epigenetic modulation of Ceratorhiza hydrophila by 5-azacytidine enhances antifungal metabolite production: Insights from antimicrobial, metabolic, genomic and computational analyses. *BMC Microbiol*. 25, (2025). https://doi.org/10.1186/s12866-025-04330-8
100. Kharwar, R.N., Mishra, A., Gond, S.K., Stierle, A., Stierle, D.: Anticancer compounds derived from fungal endophytes: Their importance and future challenges. *Natural Product Reports*. 28(7):1208-28 (2011). doi: 10.1039/c1np00008j
101. Varghese, S., Jisha, M.S., Rajeshkumar, K.C., Gajbhiye, V., Alrefaei, A.F., Jeewon, R.: Endophytic fungi: A future prospect for breast cancer therapeutics and drug development. *Heliyon*. 10(13) (2024). doi: 10.1016/j.heliyon.2024.e33995

Chapter 6

Epigenetic modulators from natural sources

Unlocking new horizons in therapeutics

Indrani Mukherjee, Neil Ferland, Katelyn Tran Nguyen, Dawn Littlefield, and Rana Chakraborty
Department of Pediatrics, University of Miami, Miami, FL

6.1 INTRODUCTION

Epigenetic regulation represents an advanced mechanism of inheritance control in biology. Living systems can rapidly adapt to environmental challenges without altering their underlying DNA sequence [1]. Through orchestrated molecular interactions involving DNA methylation, histone modification, chromatin remodeling, and non-coding RNA signaling, epigenetic mechanisms help establish cell identity, developmental timing, and long-term genome responsiveness [2,3]. The same genome can thus support a broad spectrum of cell types and phenotypes, depending on the molecular switches in each context [4]. The human epigenome is both dynamic and well-structured, and its proper regulation is essential for physiological homeostasis [5]. However, perturbations in DNA methylation patterns, histone acetylation balance, or RNA-mediated silencing often lead to aberrant transcriptional outcomes favoring various diseases [6]. Such epigenetic disturbances are often recognized as hallmarks of many complex diseases, including cancer, diabetes, neurodegenerative disorders, and autoimmune or chronic inflammatory conditions [7–11]. Recognizing and restoring homeostatic epigenetic equilibrium has thus become a major objective in curing these disorders [12].

Epigenetic drugs, or "epidrugs," emerged as an important class of therapeutic agents during the 21st century [13]. The first generation of synthetic DNA methyltransferase (DNMT) inhibitors, such as azacitidine and decitabine, and histone deacetylase (HDAC) inhibitors, such as vorinostat and romidepsin, gained regulatory approval for treating hematological malignancies [14–17]. Despite these successes, these agents can be associated with cytotoxicity, limited tissue selectivity, and acquired resistance [18]. Attention has therefore shifted toward natural compounds capable of modulating epigenetic processes [19,20]. Nature offers a vast source of chemically diverse molecules, many of which act as multi-targeted or pleiotropic modulators of the chromatin-regulatory machinery [21].

 DOI: 10.1201/9781003733010-6

The molecular scaffolds of these bioactives—including flavonoids, alkaloids, terpenoids, phenolic acids, peptides, and organosulfur compounds—embody unique structural and functional diversity [22,23]. Their ability to orchestrate simultaneous effects on multiple epigenetic targets, including DNMTs, HDACs, histone methyltransferases, and microRNAs, allows the restoration or fine-tuning of gene expression networks rather than simply the inhibition of a single enzyme [24,25].

Nutritional and epidemiological studies indicate that diet-derived bioactive compounds can significantly influence epigenetic regulation, contributing to disease prevention and the modulation of disease progression [26–28]. Phytochemicals such as curcumin, resveratrol, epigallocatechin gallate (EGCG), and sulforaphane alter DNA methylation signatures, histone acetylation status, and non-coding RNA expression in different experimental setups. Additionally, these compounds modulate inflammatory, metabolic, and oxidative stress-related signaling cascades [29–30]. Though substantial research has been conducted, variations in bioavailability, dose responsiveness, and tissue specificity remains challenges for translational application [31–33]. Recent advances in nanotechnology-based cargo systems have thus been explored to enhance the targeted delivery of these epigenetic modulators, thereby improving better stability and cellular uptake [34].

The main aim of this chapter is to examine the fundamental concepts, classes, and therapeutic significance of epigenetic modulators from natural origins. We review the molecular details of key natural regulators, discuss their mechanistic interactions within chromatin-modifying systems, and explore how emerging nanotechnology amplifies their clinical potential. This chapter highlights how the interplay between natural compounds and modern biomedical innovation may collectively move toward safer and more effective epigenetic therapies [35].

6.2 THE ARCHITECTURE OF EPIGENETIC REGULATION

Epigenetic control relies on dynamic interactions among enzymes that write, read, and erase histone and DNA modifications [36]. They can be broadly classified as follows:

6.2.1 Writers

These add marks such as methyl or acetyl groups (e.g., DNA, histone acetyltransferases, and methyltransferases) [37].

6.2.2 Erasers

These remove these marks, with examples including histone deacetylases, demethylases, or TET family enzymes [37].

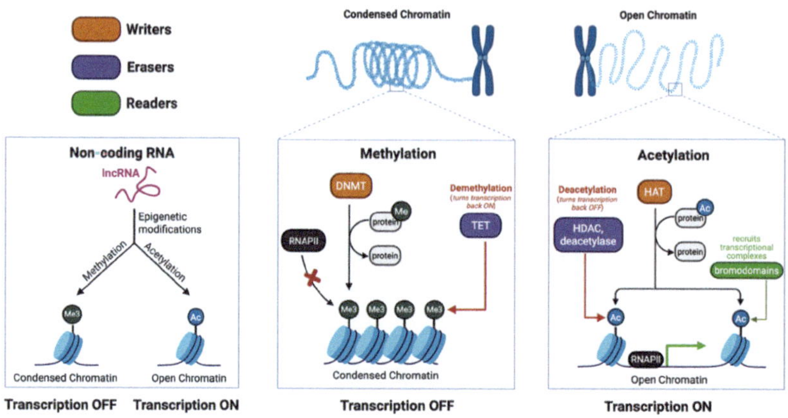

Figure 6.1 Overview of Epigenetic Mechanisms: Diagram showing the crosstalk among DNA methylation, histone acetylation/methylation, and non-coding RNA pathways, illustrating how writers, erasers, and readers influence chromatin accessibility.

6.2.3 Readers

These recognize specific epigenetic signatures (e.g., bromodomains, chromodomains) and recruit transcriptional complexes [38].

In normal physiology, such modifications coordinate cell differentiation, immune resilience, and developmental plasticity [39]. Pathologically, uncontrolled methylation or deacetylation results in gene silencing or overactivation associated with carcinogenesis, neurodegeneration, and metabolic dysregulation [9,40–42]. Therapeutic targeting of epigenetic modifiers has already proven feasible, as seen with FDA-approved drugs such as vorinostat (an HDAC inhibitor) and azacitidine (a DNMT inhibitor) [43]. Nevertheless, next-generation treatments require modulators with greater target selectivity and minimal adverse effects, gaps that natural products may address [44,45]. An overview summarizing the epigenetic mechanisms is illustrated in Figure 6.1.

6.3 PHYTOCHEMICALS AS EPIGENETIC MODULATORS

The different types of phytochemicals that act as epigenetic modulators are classified as follows:

6.3.1 Polyphenols

Polyphenols constitute a broad class of micronutrients abundant in fruits, tea, coffee, wine, and spices. These compounds exert strong antioxidant and signaling-modulatory functions across numerous cellular pathways [46].

6.3.1.1 *Resveratrol*

Obtained from grapes and berries, resveratrol modulates SIRT1, a NAD+-dependent histone deacetylase, thereby influencing chromatin remodeling and metabolic control [47]. Through the activation of SIRT1 and the inhibition of DNMT1, resveratrol re-expresses tumor suppressor genes in various cancer models [48,49].

6.3.1.2 *Curcumin*

Derived from Curcuma longa, curcumin inhibits HDAC and HAT activities, demethylates CpG islands at the promoters of genes such as p16 and MLH1, and upregulates miR-22 expression, thus reversing epigenetic silencing [50–52]. The pleiotropic activity of curcumin extends to the modulation of histone acetylation, DNA methylation, and non-coding RNA expression [53].

6.3.1.3 *Epigallocatechin gallate [EGCG]*

EGCG, a major catechin from green tea, functions as a competitive inhibitor of DNMTs, reactivating methylation-silenced genes—including GSTP1 and E-cadherin [54]. Moreover, EGCG restores the expression of tumor-suppressive miRNAs and impairs HDAC activity, inducing global chromatin relaxation [55,56]. These polyphenols demonstrate the profound interplay between diet and epigenetic homeostasis [57].

6.3.2 Terpenoids and isoprenoids

Compounds like ursolic acid and betulinic acid target histone deacetylases and histone methyltransferases, leading to epigenetically mediated apoptosis in cancer cells [58,59]. Ginsenosides, primarily extracted from *Panax ginseng*, modify histone acetylation by modulating HDAC activity and regulating the DNA methylation of genes associated with stress and immunity [60,61].

6.3.3 Alkaloids and other phytochemicals

Berberine, from Berberis species, demethylates the promoters of tumor suppressor genes and increases histone acetylation by inhibiting HDAC1 and DNMT1 [62,63]. Camptothecin, initially identified from *Camptotheca acuminata*, not only targets topoisomerase I but also induces histone modifications linked to apoptosis [64,65].

6.3.4 Sulfur-containing compounds

Organosulfur molecules such as sulforaphane (found in broccoli sprouts) act as potent HDAC inhibitors, restoring the acetylation of histones H3

Table 6.1 Representative plant-derived epigenetic modulators and their targets

Compound	*Natural source*	*Primary epigenetic target*	*Mode of action*	*Disease context*
Curcumin	Turmeric (Curcuma longa)	HDAC, HAT, DNMT	Inhibits HDACs and DNMTs; demethylates promoters	Cancer, inflammation
Resveratrol	Grapes, berries	SIRT1, DNMT1	Activates SIRT1; modifies DNA methylation	Metabolic and cardiovascular diseases
Epigallocatechin gallate	Green tea	DNMT1, HDAC	DNMT and HDAC inhibition	Prostate and breast cancers
Sulforaphane	Broccoli sprouts	HDAC	Increases histone acetylation	Chemoprevention
Berberine	Berberis species	DNMT1, HDAC1	Modulates promoter methylation	Leukemia, diabetes
Quercetin	Apples, onions	HDAC, miRNAs	HDAC inhibition; miRNA regulation	Colon and breast cancer

and H4 in the *p21* promoter region [66,67]. Allicin, from garlic, enhances histone acetylation and reduces global methylation levels, thereby improving antioxidant gene expression [68,69].

A detailed list of natural modulators, showing their major epigenetic targets and modes of action, is presented in Table 6.1.

6.4 MARINE-DERIVED METABOLITES WITH EPIGENETIC ACTIVITY

Marine ecosystems are prolific sources of novel bioactive compounds. Due to diverse ecological pressures, marine organisms synthesize secondary metabolites with unique functional groups capable of modulating chromatin-regulatory enzymes [70,71]. Psammaplin A, isolated from sponges, is a naturally occurring HDAC inhibitor possessing a disulfide bond critical for its bioactivity. This compound effectively reactivates silenced genes in melanoma and leukemia cells through the hyperacetylation of histone H3 [72]. Largazole, derived from cyanobacteria, selectively targets class I HDACs with remarkable potency, bridging marine chemistry with clinical epigenetic drug design [73,74]. Semi-synthetic analogs of largazole are currently being evaluated for anticancer applications [75]. Marine-derived depsipeptides such as romidepsin have gained regulatory approval as targeted anticancer agents, validating the marine chemical space as an invaluable resource for epigenetic modulation [76,77]. The compounds—along with their primary targets, mechanistic highlights, and therapeutic potential—are summarized in Table 6.2 and Figure 6.2.

Table 6.2 Marine and microbial-derived epigenetic compounds

Compound	*Source organism*	*Primary target*	*Mechanistic highlights*	*Therapeutic potential*
Psammaplin A	Marine sponge	HDAC, DNMT	Dual inhibition via disulfide bond mechanism	Anticancer
Largazole	Cyanobacteria	HDAC class I	Prodrug converted to thioester active form	Antitumor and anti-inflammatory
Romidepsin	Marine bacterium	HDAC1/2	Thiol-dependent target engagement	Approved for T-cell lymphoma
Trichostatin A	Soil/marine bacteria	HDACs	Pan-HDAC inhibition, apoptosis activation	Broad anticancer utility
Spiruchostatin	*Pseudomonas* sp.	HDACs	Inhibits proliferation, induces differentiation	Glioma and colon cancer models

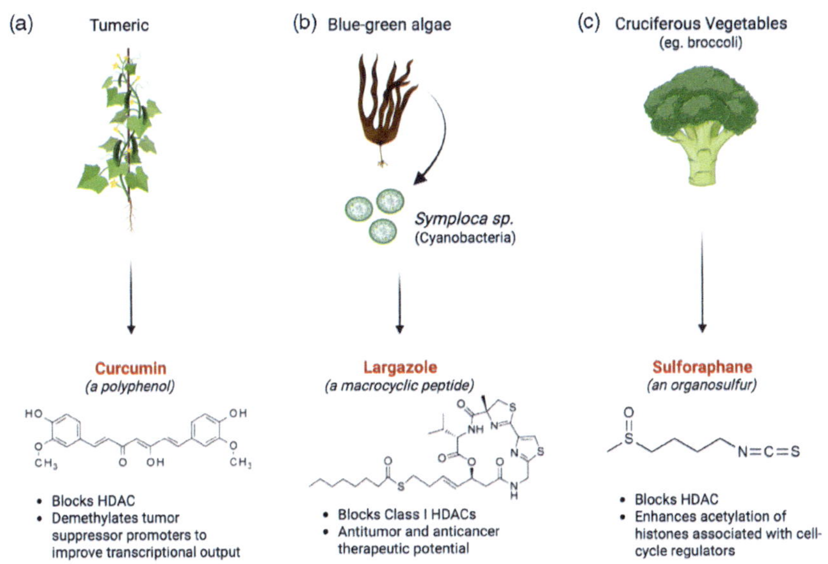

Figure 6.2 Structural Diversity of Selected Natural Epigenetic Modulators: Representative structures of (a) curcumin (polyphenol), (b) largazole (macrocyclic peptide), and (c) sulforaphane (organosulfur compound), and their links to epigenetic activity.

6.5 MECHANISTIC INSIGHTS: MOLECULAR CROSSROADS OF EPIGENETIC REGULATION

Natural compounds significantly alter the epigenome through various interlinked mechanisms that modify chromatin structure and gene expression. Rather than acting on a single enzyme, many bioactive compounds affect several epigenetic regulators simultaneously, allowing a single compound to modulate multiple transcriptional pathways simultaneously. The major mechanisms through which natural epigenetic modulators exert their effects are given below [78]:

6.5.1 DNA demethylation

EGCG competes with *S*-adenosylmethionine at the DNMT1 active site [79].

6.5.2 Histone acetylation remodeling

Curcumin blocks HDAC activity and promotes histone acetylation, which relaxes the chromatin conformation [80] (Figure 6.2a).

6.5.3 Non-coding RNA regulation

Polyphenols modulate miRNAs involved in the *p53*, Wnt, and Notch pathways [81].

6.5.4 Chromatin accessibility

Terpenoids increase the nuclear accessibility of transcription factors by altering nucleosome spacing [82].

For example, curcumin-mediated demethylation of tumor suppressor promoters increases transcriptional output, while sulforaphane-induced HDAC inhibition enhances the acetylation of histones associated with cell-cycle regulators [83,84]. These combined effects produce systemic epigenetic reprogramming, often with lower toxicity compared to synthetic drugs [85,86].

6.6 SYNERGISTIC INTERACTIONS AND EPIGENETIC CROSSTALK

Epigenetic regulation primarily involves dynamic interactions between DNA methylation, histone modifications, and non-coding RNA pathways. Natural compounds often influence more than one epigenetic layer, resulting in combined effects on gene expression [87]. For example, polyphenols can alter histone acetylation while simultaneously regulating microRNAs that target transcripts encoding epigenetic enzymes [86]. Resveratrol modulates SIRT1 activity while influencing the microRNAs such as miR-34a, thereby participating in the post-transcriptional regulation of genes involved in metabolism and cell survival [88]. Similarly, epigallocatechin-3-gallate has been reported to affect both DNA methylation and histone acetylation in cancer models [89,90]. Previous studies combining multiple phytochemicals, including curcumin and EGCG, suggest that additive or synergistic epigenetic effects may occur under very specific conditions. However, these outcomes are influenced by factors such as compound concentration, treatment duration, and cellular context, emphasizing the need for detailed experimental design and validation [91].

6.7 EMERGING ANALYTICAL TOOLS FOR DISCOVERY AND VALIDATION

Technological innovations have accelerated the identification of natural epigenetic modulators. High-throughput epigenomic profiling combined with computational analysis has advanced significantly, accelerating the identification and characterization of several such compounds. Techniques like ChIP-seq [92], DNA methylation profiling [93], and transcriptomic analysis [94] allow detailed mapping of chromatin and gene expression changes primarily induced by specific bioactive compounds [92,94]. Similarly, computational modeling, molecular docking, and machine learning approaches are extensively used to predict compound–target interactions [95]. However, further experimental validation is required to confirm the findings further [96,97] (Figure 6.3).

6.8 NANOTECHNOLOGY-ENABLED DELIVERY OF NATURAL EPIGENETIC COMPOUNDS

A major limitation in the clinical translation of many natural epigenetic modulators includes poor solubility, instability, and insufficient cellular uptake. Nanotechnology-based drug delivery has been widely applied to improve pharmacokinetic properties and enhance intracellular delivery. Specifically, liposomal formulations, polymeric nanoparticles, and organic nanocarriers have been shown to increase the bioavailability and sustained release of different phytochemicals [98] into the systemic circulation. Some of the well-known cargoes used in these studies have been listed below:

8.1 **Liposomal Curcumin**: Improves tissue distribution and enhances engagement with nuclear targets [99].

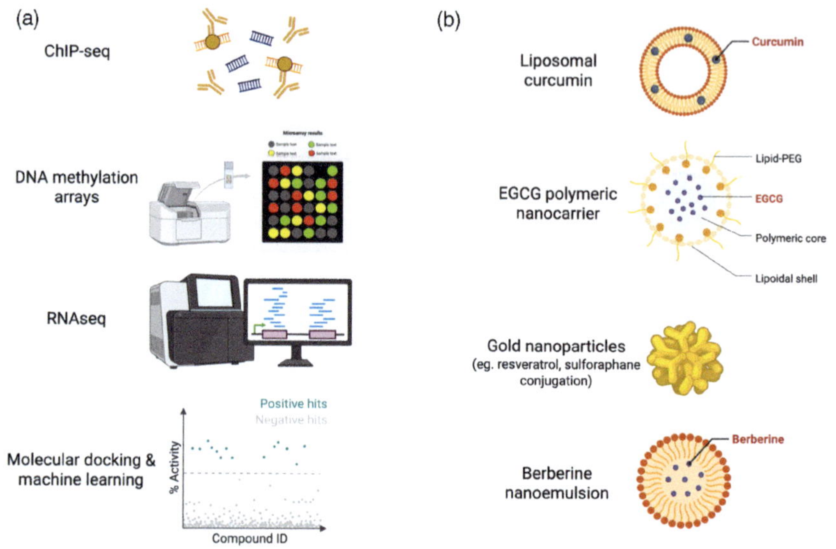

Figure 6.3 (a) Emerging analytical tools to identify natural epigenetic modulators and (b) nanotechnology-enabled delivery of an epigenetic compound.

8.2 **PLGA-encapsulated EGCG**: Increases serum stability and cellular uptake [100].
8.3 **Gold Nanoparticles**: When conjugated with resveratrol or sulforaphane, they enable targeted epigenetic modulation in tumor microenvironments via pH-triggered release [101].
8.4 **Nanoemulsions**: Enhance brain delivery of berberine, crossing the blood–brain barrier for neuroepigenetic therapy [102].

Advanced systems integrate stimuli-responsive nanocarriers that release epigenetic modulators upon exposure to pH gradients, enzymes, or oxidative stress [103]. Coupling them with cell-penetrating peptides can further enhance nuclear localization, directly impacting chromatin-associated targets [104].

6.9 CLINICAL AND TRANSLATIONAL PROSPECTS

Many natural epigenetic compounds have advanced to preclinical evaluation or early-phase clinical trials. Curcumin analogs such as EF24 and GO-Y030 show selective inhibition of HDACs and exhibit higher potency [105]. Sulforaphane-rich formulations have been tested in chemoprevention trials for prostate and breast cancers, showing the modulation of epigenetic biomarkers [106,107]. However, natural epigenetic drug development faces challenges, including variable purity, poor pharmacokinetic characterization, and off-target effects [108]. Implementing nanocarriers, prodrug strategies, and combined regimens with approved drugs can enhance translational feasibility [109,110]. Regulatory agencies are increasingly urging research that incorporates dietary epigenetics into preventive and therapeutic medical approaches [111]. However, despite progress, challenges remain, including variability in compound purity, inconsistent *in vivo* responses, and a lack of long-term efficacy data. Nano-delivery systems and standardized phytopharmaceutical formulations are potential solutions to these translational gaps [112].

6.10 INTEGRATING MULTI-OMICS AND PRECISION EPIGENETIC HEALTHCARE

Epigenetics operates at the center of genomics, transcriptomics, proteomics, and metabolomics [113]. Integrative multi-omic profiling enables the personalized prediction of which patients may benefit most from specific natural compounds [114]. Precision healthcare leverages this data-driven understanding to guide optimized interventions, such as tailoring resveratrol therapy for metabolic syndrome patients with SIRT1 mutations, or providing curcumin supplementation for inflammatory epigenotypes [115]. Machine learning models analyzing methylation signatures and metabolite networks are now informing individualized nutriepigenomic recommendations [116,117].

6.11 FUTURE CHALLENGES AND OPPORTUNITIES

Although enthusiasm for natural epigenetic modulators remains high, significant gaps persist, including:

a. The complexity of multi-target actions, which complicates pharmacodynamic evaluation.
b. Standardization and quality control across extracts are inconsistent.
c. Limited clinical data linking epigenetic modulation to long-term outcomes.
d. Bioavailability that remains low for many polyphenolic compounds.

Emerging solutions include artificial intelligence-guided molecular design of semi-synthetic analogs, CRISPR-based readouts for functional epigenetic screening, and next-generation nanoplatforms enabling precision delivery. Further collaboration among phytochemistry, systems biology, clinical pharmacology, and nanomedicine will be essential to turn these multi-target epigenetic modulators into safer interventions for patients.

6.12 CONCLUSION

In this chapter, we discussed the evidence indicating that natural compounds derived from plants and marine organisms possess the capacity to modulate epigenetic regulation, primarily affecting DNA methylation, histone modifications, and non-coding RNA pathways. Several studies have shown that these compounds alter gene expression associated with cancer, metabolic disorders, inflammation, and neurodegenerative diseases; however, their regulatory extent varies across experimental groups. At this stage, this variability poses a major challenge, preventing natural modulators from advancing to clinical trials due to inconsistent compound composition, limited bioavailability, and an incomplete understanding of their long-term effects *in vivo*. Nanotechnology-based systems could mitigate certain limitations; however, substantial future investigation is required for their clinical validation. Future research will enable the exploration of the full potential of these compounds in therapeutic strategies and patient management.

REFERENCES

1. Bird A. Perceptions of epigenetics. *Nature*. 2007 May 24;447[7143]:396–398. doi: 10.1038/nature05913. PMID: 17522671.
2. Cedar H, Bergman Y. Linking DNA methylation and histone modification: patterns and paradigms. *Nat Rev Genet*. 2009 May;10[5]:295–304. doi: 10.1038/nrg2540. PMID: 19308066.
3. Allis CD, Jenuwein T. The molecular hallmarks of epigenetic control. *Nat Rev Genet*. 2016 Aug;17[8]:487–500. doi: 10.1038/nrg.2016.59. Epub 2016 Jun 27. PMID: 27346641.
4. Portela A, Esteller M. Epigenetic modifications and human disease. *Nat Biotechnol*. 2010 Oct;28[10]:1057–68. doi: 10.1038/nbt.1685. PMID: 20944598.
5. Wu YL, Lin ZJ, Li CC, Lin X, Shan SK, Guo B, Zheng MH, Li F, Yuan LQ, Li ZH. Epigenetic regulation in metabolic diseases: mechanisms and advances in clinical study. *Signal Transduct Target Ther*. 2023 Mar 2;8[1]:98. doi: 10.1038/s41392-023-01333-7. PMID: 36864020; PMCID: PMC9981733.
6. Jones PA, Issa JP, Baylin S. Targeting the cancer epigenome for therapy. *Nat Rev Genet*. 2016 Sep 15;17[10]:630–641. doi: 10.1038/nrg.2016.93. PMID: 27629931.
7. Baylin SB, Jones PA. Epigenetic determinants of cancer. *Cold Spring Harb Perspect Biol*. 2016 Sep 1;8[9]:a019505. doi: 10.1101/cshperspect.a019505. PMID: 27194046; PMCID: PMC5008069.
8. Fischer A, Sananbenesi F, Mungenast A, Tsai LH. Targeting the correct HDAC[s] to treat cognitive disorders. *Trends Pharmacol Sci*. 2010 Dec;31[12]:605–617. doi: 10.1016/j.tips.2010.09.003. Epub 2010 Oct 25. PMID: 20980063.
9. Berdasco M, Esteller M. Clinical epigenetics: seizing opportunities for translation. *Nat Rev Genet*. 2019 Feb;20[2]:109–127. doi: 10.1038/s41576-018-0074-2. PMID: 30479381.

10. Choi SW, Friso S. Epigenetics: A new bridge between nutrition and health. *Adv Nutr.* 2010 Nov;1[1]:8–16. doi: 10.3945/an.110.1004. Epub 2010 Nov 16. PMID: 22043447; PMCID: PMC3042783.
11. Ling C, Rönn T. Epigenetics in human obesity and type 2 diabetes. *Cell Metab.* 2019 May 7;29[5]:1028–1044. doi: 10.1016/j.cmet.2019.03.009. Epub 2019 Apr 11. PMID: 30982733; PMCID: PMC6509280.
12. Nebbioso A, Tambaro FP, Dell'Aversana C, Altucci L. Cancer epigenetics: moving forward. *PLoS Genet.* 2018 Jun 7;14[6]:e1007362. doi: 10.1371/journal.pgen.1007362. PMID: 29879107; PMCID: PMC5991666.
13. Biswas S, Rao CM. Epigenetic tools [The Writers, The Readers and The Erasers] and their implications in cancer therapy. *Eur J Pharmacol.* 2018 Oct 15;837:8–24. doi: 10.1016/j.ejphar.2018.08.021. Epub 2018 Aug 18. PMID: 30125562.
14. Grant C, Rahman F, Piekarz R, Peer C, Frye R, Robey RW, Gardner ER, Figg WD, Bates SE. Romidepsin: a new therapy for cutaneous T-cell lymphoma and a potential therapy for solid tumors. *Expert Rev Anticancer Ther.* 2010 Jul;10[7]:997–1008. doi: 10.1586/era.10.88. Erratum in: *Expert Rev Anticancer Ther.* 2011 Oct;11[10]:1622. PMID: 20645688; PMCID: PMC6361116.
15. Ma J, Ge Z. Comparison between decitabine and azacitidine for patients with acute myeloid leukemia and higher-risk myelodysplastic syndrome: a systematic review and network meta-analysis. *Front Pharmacol.* 2021 Aug 17;12:701690. doi: 10.3389/fphar.2021.701690. Erratum in: *Front Pharmacol.* 2023 May 05;14:1213053. doi: 10.3389/fphar.2023.1213053. PMID: 34483903; PMCID: PMC8416074.
16. Kelly AD, Issa JJ. The promise of epigenetic therapy: reprogramming the cancer epigenome. *Curr Opin Genet Dev.* 2017 Feb;42:68–77. doi: 10.1016/j.gde.2017.03.015. Epub 2017 Apr 14. PMID: 28412585.
17. Marks PA, Breslow R. Dimethyl sulfoxide to vorinostat: development of this histone deacetylase inhibitor as an anticancer drug. *Nat Biotechnol.* 2007 Jan;25[1]:84–90. doi: 10.1038/nbt1272. PMID: 17211407.
18. Sharma S, Kelly TK, Jones PA. Epigenetics in cancer. *Carcinogenesis.* 2010 Jan;31[1]:27–36. doi: 10.1093/carcin/bgp220. Epub 2009 Sep 13. PMID: 19752007; PMCID: PMC2802667.
19. Link A, Balaguer F, Goel A. Cancer chemoprevention by dietary polyphenols: promising role for epigenetics. *Biochem Pharmacol.* 2010 Dec 15;80[12]:1771–92. doi: 10.1016/j.bcp.2010.06.036. Epub 2010 Jun 26. PMID: 20599773; PMCID: PMC2974019.
20. Roy S, Deka D, Kondaveeti SB, Ayyadurai P, Siripragada S, Philip N, Pathak S, Duttaroy AK, Banerjee A. An overview of potential of natural compounds to regulate epigenetic modifications in colorectal cancer: a recent update. *Epigenetics.* 2025 Dec;20[1]:2491316. doi: 10.1080/15592294.2025.2491316. Epub 2025 Apr 16. PMID: 40239010; PMCID: PMC12005453.
21. Carlos-Reyes Á, López-González JS, Meneses-Flores M, Gallardo-Rincón D, Ruíz-García E, Marchat LA, Astudillo-de la Vega H, Hernández de la Cruz ON, López-Camarillo C. Dietary compounds as epigenetic modulating agents in cancer. *Front Genet.* 2019 Mar 1;10:79. doi: 10.3389/fgene.2019.00079. PMID: 30881375; PMCID: PMC6406035.

22. Thakur VS, Deb G, Babcook MA, Gupta S. Plant phytochemicals as epigenetic modulators: role in cancer chemoprevention. AAPS J. 2014 Jan;16[1]:151–63. doi: 10.1208/s12248-013-9548-5. Epub 2013 Dec 5. PMID: 24307610; PMCID: PMC3889536.
23. Tollefsbol TO, ed. *Epigenetics in Human Disease*, 2nd ed. Academic Press, 2022.
24. Bhat KP, Pezzuto JM. Cancer chemopreventive activity of resveratrol. *Ann N Y Acad Sci*. 2002 May;957:210–29. doi: 10.1111/j.1749-6632.2002.tb02918.x. PMID: 12074974.
25. van der Maarel SM. Epigenetic mechanisms in health and disease. *Ann Rheum Dis*. 2008 Dec;67 Suppl 3:iii97–100. doi: 10.1136/ard.2008.098392. PMID: 19022824.
26. Hardy TM, Tollefsbol TO. Epigenetic diet: impact on the epigenome and cancer. *Epigenomics*. 2011 Aug;3[4]:503–518. doi: 10.2217/epi.11.71. PMID: 22022340; PMCID: PMC3197720.
27. Liu RH. Health benefits of fruit and vegetables are from additive and synergistic combinations of phytochemicals. *Am J Clin Nutr*. 2003 Sep;78[3 Suppl]:517S–520S. doi: 10.1093/ajcn/78.3.517S. PMID: 12936943.
28. Vanden Berghe W. Epigenetic impact of dietary polyphenols in cancer chemoprevention: lifelong remodeling of our epigenomes. *Pharmacol Res*. 2012 Jun;65[6]:565–576. doi: 10.1016/j.phrs.2012.03.007. Epub 2012 Mar 22. PMID: 22465217.
29. Pandey KB, Rizvi SI. Plant polyphenols as dietary antioxidants in human health and disease. *Oxid Med Cell Longev*. 2009 Nov-Dec;2[5]:270–278. doi: 10.4161/oxim.2.5.9498. PMID: 20716914; PMCID: PMC2835915.
30. Reuter S, Gupta SC, Chaturvedi MM, Aggarwal BB. Oxidative stress, inflammation, and cancer: how are they linked? *Free Radic Biol Med*. 2010 Dec 1;49[11]:1603–1616. doi: 10.1016/j.freeradbiomed.2010.09.006. Epub 2010 Sep 16. PMID: 20840865; PMCID: PMC2990475.
31. Lorenzo PM, Izquierdo AG, Rodriguez-Carnero G, Fernández-Pombo A, Iglesias A, Carreira MC, Tejera C, Bellido D, Martinez-Olmos MA, Leis R, Casanueva FF, Crujeiras AB. Epigenetic effects of healthy foods and lifestyle habits from the Southern European Atlantic diet pattern: a narrative review. *Adv Nutr*. 2022 Oct 2;13[5]:1725–1747. doi: 10.1093/advances/nmac038. PMID: 35421213; PMCID: PMC9526853.
32. Dwivedi K, Sahoo A, Almalki WH, Almujri SS, Aodah A, Alruwaili NK, Rab SO, Alanezi AA, Haji EM, Barkat MA, Singh T, Rahman M. Innovative nanocarrier systems for enhanced delivery of phyto-active compounds in cancer therapy. *Nanomedicine [Lond]*. 2025 Jan;20[1]:91–116. doi: 10.1080/17435889.2024.2440301. Epub 2024 Dec 20. PMID: 39703154; PMCID: PMC11703466.
33. Jain KK. The role of nanobiotechnology in drug discovery. *Drug Discov Today*. 2005 Nov 1;10[21]:1435–1442. doi: 10.1016/S1359-6446[05]03573-7. PMID: 16243263.
34. Kopelovich L, Crowell JA, Fay JR. The epigenome as a target for cancer chemoprevention. *J Natl Cancer Inst*. 2003 Dec 3;95[23]:1747–1757. doi: 10.1093/jnci/djg109. PMID: 14652236.

35. Aggarwal R, Jha M, Shrivastava A, Jha AK. Natural compounds: role in reversal of epigenetic changes. *Biochemistry [Mosc]*. 2015 Aug;80[8]: 972–989. doi: 10.1134/S0006297915080027. PMID: 26547065.
36. Zhang Y, Reinberg D. Transcription regulation by histone methylation: interplay between different covalent modifications of the core histone tails. *Genes Dev.* 2001 Sep 15;15[18]:2343–2360. doi: 10.1101/gad.927301. PMID: 11562345.
37. Jones PA. Epigenetics in carcinogenesis and cancer prevention. *Ann N Y Acad Sci.* 2003 Mar;983:213–219. doi: 10.1111/j.1749-6632.2003.tb05976.x. PMID: 12724226.
38. Bottomley MJ. Structures of protein domains that create or recognize histone modifications. *EMBO Rep.* 2004 May;5[5]:464–469. doi: 10.1038/sj.embor.7400146. PMID: 15184976; PMCID: PMC1299057.
39. Vasanthi D, Mishra RK. Epigenetic regulation of genes during development: a conserved theme from flies to mammals. *J Genet Genomics.* 2008 Jul;35[7]:413–429. doi: 10.1016/S1673-8527[08]60059-4. PMID: 18640621.
40. Babar Q, Saeed A, Tabish TA, Pricl S, Townley H, Thorat N. Novel epigenetic therapeutic strategies and targets in cancer. *Biochim Biophys Acta Mol Basis Dis.* 2022 Dec 1;1868[12]:166552. doi: 10.1016/j.bbadis.2022.166552. Epub 2022 Sep 17. PMID: 36126898.
41. Benedetti R, Conte M, Altucci L. Targeting histone deacetylases in diseases: where are we? *Antioxid Redox Signal.* 2015 Jul 1;23[1]:99–126. doi: 10.1089/ars.2013.5776. Epub 2014 Mar 6. PMID: 24382114; PMCID: PMC4492558.
42. Suárez R, Chapela SP, Álvarez-Córdova L, Bautista-Valarezo E, Sarmiento-Andrade Y, Verde L, Frias-Toral E, Sarno G. Epigenetics in obesity and diabetes mellitus: new insights. *Nutrients.* 2023 Feb 4;15[4]:811. doi: 10.3390/nu15040811. PMID: 36839169; PMCID: PMC9963127.
43. Clarke K, Young C, Liberante F, McMullin MF, Thompson A, Mills K. The histone deacetylase inhibitor Romidepsin induces as a cascade of differential gene expression and altered histone H3K9 marks in myeloid leukaemia cells. *Oncotarget.* 2019 May 28;10[37]:3462–3471. doi: 10.18632/oncotarget.26877. PMID: 31191819; PMCID: PMC6544403.
44. Tieu S, Charchoglyan A, Wagter-Lesperance L, Karimi K, Bridle BW, Karrow NA, Mallard BA. Immunoceuticals: harnessing their immunomodulatory potential to promote health and wellness. *Nutrients.* 2022 Sep 30;14[19]:4075. doi: 10.3390/nu14194075. PMID: 36235727; PMCID: PMC9571036.
45. Zoghbi HY, Beaudet AL. Epigenetics and human disease. *Cold Spring Harb Perspect Biol.* 2016 Feb 1;8[2]:a019497. doi: 10.1101/cshperspect.a019497. PMID: 26834142; PMCID: PMC4743078.
46. Vauzour D, Rodriguez-Mateos A, Corona G, Oruna-Concha MJ, Spencer JP. Polyphenols and human health: prevention of disease and mechanisms of action. *Nutrients.* 2010 Nov;2[11]:1106–1131. doi: 10.3390/nu2111106. Epub 2010 Nov 8. PMID: 22254000; PMCID: PMC3257622.
47. Howitz KT, Bitterman KJ, Cohen HY, Lamming DW, Lavu S, Wood JG, Zipkin RE, Chung P, Kisielewski A, Zhang LL, Scherer B, Sinclair DA. Small molecule activators of sirtuins extend Saccharomyces cerevisiae lifespan. *Nature.* 2003 Sep 11;425[6954]:191–196. doi: 10.1038/nature01960. Epub 2003 Aug 24. PMID: 12939617.

48. Hubbard BP, Sinclair DA. Small molecule SIRT1 activators for the treatment of aging and age-related diseases. *Trends Pharmacol Sci.* 2014 Mar;35[3]: 146–154. doi: 10.1016/j.tips.2013.12.004. Epub 2014 Jan 16. PMID: 24439680; PMCID: PMC3970218.
49. Fernandes GFS, Silva GDB, Pavan AR, Chiba DE, Chin CM, Dos Santos JL. Epigenetic regulatory mechanisms induced by resveratrol. *Nutrients.* 2017 Nov 1;9[11]:1201. doi: 10.3390/nu9111201. PMID: 29104258; PMCID: PMC5707673.
50. Teiten MH, Dicato M, Diederich M. Curcumin as a regulator of epigenetic events. *Mol Nutr Food Res.* 2013 Sep;57[9]:1619–1629. doi: 10.1002/mnfr.201300201. Epub 2013 Jun 11. PMID: 23754571.
51. Hong SM, Choi J, Ryu K, Ro JY, Yu E. Promoter hypermethylation of the p16 gene and loss of its protein expression is correlated with tumor progression in extrahepatic bile duct carcinomas. *Arch Pathol Lab Med.* 2006 Jan;130[1]:33–38. doi: 10.5858/2006-130-33-PHOTPG. PMID: 16390235.
52. Hassan FU, Rehman MS, Khan MS, Ali MA, Javed A, Nawaz A, Yang C. Curcumin as an alternative epigenetic modulator: mechanism of action and potential effects. *Front Genet.* 2019 Jun 4;10:514. doi: 10.3389/fgene.2019.00514. PMID: 31214247; PMCID: PMC6557992.
53. Gupta SC, Kim JH, Prasad S, Aggarwal BB. Regulation of survival, proliferation, invasion, angiogenesis, and metastasis of tumor cells through modulation of inflammatory pathways by nutraceuticals. *Cancer Metastasis Rev.* 2010 Sep;29[3]:405–434. doi: 10.1007/s10555-010-9235-2. PMID: 20737283; PMCID: PMC2996866.
54. Fang MZ, Wang Y, Ai N, Hou Z, Sun Y, Lu H, Welsh W, Yang CS. Tea polyphenol [-]-epigallocatechin-3-gallate inhibits DNA methyltransferase and reactivates methylation-silenced genes in cancer cell lines. *Cancer Res.* 2003 Nov 15;63[22]:7563–7570. PMID: 14633667.
55. Lecumberri E, Dupertuis YM, Miralbell R, Pichard C. Green tea polyphenol epigallocatechin-3-gallate [EGCG] as adjuvant in cancer therapy. *Clin Nutr.* 2013 Dec;32[6]:894–903. doi: 10.1016/j.clnu.2013.03.008. Epub 2013 Mar 15. PMID: 23582951.
56. Henning SM, Wang P, Carpenter CL, Heber D. Epigenetic effects of green tea polyphenols in cancer. *Epigenomics.* 2013 Dec;5[6]:729–741. doi: 10.2217/epi.13.57. PMID: 24283885; PMCID: PMC3970408.
57. Carlberg C, Velleuer E. Nutrition and epigenetic programming. *Curr Opin Clin Nutr Metab Care.* 2023 May 1;26[3]:259–265. doi: 10.1097/MCO.0000000000000900. Epub 2022 Dec 23. PMID: 36728887.
58. Galgon T, Wohlrab W, Dräger B. Betulinic acid induces apoptosis in skin cancer cells and differentiation in normal human keratinocytes. *Exp Dermatol.* 2005 Oct;14[10]:736–743. doi: 10.1111/j.1600-0625.2005.00352.x. PMID: 16176281.
59. Limami Y, Pinon A, Wahnou H, Oudghiri M, Liagre B, Simon A, Duval RE. Ursolic acid's alluring journey: one triterpenoid vs. cancer hallmarks. *Molecules.* 2023 Dec 1;28[23]:7897. doi: 10.3390/molecules28237897. PMID: 38067626; PMCID: PMC10707789.
60. Ratan ZA, Haidere MF, Hong YH, Park SH, Lee JO, Lee J, Cho JY. Pharmacological potential of ginseng and its major component ginsenosides. *J Ginseng Res.* 2021 Mar;45[2]:199–210. doi: 10.1016/j.jgr.2020.02.004. Epub 2020 Mar 25. PMID: 33841000; PMCID: PMC8020288.

61. Lü JM, Yao Q, Chen C. Ginseng compounds: an update on their molecular mechanisms and medical applications. *Curr Vasc Pharmacol.* 2009 Jul;7[3]:293–302. doi: 10.2174/157016109788340767. PMID: 19601854; PMCID: PMC2928028.
62. Saha SK, Khuda-Bukhsh AR. Berberine alters epigenetic modifications, disrupts microtubule network, and modulates HPV-18 E6-E7 oncoproteins by targeting p53 in cervical cancer cell HeLa: a mechanistic study including molecular docking. *Eur J Pharmacol.* 2014 Dec 5;744:132–146. doi: 10.1016/j.ejphar.2014.09.048. Epub 2014 Oct 18. PMID: 25448308.
63. Brocks D, Schmidt CR, Daskalakis M, Jang HS, Shah NM, Li D, Li J, Zhang B, Hou Y, Laudato S, Lipka DB, Schott J, Bierhoff H, Assenov Y, Helf M, Ressnerova A, Islam MS, Lindroth AM, Haas S, Essers M, Imbusch CD, Brors B, Oehme I, Witt O, Lübbert M, Mallm JP, Rippe K, Will R, Weichenhan D, Stoecklin G, Gerhäuser C, Oakes CC, Wang T, Plass C. DNMT and HDAC inhibitors induce cryptic transcription start sites encoded in long terminal repeats. *Nat Genet.* 2017 Jul;49[7]:1052–1060. doi: 10.1038/ng.3889. Epub 2017 Jun 12. Erratum in: *Nat Genet.* 2017 Oct 27;49[11]:1661. doi: 10.1038/ng1117-1661c. PMID: 28604729; PMCID: PMC6005702.
64. Sriram D, Yogeeswari P, Thirumurugan R, Bal TR. Camptothecin and its analogues: a review on their chemotherapeutic potential. *Nat Prod Res.* 2005 Jun;19[4]:393–412. doi: 10.1080/14786410412331299005. PMID: 15938148.
65. Farhana A, Koh AE, Ling Mok P, Alsrhani A, Khan YS, Subbiah SK. Camptothecin encapsulated in β-cyclodextrin-EDTA-Fe_3O_4 nanoparticles induce metabolic reprogramming repair in HT29 cancer cells through epigenetic modulation: a bioinformatics approach. *Nanomaterials [Basel].* 2021 Nov 23;11[12]:3163. doi: 10.3390/nano11123163. PMID: 34947512; PMCID: PMC8705212.
66. Myzak MC, Dashwood WM, Orner GA, Ho E, Dashwood RH. Sulforaphane inhibits histone deacetylase in vivo and suppresses tumorigenesis in Apc-minus mice. *FASEB J.* 2006 Mar;20[3]:506–508. doi: 10.1096/fj.05-4785fje. Epub 2006 Jan 11. PMID: 16407454; PMCID: PMC2373266.
67. Dashwood RH, Ho E. Dietary histone deacetylase inhibitors: from cells to mice to man. *Semin Cancer Biol.* 2007 Oct;17[5]:363–369. doi: 10.1016/j.semcancer.2007.04.001. Epub 2007 May 5. PMID: 17555985; PMCID: PMC2737738.
68. Zhang H, Wang H, Qin L, Lin S. Garlic-derived compounds: epigenetic modulators and their antitumor effects. *Phytother Res.* 2024 Mar;38[3]:1329–1344. doi: 10.1002/ptr.8108. Epub 2024 Jan 9. PMID: 38194996.
69. Iciek M, Kwiecień I, Włodek L. Biological properties of garlic and garlic-derived organosulfur compounds. *Environ Mol Mutagen.* 2009 Apr;50[3]:247–265. doi: 10.1002/em.20474. PMID: 1925333.
70. Conte M, Fontana E, Nebbioso A, Altucci L. Marine-derived secondary metabolites as promising epigenetic bio-compounds for anticancer therapy. *Mar Drugs.* 2020 Dec 31;19[1]:15. doi: 10.3390/md19010015. PMID: 33396307; PMCID: PMC7824531.
71. Newman DJ, Cragg GM. Marine-sourced anti-cancer and cancer pain control agents in clinical and late preclinical development. *Mar Drugs.* 2014 Jan 14;12[1]:255–278. doi: 10.3390/md12010255. PMID: 24424355; PMCID: PMC3917273.

72. Kim DH, Shin J, Kwon HJ. Psammaplin A is a natural prodrug that inhibits class I histone deacetylase. *Exp Mol Med.* 2007 Feb 28;39[1]:47–55. doi: 10.1038/emm.2007.6. PMID: 17334228.
73. Taori K, Paul VJ, Luesch H. Structure and activity of largazole, a potent antiproliferative agent from the Floridian marine cyanobacterium Symploca sp. *J Am Chem Soc.* 2008 Feb 13;130[6]:1806–1807. doi: 10.1021/ja7110064. Epub 2008 Jan 19. Erratum in: *J Am Chem Soc.*2008 Oct 8;130[40]:13506. PMID: 18205365.
74. Kim B, Ratnayake R, Lee H, Shi G, Zeller SL, Li C, Luesch H, Hong J. Synthesis and biological evaluation of largazole zinc-binding group analogs. *Bioorg Med Chem.* 2017 Jun 15;25[12]:3077–3086. doi: 10.1016/j.bmc.2017.03.071. Epub 2017 Apr 4. PMID: 28416100; PMCID: PMC5494147.
75. Schwartsmann G, Da Rocha AB, Mattei J, Lopes R. Marine-derived anticancer agents in clinical trials. *Expert Opin Investig Drugs.* 2003 Aug;12[8]: 1367–1383. doi: 10.1517/13543784.12.8.1367. PMID: 12882622.
76. Barbarotta L, Hurley K. Romidepsin for the treatment of peripheral T-cell lymphoma. *J Adv Pract Oncol.* 2015 Jan-Feb;6[1]:22–36. PMID: 26413372; PMCID: PMC4577031.
77. Whittaker SJ, Demierre MF, Kim EJ, Rook AH, Lerner A, Duvic M, Scarisbrick J, Reddy S, Robak T, Becker JC, Samtsov A, McCulloch W, Kim YH. Final results from a multicenter, international, pivotal study of romidepsin in refractory cutaneous T-cell lymphoma. *J Clin Oncol.* 2010 Oct 10;28[29]:4485–4491. doi: 10.1200/JCO.2010.28.9066. Epub 2010 Aug 9. PMID: 20697094.
78. Ratovitski EA. Anticancer natural compounds as epigenetic modulators of gene expression. *Curr Genomics.* 2017 Apr;18[2]:175–205. doi: 10.2174/1389202917666160803165229. PMID: 28367075; PMCID: PMC5345332.
79. Agarwal A, Kansal V, Farooqi H, Prasad R, Singh VK. Epigallocatechin gallate [EGCG], an active phenolic compound of green tea, inhibits tumor growth of head and neck cancer cells by targeting DNA hypermethylation. *Biomedicines.* 2023 Mar 5;11[3]:789. doi: 10.3390/biomedicines11030789. PMID: 36979768; PMCID: PMC10045148.
80. Link A, Balaguer F, Shen Y, Lozano JJ, Leung HC, Boland CR, Goel A. Curcumin modulates DNA methylation in colorectal cancer cells. *PLoS One.* 2013;8[2]:e57709. doi: 10.1371/journal.pone.0057709. Epub 2013 Feb 27. PMID: 23460897; PMCID: PMC3584082.
81. Phuah NH, Nagoor NH. Regulation of microRNAs by natural agents: new strategies in cancer therapies. *Biomed Res Int.* 2014;2014:804510. doi: 10.1155/2014/804510. Epub 2014 Sep 1. PMID: 25254214; PMCID: PMC4165563.
82. Huang M, Lu JJ, Huang MQ, Bao JL, Chen XP, Wang YT. Terpenoids: natural products for cancer therapy. *Expert Opin Investig Drugs.* 2012 Dec;21[12]:1801–1818. doi: 10.1517/13543784.2012.727395. Epub 2012 Oct 23. PMID: 23092199.
83. Rajendran P, Delage B, Dashwood WM, Yu TW, Wuth B, Williams DE, Ho E, Dashwood RH. Histone deacetylase turnover and recovery in sulforaphane-treated colon cancer cells: competing actions of 14-3-3 and Pin1 in HDAC3/SMRT corepressor complex dissociation/reassembly. *Mol Cancer.* 2011 May 30;10:68. doi: 10.1186/1476-4598-10-68. PMID: 21624135; PMCID: PMC3127849.

84. Ho E, Clarke JD, Dashwood RH. Dietary sulforaphane, a histone deacetylase inhibitor for cancer prevention. *J Nutr.* 2009 Dec;139[12]:2393–2396. doi: 10.3945/jn.109.113332. Epub 2009 Oct 7. PMID: 19812222; PMCID: PMC2777483.
85. Shankar E, Kanwal R, Candamo M, Gupta S. Dietary phytochemicals as epigenetic modifiers in cancer: promise and challenges. *Semin Cancer Biol.* 2016 Oct;40–41:82–99. doi: 10.1016/j.semcancer.2016.04.002. Epub 2016 Apr 23. PMID: 27117759; PMCID: PMC5067170.
86. Chou PJ, Peter RM, Shannar A, Pan Y, Dave PD, Xu J, Sarwar MS, Kong AN. Epigenetics of dietary phytochemicals in cancer prevention: fact or fiction. *Cancer J.* 2024 Sep-Oct 01;30[5]:320–328. doi: 10.1097/PPO.0000000000000742. PMID: 39312452; PMCID: PMC11573353.
87. Arora I, Sharma M, Tollefsbol TO. Combinatorial epigenetics impact of polyphenols and phytochemicals in cancer prevention and therapy. *Int J Mol Sci.* 2019 Sep 14;20[18]:4567. doi: 10.3390/ijms20184567. PMID: 31540128; PMCID: PMC6769666.
88. Choi SE, Kemper JK. Regulation of SIRT1 by microRNAs. *Mol Cells.* 2013 Nov;36[5]:385–392. doi: 10.1007/s10059-013-0297-1. Epub 2013 Nov 6. PMID: 24213676; PMCID: PMC3887935. Wu W, Gou H, Xiang B, Geng R, Dong J, Yang X, Chen D, Dai R, Chen L, Liu J. EGCG enhances the chemosensitivity of colorectal cancer to irinotecan through GRP78-mediated endoplasmic reticulum stress. *J Oncol.* 2022 Sep 13;2022:7099589. doi: 10.1155/2022/7099589. PMID: 36147440; PMCID: PMC9489388.
89. Thakur VS, Deb G, Babcook MA, Gupta S. Green tea catechins and other flavonoids enhance the reactivation potential of epigenetically silenced genes in cancer. *Mol Carcinog.* 2014;53[7]:498–508.
90. Costea T, Hudiță A, Ciolac OA, Gălățeanu B, Ginghină O, Costache M, Ganea C, Mocanu MM. Chemoprevention of colorectal cancer by dietary compounds. *Int J Mol Sci.* 2018 Nov 28;19[12]:3787. doi: 10.3390/ijms19123787. PMID: 30487390; PMCID: PMC6321468.
91. Eom DW, Lee JH, Kim YJ, Hwang GS, Kim SN, Kwak JH, Cheon GJ, Kim KH, Jang HJ, Ham J, Kang KS, Yamabe N. Synergistic effect of curcumin on epigallocatechin gallate-induced anticancer action in PC3 prostate cancer cells. *BMB Rep.* 2015 Aug;48[8]:461–466. doi: 10.5483/bmbrep.2015.48.8.216. PMID: 25441423; PMCID: PMC4576954.
92. Park PJ. ChIP-seq: advantages and challenges of a maturing technology. *Nat Rev Genet.* 2009 Oct;10[10]:669–680. doi: 10.1038/nrg2641. Epub 2009 Sep 8. PMID: 19736561; PMCID: PMC3191340.
93. Laird PW. Principles and challenges of genomewide DNA methylation analysis. *Nat Rev Genet.* 2010 Mar;11[3]:191–203. doi: 10.1038/nrg2732. PMID: 20125086.
94. Wang Z, Gerstein M, Snyder M. RNA-Seq: a revolutionary tool for transcriptomics. *Nat Rev Genet.* 2009 Jan;10[1]:57–63. doi: 10.1038/nrg2484. PMID: 19015660; PMCID: PMC2949280.
95. Zhavoronkov A, Mamoshina P, Vanhaelen Q, Scheibye-Knudsen M, Moskalev A, Aliper A. Artificial intelligence for aging and longevity research: recent advances and perspectives. *Ageing Res Rev.* 2019 Jan;49:49–66. doi: 10.1016/j.arr.2018.11.003. Epub 2018 Nov 22. PMID: 30472217.

96. Zhang X, Jonassen I, Goksøyr A. Machine Learning Approaches for Biomarker Discovery Using Gene Expression Data. In: Helder IN, editor. *Bioinformatics* [Internet]. Brisbane [AU]: Exon Publications; 2021 Mar 20. Chapter 4. PMID: 33877765.
97. Ocana A, Pandiella A, Privat C, Bravo I, Luengo-Oroz M, Amir E, Gyorffy B. Integrating artificial intelligence in drug discovery and early drug development: a transformative approach. *Biomark Res.* 2025 Mar 14;13[1]:45. doi: 10.1186/s40364-025-00758-2. PMID: 40087789; PMCID: PMC11909971.
98. Serri C. Editorial on Special Issue: "Advances in nanotechnology-based drug delivery systems". *Pharmaceutics.* 2025 Aug 10;17[8]:1038. doi: 10.3390/pharmaceutics17081038. PMID: 40871059; PMCID: PMC12389492.
99. Radha R, Paul V, Anjum S, Bouakaz A, Pitt WG, Husseini GA. Enhancing Curcumin's therapeutic potential in cancer treatment through ultrasound mediated liposomal delivery. *Sci Rep.* 2024 May 7;14[1]:10499. doi: 10.1038/s41598-024-61278-x. PMID: 38714740; PMCID: PMC11076529.
100. Granja A, Pinheiro M, Reis S. Epigallocatechin gallate nanodelivery systems for cancer therapy. *Nutrients.* 2016 May 20;8[5]:307. doi: 10.3390/nu8050307. PMID: 27213442; PMCID: PMC4882719.
101. Kanu GA, Parambath JBM, Abu Odeh RO, Mohamed AA. Gold nanoparticle-mediated gene therapy. *Cancers [Basel].* 2022 Oct 31;14[21]:5366. doi: 10.3390/cancers14215366. PMID: 36358785; PMCID: PMC9653658.
102. Dhariwal R, Jain M, Mir YR, Singh A, Jain B, Kumar P, Tariq M, Verma D, Deshmukh K, Yadav VK, Malik T. Targeted drug delivery in neurodegenerative diseases: the role of nanotechnology. *Front Med [Lausanne].* 2025 Jan 29;12:1522223. doi: 10.3389/fmed.2025.1522223. PMID: 39963432; PMCID: PMC11831571.
103. Zhao X, Bai J, Yang W. Stimuli-responsive nanocarriers for therapeutic applications in cancer. *Cancer Biol Med.* 2021 Mar 25;18[2]:319–335. doi: 10.20892/j.issn.2095-3941.2020.0496. PMID: 33764711; PMCID: PMC8185873.
104. Deshayes S, Morris MC, Divita G, Heitz F. Cell-penetrating peptides: tools for intracellular delivery of therapeutics. *Cell Mol Life Sci.* 2005 Aug;62[16]:1839–1849. doi: 10.1007/s00018-005-5109-0. PMID: 15968462; PMCID: PMC11139131.
105. Wang H, Shi L, Wang Z. A novel hydroxamic acid-based curcumin derivative as potent histone deacetylase inhibitor for the treatment of glioblastoma. *Front Oncol.* 2021 Nov 5;11:756817. doi: 10.3389/fonc.2021.756817. PMID: 34804949; PMCID: PMC8602566.
106. Alumkal JJ, Slottke R, Schwartzman J, Cherala G, Munar M, Graff JN, Beer TM, Ryan CW, Koop DR, Gibbs A, Gao L, Flamiatos JF, Tucker E, Kleinschmidt R, Mori M. A phase II study of sulforaphane-rich broccoli sprout extracts in men with recurrent prostate cancer. *Invest New Drugs.* 2015 Apr;33[2]:480–489. doi: 10.1007/s10637-014-0189-z. Epub 2014 Nov 29. PMID: 25431127; PMCID: PMC4390425.
107. Atwell LL, Beaver LM, Shannon J, Williams DE, Dashwood RH, Ho E. Epigenetic regulation by sulforaphane: opportunities for breast and prostate cancer chemoprevention. *Curr Pharmacol Rep.* 2015 Apr 1;1[2]:102–111. doi: 10.1007/s40495-014-0002-x. PMID: 26042194; PMCID: PMC4450146.

108. Dai W, Qiao X, Fang Y, Guo R, Bai P, Liu S, Li T, Jiang Y, Wei S, Na Z, Xiao X, Li D. Epigenetics-targeted drugs: current paradigms and future challenges. *Signal Transduct Target Ther.* 2024 Nov 26;9[1]:332. doi: 10.1038/s41392-024-02039-0. PMID: 39592582; PMCID: PMC11627502.
109. Park S, Lu GL, Zheng YC, Davison EK, Li Y. Nanoparticle-based delivery strategies for combating drug resistance in cancer therapeutics. *Cancers [Basel].* 2025 Aug 11;17[16]:2628. doi: 10.3390/cancers17162628. PMID: 40867257; PMCID: PMC12384834.
110. Honer MA, Ferman BI, Gray ZH, Bondarenko EA, Whetstine JR. Epigenetic modulators provide a path to understanding disease and therapeutic opportunity. *Genes Dev.* 2024 Jul 19;38[11–12]:473–503. doi: 10.1101/gad.351444.123. PMID: 38914477; PMCID: PMC11293403.
111. Ostaiza-Cardenas J, Tobar AC, Costa SC, Calero DS, López-Carrera A, Bermúdez FG, Orellana-Manzano A. Epigenetic modulation by life-style: advances in diet, exercise, and mindfulness for disease prevention and health optimization. *Front Nutr.* 2025 Aug 21;12:1632999. doi: 10.3389/fnut.2025.1632999. PMID: 40917096; PMCID: PMC12408607.
112. Ahmad U, Islam A, Khan MM, Akhtar J. Nanotechnology-driven epigenetic cancer therapy: precision delivery and sustained release of DNA methylation modulators. *Yale J Biol Med.* 2025 Jun 30;98[2]:227–235. doi: 10.59249/GVNM8843. PMID: 40589931; PMCID: PMC12204028.
113. Hasin Y, Seldin M, Lusis A. Multi-omics approaches to disease. *Genome Biol.* 2017 May 5;18[1]:83. doi: 10.1186/s13059-017-1215-1. PMID: 28476144; PMCID: PMC5418815.
114. Abdelaziz EH, Ismail R, Mabrouk MS, Amin E. Multi-omics data integration and analysis pipeline for precision medicine: systematic review. *Comput Biol Chem.* 2024 Dec;113:108254. doi: 10.1016/j.compbiolchem.2024.108254. Epub 2024 Oct 16. PMID: 3944740.
115. Ahluwalia MK. Nutrigenetics and nutrigenomics: a personalized approach to nutrition. *Adv Genet.* 2021;108:277–340. doi: 10.1016/bs.adgen.2021.08.005. Epub 2021 Oct 28. PMID: 34844714.
116. Burdge GC, Hoile SP, Lillycrop KA. Epigenetics: are there implications for personalised nutrition? *Curr Opin Clin Nutr Metab Care.* 2012 Sep;15[5]:442–447. doi: 10.1097/MCO.0b013e3283567dd2. PMID: 22878237.
117. Kirk D, Kok E, Tufano M, Tekinerdogan B, Feskens EJM, Camps G. Machine learning in nutrition research. *Adv Nutr.* 2022 Dec 22;13[6]:2573–2589. doi: 10.1093/advances/nmac103. Erratum in: *Adv Nutr.* 2023 May;14[3]:584. doi: 10.1016/j.advnut.2023.03.001. Erratum in: *Adv Nutr.* 2023 Apr 1:S2161-8313[23]00281-8. doi: 10.1016/j.advnut.2023.03.012. PMID: 36166846; PMCID: PMC9776646.

Chapter 7

Targeting cancer stem cells

Novel natural compounds in precision oncology

Veena Vishwakarma[1], *Shajda Khatoon*[2], *Ajay K. Ade*[3], *and Roma Pandey*[1]

[1]Department of Biotechnology, School of Engineering, IILM University, Greater Noida, Uttar Pradesh, India

[2]Mylab Discovery Solution Pvt Ltd., Pune, India

[3]Molecular Diagnostic Laboratory, National Institute of Biologicals, Noida, Uttar Pradesh, India

7.1 INTRODUCTION

Cancer stem cells (CSCs) are subpopulations within a tumour that have a characteristic property of self-renewal, multipotency, and therapeutic tolerance [1].

These properties are increasingly recognized as driving initiation, progression, relapse, and metastasis in cancer. Unlike the majority of other tumour cells that can be treated with cytotoxic or targeted therapies, CSCs frequently persist. They survive through their intrinsic defenses such as enhanced DNA repair, aldehyde detoxification, and efflux transporters. They also rely on extrinsic defences, which include specialized niches such as hypoxia, stroma crosstalk, and immune evasion [2]. Due to their functional plasticity, CSCs can shift between stem cell-like and differentiated states in response to environmental conditions. This phenomenon contributes to tumour heterogeneity and therapy resistance. Because of these properties CSCs have become a high-priority target in precision oncology. It is important that cancer therapy not only reduces the tumour bulk but also silences or eradicates the self-renewing reservoirs of CSCs to avoid relapse.

At the molecular level, CSC phenotypes are regulated by a dense network of developmental and stress-response pathways. The key regulators include Wnt/β-catenin, Notch, Hedgehog, TGF-β, JAK/STAT, NF-κB, and PI3K/AKT/mTOR [1]. These signalling pathways interact with the epigenetic regulators including DNA methyltransferases (DNMTs), and histone deacetylases/histone acetyltransferases (HDACs/HATs), histone methyltransferases/demethylases, and non-coding RNAs. These regulators can stabilize stemness and epithelial-mesenchymal transition/mesenchymal-epithelial transition (EMT/MET) in CSCs [3].

DOI: 10.1201/9781003733010-7

Furthermore, CSCs can efficiently switch between glycolysis and oxidative phosphorylation. They show higher levels of lipid synthesis/β-oxidation and have finely regulated redox homeostasis. This unique metabolic feature supports their survival under hypoxic and nutrient-deficient conditions [4]. However, it is important to note that these pathways rarely act separately. Even if one pathway is inhibited in CSCs, alternative pathways in the network keep their stemness intact. Therefore, this network redundancy argues for multi-target strategies to explore the approaches capable of targeting multiple pathways simultaneously with acceptable safety.

Many phytochemicals and other natural products target multiple CSC-relevant pathways at once. For instance, polyphenols (e.g., curcumin, resveratrol, Epigallocatchein gallate (EGCG), and quercetin) can attenuate Wnt/Notch signalling, modify chromatin, and normalize redox balance [5]. Many terpenoids and withanolides (e.g., ursolic acid, withaferin A) can interfere with NF-κB and cytoskeletal dynamics, restricting invasion and EMT [6,7]. Organosulfur compounds (e.g., sulforaphane) inhibit HDACs and reactivate tumour suppressors. Microbial agents (e.g., salinomycin) selectively reduce CSC viability by disrupting ion homeostasis and mitochondrial function [8–10].

Though the multi-targeted approach of natural products complicates the understanding of their mechanism of action, it can be beneficial against the networked biology of stemness. Moreover, many natural molecules show safe therapeutic indices and have been used for decades as dietary components or traditional remedies. Thus, they provide a feasible starting point for therapy.

The potential of natural products seems promising; however, it should be balanced with the careful approach in precision oncology. While formulating drug combinations, biomarker-guided selection is crucial for patient treatment. CSC marker molecules (e.g., CD44 high/ALDH1+), signalling pathway patterns (Wnt/Notch/Hedgehog), EMT status, and metabolic profiles can inform about compound selection and dosage. Companion diagnostics can detect early pharmacodynamic effects, such as increased histone acetylation with HDAC-modulating nutraceuticals or decreased ATP-binding cassette (ABC) transporter activity. These pharmacokinetic realities are equally crucial in drug design [5,11,12].

One of the challenges with natural products is that many phytochemicals dissolve poorly, are quickly metabolized, and thus have low bioavailability. Advances in drug delivery systems such as nanoparticles, liposomes, polymeric micelles, solid lipid carriers, and ligand-directed systems (e.g., hyaluronic acid for CD44, aptamers for CD133) can enhance bioavailability and target CSCs more precisely. Strategies like prodrug design, co-crystallization, and metabolic engineering can further enhance drug exposure while reducing off-target effects [11,13].

Combination therapy is another aspect of natural product-based precision care. Natural products can make CSCs more sensitive to therapies by reducing the apoptotic threshold, inhibiting ABC transporters, or reversing EMT. They may also help targeted inhibitors by preventing alternate pathway activation and altering the tumour microenvironment to aid immune surveillance via attenuation of NF-κB-driven inflammation and modulation of myeloid cells. Strategies for scheduled administrations of natural epigenetic modulators for pre-conditioning of CSCs before cytotoxic therapy, or metronomic maintenance, can be as important as agent selection to prevent tumour regrowth [8].

For clinical adoption, quality, standardization, and regulatory frameworks are essential considerations. The composition of bioactive compounds in plants varies by species, environmental conditions, and post-harvest processing. Ensuring batch-to-batch consistency and comprehensive chemical/metabolite characterization is mandatory for reproducibility. Chemistry, Manufacturing, and Controls (CMC) frameworks must be up to the regulatory standards, along with critical assessment of natural product and drug interactions to avoid compromising standard treatments. Sustainability and responsible sourcing are ethical and practically essential as the demand grows [14,15].

This chapter frames natural products in the evolution of the modern concept of CSCs and provides an overview from mechanism to medicine. First, the chapter discusses CSC biology, focusing on targetable vulnerabilities in signalling pathways, epigenetics, metabolism, and the tumour niche. We next list the major categories of natural products associated with anti-CSC activity, emphasizing the representative compounds, their key targets, and the level of evidence in different tumour models. Mechanistic subsections address the action of these agents on self-renewal pathways as well as their EMT-reversing, drug-resistance-limiting, and metabolic- and redox-programming properties.

Later sections cover formulations and targeted delivery strategies that break through pharmacokinetic barriers; combination regimes with chemotherapy, targeted therapies, and immuno-oncology; as well as the clinical translation landscape including safety, pharmacokinetics, biomarker development, and trial design.

Finally, the chapter concludes with a discussion of discovery engines, from omics technologies to *in silico* approaches, that can drive the transformation of natural compounds into precisely targeted CSC-directed drugs. To combine traditional ethnopharmacological knowledge with modern systems biology and drug delivery science, we hope to establish an informed clinical foundation for launching CSC-active natural compounds as useful strategies to achieve durable responses and better outcomes in precision oncology.

7.2 BIOLOGY OF CANCER STEM CELLS

7.2.1 Defining features: self-renewal, differentiation, quiescence, and plasticity

CSC biology is a crucial area of cancer research due to its resistance to conventional cancer therapies. CSCs are rare subsets of cells in tumours characterized by self-renewal, differentiation, continuous proliferation, tumorigenesis, metastasis, and the maintenance of tumour heterogeneity. These CSCs may arise from normal stem cells that accumulate mutations, progenitor cells that reacquire stem cell-like properties, and cancer cells that undergo dedifferentiation via the EMT pathway [1].

The resistance of CSCs to chemotherapy lies in their two unique abilities: quiescence and plasticity. Quiescence is their ability to remain in a dormant, non-proliferative stage of the cell cycle with low metabolic activity. This allows the CSCs to remain hidden in the tumour microenvironment and re-enter the cell cycle, thus evading the traditional cancer therapies which target rapidly dividing cells. On the other hand, plasticity is the ability of CSCs to switch between the CSC and non-CSC states and vice versa. It is now evident that CSCs exist in a stochastic state in which they can switch between stem cell and non-stem-like phenotypes [16–18]. A major challenge in targeted cancer therapy lies in the inability to differentiate CSCs from non-CSCs due to absence of unique/universal biomarkers. Several surface biomarkers such as CD44, CD133, ALDH1, CXCR4, and EpCAM are being utilized to enrich CSC populations, but these markers are not exclusive to only CSCs. Non-CSCs and even normal stem cells also express these molecules, and thus there is an overlap between these cells. The phenotypic plasticity further diminishes the distinctions between these cells [2]. As discussed in the introduction section, these CSC phenotypes are regulated by a complex network of various pathways, which include pathways such as Wnt/β-catenin, Notch, Hedgehog, TGF-β, JAK/STAT, NF-κB, and PI3K/AKT/mTOR (Figure 7.1).

7.2.1.1 Wnt/β-catenin signalling (targeting self-renewal)

The Wnt/β-catenin pathway is a conserved developmental programme that cancer cells hijack to control CSC properties. It preserves stem cell fate and tissue homeostasis in normal tissues. Upregulation of this pathway is associated with CSC phenotypes and favours self-renewal, tumour initiation, and therapy resistance [19].

Wnt ligands attach to Frizzled/LRP5/6 receptors and block the associated degradation complex, allowing β-catenin to accumulate. Now, active β-catenin can translocate into the nucleus and activate stemness genes (e.g., *CD44, SOX2, LGR5, NANOG, MYC, CCND1*) in conjunction

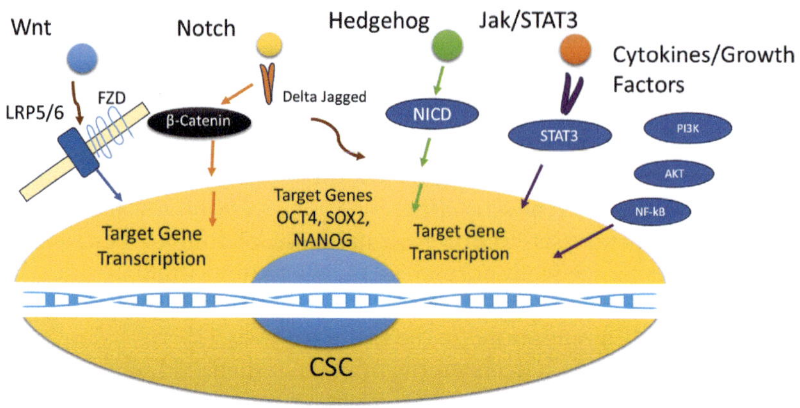

Figure 7.1 Major signalling pathways targeted in cancer stem cells (CSCs).

with TCF/LEF transcription factors [19,20]. High Wnt activity in CSCs is associated with enhanced sphere formation, tumour initiation, EMT, and metastatic potential [19–21]. By promoting DNA repair, anti-apoptotic signalling, quiescence, and drug efflux (ABCB1/ABCG2), β-catenin also contributes to therapeutic resistance by allowing CSC-mediated relapse [20,21].

Porcupine inhibitors (LGK974), Frizzled antagonists, tankyrase inhibitors, and β-catenin/TCF disruptors (CWP232291) are among the therapeutic approaches that target Wnt. These methods lower the frequency of CSCs and the initiation of tumours, but they must be carefully designed to prevent toxicity in normal stem cells [22,23].

7.2.1.2 *Notch signalling (involvement in differentiation and survival)*

Another conserved signalling system that regulates CSC survival, maintenance, and the undifferentiated state is the Notch pathway. The proteolytic release of Notch intracellular domain (NICD), which is translocated to the nucleus and promotes the transcription of target genes (p21, Cyclin D1, HES, HEY, c-MYC), is triggered by the binding of ligands (Jagged/Delta), encouraging self-renewal and stress tolerance. While Notch inhibition with γ-secretase inhibitors or receptor/ligand blocking decreases CSC frequency and tumour heterogeneity, Notch-active CSCs exhibit increased sphere/colony formation, tumour initiation, and therapy resistance [24,25]. Therapeutic targeting necessitates selective or context-specific methods to eradicate CSCs while preserving healthy tissue because of the pathway's function in regular stem cells.

7.2.1.3 Hedgehog (Hh) signalling (role in proliferation and invasiveness)

An important developmental signalling cascade, the Hedgehog (Hh) pathway is aberrantly reactivated in many malignancies, where it is essential for maintaining CSC populations. When a ligand such as Sonic hedgehog (SHH), Indian hedgehog (IHH), or Desert hedgehog (DHH) binds to patched 1 (PTCH1), this leads to the inhibition of Smoothened (SMO) and activates GLI (glioma) family zinc finger transcription factors, which drive the expression of genes related to invasion, proliferation, and stemness. This initiates canonical Hh signalling [26].

Hh signalling keeps CSCs in a dedifferentiated state, encourages tumour-initiating potential, and boosts self-renewal. Increased clonogenicity and chemoresistance are correlated with GLI1/GLI2 overexpression in several CSC models, including pancreatic, breast, and glioma stem-like cells, where elevated Hh–GLI activity has been reported [25]. For instance, by triggering EMT-associated transcription factors like Snail and Twist, Hh activation in pancreatic CSCs encourages sphere formation and supports metastatic characteristics [27].

By controlling matrix-remodelling proteins (such as MMP-2 and MMP-9) and improving regulation of the tumour, Hh signalling also promotes cellular invasiveness by promoting CSC migration and metastatic dissemination [28]. Pathway inhibition by SMO blockers (such as vismodegib and sonidegib) or GLI inhibitors has shown effectiveness in lowering CSC frequency and making tumours more susceptible to chemotherapy, as CSCs depend on prolonged Hh activity for proliferation and survival [29]. The Hh pathway is a desirable therapeutic target for stopping metastasis and recurrence because it functions as a key regulator of CSC-driven tumour growth and invasion.

7.2.1.4 STAT3 and NF-κB pathways (role in inflammation and survival)

The STAT3 and NF-κB pathways create inflammatory signalling networks that affect the regulation of the CSCs in cancers. The genes SOX2, NANOG, and c-MYC help maintain self-renewal, and the system is activated constantly by STAT3, which is typically regulated by the IL-6/JAK pathway [30].

Phosphorylated STAT3 also encourages the formation of tumour spheres that resist apoptosis and aid in metastasis, and these spheres are almost always more abundant in populations enriched with CSCs in breast, colorectal, and liver cancers. Similarly, the NF-κB pathway, which is triggered by cytokines, Toll-like receptors, and oncogenic stress, maintains CSC survival by triggering inflammatory mediators, EMT drivers, and anti-apoptotic factors, including BCL-XL and XIAP [31,32]. By upregulating

SNAIL, TWIST, and MMPs, NF-κB signalling strengthens the metastatic potential of CSCs and increases their invasiveness.

Importantly, STAT3 and NF-κB support one another via reciprocal activation loops: NF-κB increases IL-6 transcription, which activates STAT3, and STAT3 maintains NF-κB activity by maintaining RELA [33]. Therapy-resistant CSCs are characterized by this STAT3–NF-κB inflammatory circuit, which allows them to endure oxidative, metabolic, and chemotherapeutic stress [34].

In various types of tumours, pharmacological targeting of JAK/STAT3 or NF-κB signalling results in CSC depletion, decreased sphere formation, and enhanced chemosensitivity. Together, STAT3 and NF-κB form a central inflammatory axis that drives CSC survival, immune evasion, and relapse, representing a high-value therapeutic target.

7.2.2 Canonical markers and assays

Gene expression analysis of gastric cardiac tumour samples, such as Barrett's oesophagus (BE) and its advanced oesophageal adenocarcinoma (EAC) stages, showed a significant increase in both EpCAM and CD133 during the progression to EAC. Along with this expression of Musashi-1 (Absent in normal cells) an RNA binding protein also increased significantly in BE and EAC samples [30].

Ovarian CSCs (OC18) showed increased expression of multiple CSC markers, such as aldehyde dehydrogenase (ALDH), CD133, and ATP-binding cassette transporters (ABCG2), but negligible expression of CD44 and CD117, as compared to normal cells. It was observed that ALDH+ and CD133+ cells from fresh tumours formed tumours at a faster rate and lead to development of larger tumours than their negative counterparts [35].

CD44 is one of the most frequently studied biomarkers in human cancers. A significant correlation between CD44 expression and expression of drug-resistance, EMT-, and metastasis-associated genes has been observed. CD44 tumour cells exhibit higher mutation rates, and CD44 contributes to chemoresistance by activating cell survival pathways, enhancing DNA repair, and promoting drug efflux, thus making treatment less effective [36,37]. Another cell surface marker, CXCR4, also known as CD184 (a chemokine receptor type 4 involved in cell migration, growth, and survival of immune cells), is frequently overexpressed on CSCs. CXCR4 has emerged as a key molecular target as it has been found to be overexpressed in more than 20 human malignancies, including breast, prostate, and pancreatic, glioblastoma, and leukaemia [38,39].

Cancer cell identification does not rely only on the expression of characteristic markers but also on intracellular molecular markers. Many molecular markers such as OCT4, SOX2, Nanong, SaLL4, c-Myc, and BMI-1 are aberrantly expressed in different cancers. These genes are directly linked in maintaining the cell renewal, tumour initiation, drug resistance, and survival of CSCs [40–44]. Different cancer-specific canonical markers, their functions, and their respective detection assays are given in Table 7.1. An illustration of different molecular markers found in specific types of cancers is shown in Figure 7.2.

However, no single marker is found to be universally specific. Thus, a combination of marker-based profiling with functional assays is carried out

Table 7.1 Key cancer stem cell biomarkers, associated tumour types, and detection

Tumour type	*Most associated CSC markers*	*Function of the biomarker*	*Assay*	*References*
Breast cancer	$CD44^+/CD24^-/CXCR4^+$	Self-renewal, migration/ metastasis, therapy resistance	Flow cytometry, IHC, qPCR, sphere formation	[47–49]
Pancreatic cancer	$CD44^+/CD24^+$/ ESA $(EpCAM)^+/CXCR4^+$	Tumour initiation, metastasis, chemoresistance	Flow cytometry, FACS, IHC, sphere assay, Xenotransplantation	[50–54]
Colorectal cancer	CD133, CD44, Lgr5, ALDH1	Tumour initiation, Wnt signalling/ stemness, metastatic capacity	Flow cytometry, IHC, organoid culture, sphere formation	[55–60]
Glioblastoma	CD133 (Prominin-1), Nestin, SOX2	Maintenance of neural stem-like state, tumour propagation, radio/ chemoresistance	Flow cytometry, IHC, neurosphere assay, Xenograft models	[61–64]
Leukaemia (AML/CML)	$CD34^+/CD38^-$, CD123, CD96, (BCR-ABL in CML context)	Self-renewal, long-term repopulating ability, drug resistance	Flow cytometry, colony-forming assays (CFU), LTC-IC, Xenotransplantation	[65–70]
Prostate cancer	CD44, CD133, $\alpha 2\beta 1$ integrin (high)	Tumour initiation, adhesion/invasion, maintenance of stemness	Flow cytometry, prostasphere/colony formation, IHC, Xenograft assays	[71–76]
Ovarian cancer	CD133, ALDH1, CD44, EpCAM	Chemoresistance, tumour initiation, metastatic dissemination	Flow cytometry, aldefluor (ALDH) assay, sphere formation, IHC	[77–81]
Lung cancer	CD133, ALDH, CD44, ABCG2	Drug efflux/ chemoresistance, self-renewal, metastatic potential	Side population (Hoechst efflux), flow cytometry, sphere formation, IHC	[82–86]

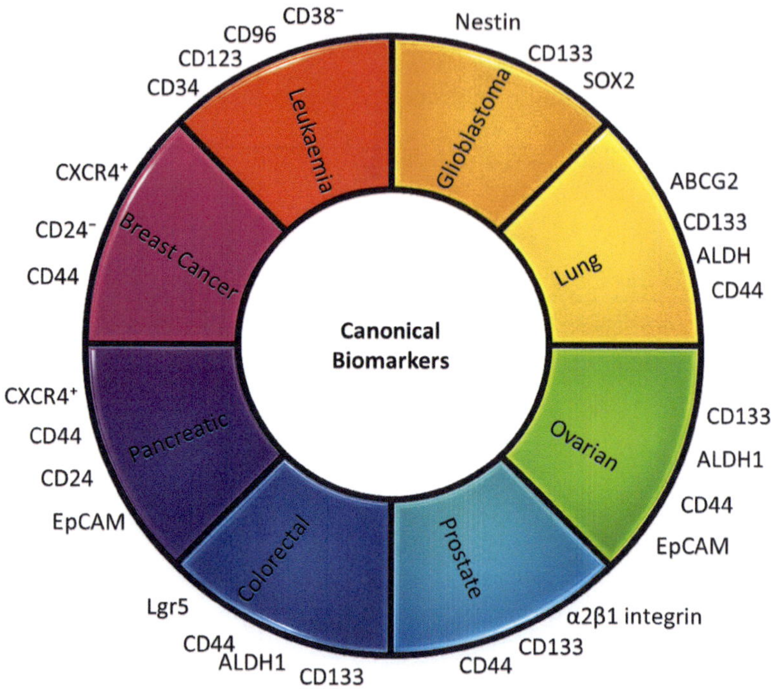

Figure 7.2 Different cancer-specific canonical markers.

for a more accurate identification and isolation of CSCs. These functional assays commonly include the in vitro measurement of ALDH activity, sphere formation assay (SFA), and side population analysis (SPA).

SFA and SPA are commonly used to characterize and enrich stem-like cells in a cell population. Sphere or spheroid formation happens in cells that can grow in anchorage-independent culture conditions, forming three-dimensional spheroids, which are called tumour spheres or neurospheres. It indicates the ability of self-renewal and differentiation, which are characteristic features of stem or progenitor cells [45]. Sphere-forming cells usually show enhanced tumour heterogeneity, drug resistance, and metastatic potential.

The SPA is rooted in the differentiation of SP cells based on their Hoechst 33342 dye efflux by active ABC transporters, especially ABCG2/BCRP1. They are easy to separate through flow cytometry because they have a unique profile of low Hoechst fluorescence [46]. The SPA is useful in the enrichment of cells that can self-renew, divide rapidly (possess high proliferative capacity), and differentiate into multiple cell types. Both SFA and SPA are valuable for studying stemness, tumour heterogeneity, and therapeutic resistance.

7.2.3 CSC niche

The CSC niche is a specialized microenvironment that sustains CSC survival, self-renewal, and therapy resistance. Hypoxia is a key characteristic of this niche, where low oxygen stabilizes HIF-1α/HIF-2α, activating angiogenic, glycolytic, and stemness programmes (including Notch, Wnt, and Hedgehog pathways), thereby reinforcing CSC maintenance and therapeutic resistance [87]. Hypoxia-driven transcriptional changes also promote chemoresistance and enhance CSC phenotypes across multiple tumours [88]. The extracellular matrix (ECM) and cancer-associated fibroblasts (CAFs) further regulate CSC behaviour. ECM components such as collagen, laminin, and fibronectin interact with CSC receptors, including integrins and CD44, to modulate adhesion, migration, and survival. CAFs can further alter the ECM and release pro-stemness mediators such as TGF-β, HGF, IL-6, and IL-8, which reinforce stem-like features and promote cancer growth [89].

Immune interactions also shape the CSC niche. CSCs evade immune surveillance by expressing PD-L1 and secreting immunosuppressive cytokines such as TGF-β and IL-10, thereby suppressing T-cell and NK-cell activity [90]. Chronic inflammation recruits tumour-associated macrophages (TAMs) and myeloid-derived suppressor cells (MDSCs), which create a pro-tumorigenic, stemness-supporting microenvironment [91]. These hypoxic, stromal, and immune cues establish a protective CSC niche that maintains plasticity and promotes resistance, thereby positioning the niche as a critical therapeutic target [87].

7.2.4 Metabolic adaptations and redox balance

CSCs are able to change the way they process oxygen and glucose to generate energy. In some CSC models, cells switch between aerobic glycolysis and oxidative phosphorylation (OXPHOS) according to oxygen or nutrient levels. Under conditions of hypoxia and nutrient limitation, CSCs predominantly use glycolysis (a "Warburg-like" metabolism) to support their rapid growth and survival [92,93]. However, when situated in well-oxygenated or nutrient-rich niches, CSCs frequently switch to OXPHOS, which is a more efficient mode of energy production. This is conducive to long-term self-renewal, quiescence, and maintenance of stemness [93,94].

In addition to central energy metabolism, lipid metabolism is also commonly upregulated in CSCs. Activated fatty acid (FA) synthesis and oxidation (FAO) provide not only energy but also lipid precursors and signalling lipids. This is necessary for membrane biosynthesis, redox homeostasis, and the preservation of stem-like features. Increased expression of enzymes, including FASN, CPT1A, and SCD1, has been described in CSC-enriched populations, providing a proliferative/survival advantage and protection against oxidative damage [93].

Importantly, to control reactive oxygen species (ROS), CSCs strictly regulate their redox balance. Although moderate ROS can encourage proliferation and signalling, excessive ROS endangers cell viability. CSCs maintain a high antioxidant capacity to avoid oxidative damage – for example, through the increased production of detoxifying enzymes like superoxide dismutases (SODs), the activation of antioxidant response regulators like NRF2, and elevated glutathione (GSH) levels [93,95]. This modulation of redox balance ensures CSC survival under diverse stress conditions – such as metabolic, hypoxic, or therapeutic stress – that contribute to therapy resistance and cancer recurrence.

Together, these metabolic adaptations give CSCs the resilience and plasticity required to survive in harsh microenvironments, evade therapies, and regenerate tumours. Therefore, addressing these metabolic and redox vulnerabilities is a viable approach to eliminate CSCs and prevent relapse.

7.3 LANDSCAPE OF NATURAL COMPOUNDS WITH ANTI-CSC ACTIVITY

Natural compounds or phytochemicals have multiple health benefits and have been used in the treatment of serious conditions, such as cancer. Plant components are characterized by their ability to prevent the development of certain diseases [96]. Medicinal plants have gained popularity due to their multi-target action [97].

The medicinal properties of a plant are due to the presence of metabolites. Plant metabolites are organic compounds that can be classified into primary metabolites and secondary metabolites. Primary metabolites are organic compounds – including glucose, starch, polysaccharides, proteins, lipids, and nucleic acids – which serve the function of growth and development in the plant system, as well as food for human beings [98,99]. In contrast, secondary metabolites play a role in the interaction of the plant with its surroundings, ensuring the continued existence of the organism in its ecosystem and its adaptation [100]. There are over 2,140,000 secondary metabolites reported [101].

The abundance of these secondary metabolites, such as phenolics, terpenoids, and alkaloids, makes medicinal plants rich in nutraceutical and pharmaceutical properties including antidiabetic, anticancer, galactagogic, digestive, hepatoprotective, regulatory, and antioxidant activities, while also working against anorexic, lithogenesis, and pathogens [102]. The major categories of secondary metabolites, their sources, and examples have been shown in Table 7.2.

The secondary metabolites extracted from plants can be further subdivided into three major classes: terpenoids, alkaloids, and phenolics [103,104]. More specifically, these consist of 29,000 terpenoids, 12,000

Table 7.2 Major categories of secondary metabolites

Categories	*Major classes*	*Common sources*	*Examples*
Phenolics	Flavonoids	Onion, turnip, apple, olive, lettuce, tea leaves, tomato, cherry, broccoli, beans	Quercetin, kaempferol, myricetin, catechin, epicatechin, genistein, daidzein, apigenin, luteolin
	Phenolic acid	Potato, apple, orange, pear, cherry, blueberry, coffee, strawberry, raspberry, grapes	Neochlorogenic, caffeic, ferulic, chlorogenic acid ellagic, gallic acids
	Polyphenols (tannins, lignans)	Lentils, peas, grape, wine, apple	Catechin, epicatechin polymers
Alkaloids	Indole	*Catharanthus roseus*	Catharanthine
		Pausinystalia johimbe	Yohimbine
		Achnatherum robustum, Claviceps purpurea	Ergot alkaloids (e.g., ergine)
		Aconitum delphinium	Terpenoid-alkaloids (e.g., aconite and actinitine)
		Vinca rosea	Vinca
		Strychnos nux-vomica	Strychnine
	Steroids	Solanum plants, nightshades, Jerusalem cherries, and tomatoes	Solanine, chaconine, aglycone
	Quinolines	*Cinchona*	Quinine, quinidine, cinchonidine, cinchonine
	Pyrrolidine	*Meloidogyne hapla, Broussonetia kazinoki*	Broussonetia, DMDP
	Pyridine	*Nicotiana tobacum*	Nicotine, anabasine
	Pyrrolizidine	*Broussonetia kazinoki, Crotalaria spectabilis Ageratum* sp., *Chromolaena odorata Senecio jacobaea*	Monocrotaline, heliotrine, lasiocarpine, senecionine
Terpenoids	Monoterpene	Oil of mint plants, lemongrass, pine tree, citrus fruits, *Cinnamomum camphora*	Menthol, nerol, pinene, limonene, citral camphor
	Sesquiterpene	*Vachellia farnesiana, Ginkgo biloba*	Farnesol, nerolidol
	Diterpenes	Green vegetables, *Andrographis paniculata*	Phytol, vitamin a1, andrographolide

(*Continued*)

Table 7.2 (Continued) Major categories of secondary metabolites

Categories	*Major classes*	*Common sources*	*Examples*
	Triterpenes	Tomato, cabbage, cucumber, pepper, figs, mango, tamarind, ginseng, shea, strawberry, lavender, rosemary, mirabilis	Lupeol, ursolic corrosive, cycloartane
	Tetraterpenes	Carrots, oranges, turmeric, rice bran oil	Carotenoids, curcumin, oryzanol
	Polyterpenes	*Hevea brasiliensis, Palaquium and Payena*	Rubber, gutta-percha
Carbohydrates containing phytochemicals	Saccharides, sugar alcohols	Sugar cane, *Ganoderma lucidum*	Policosanols
Lipids	Phytosterols, PUFAs	Nuts and oils	Stigmasterol, sitosterols
Other		Oil seeds, whole grains, vegetables, fruits	Bromelain, capsaicin, phytoestrogens

alkaloids, and 8,000 phenolics (including steroids, flavonoids, and tannins), while more are in the process of identification [104,105].

In terms of their biosynthetic pathways, secondary metabolites can be classified into three main groups: phenolic compounds synthesized via the shikimate pathway, terpenes synthesized via the mevalonic acid pathway, and nitrogen-containing compounds synthesized via the tricarboxylic acid cycle pathway [106,107].

Alkaloids are an additional class of secondary metabolites, which are nitrogenous organic molecules biosynthesized mainly from amino acids – such as tryptophan, tyrosine, phenylalanine, lysine, and arginine – using many unique enzymes [108]. Many of the most important therapeutic agents are alkaloids. The sites of biosynthesis are compartmentalized at the cellular or sub-cellular level. However, secondary metabolites can be transported over long distances and accumulate away from their location of synthesis [109].

Plant-based natural products such as curcumin have demonstrated promising immunomodulatory activity and can be used as an adjunctive therapy for cerebral malaria [110]. A representation of the general immunomodulation mechanisms exhibited by medicinal plants is shown in Figure 7.3.

In mosquito-borne viral diseases like dengue, which are caused by high levels of pro-inflammatory cytokine production, natural immunomodulating agents from plant sources – such as antioxidants – have demonstrated promising results [112].

Colchicine an alkaloid found in significant quantities only in two plants – namely, *Colchicum autumnale* and *Gloriosa* sp. [113] – is also

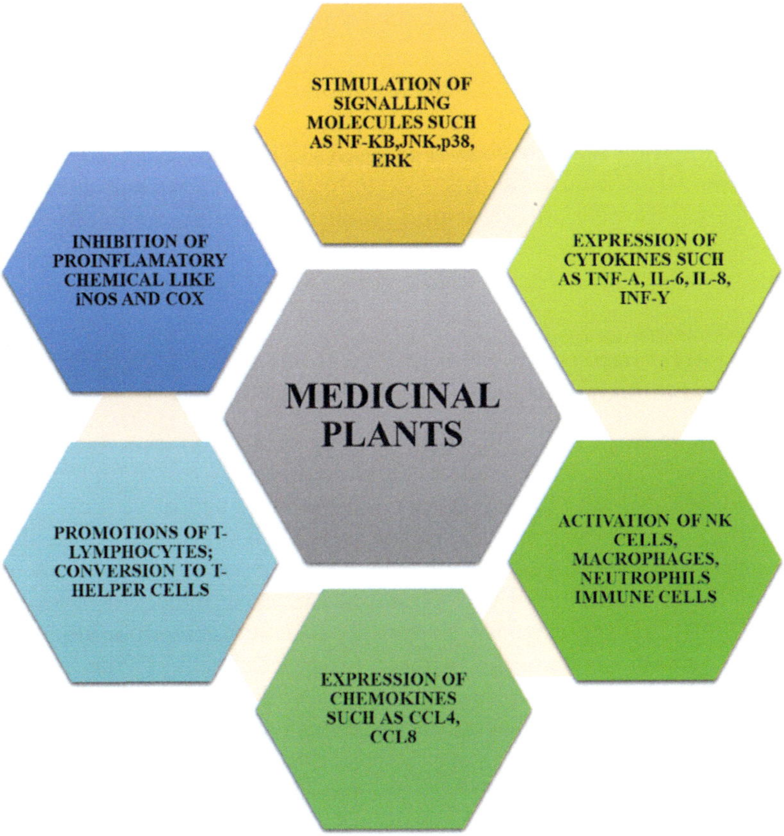

Figure 7.3 General immunomodulation mechanisms shown by medicinal plants [111].

widely utilized to inhibit the polymerization of microtubules and to inhibit fibrosis in experimental animal models [114–116]. It is also used to treat cancer [117,118].

7.3.1 Polyphenols and flavonoids

7.3.1.1 Curcumin (from Curcuma longa): targeting NF-κB and Wnt/β-Catenin in breast and colon CSCs

Polyphenols and flavonoids constitute one of the most extensively studied classes of natural compounds with potent anti-CSC activity. Characterized by their distinct chemical structures featuring multiple phenolic rings, these molecules exhibit remarkable pleiotropy, enabling them to simultaneously engage several critical survival and self-renewal pathways in CSCs. Curcumin is the primary yellow pigment of the spice *Curcuma longa* (turmeric) that has a known impact on chronic inflammation and oncogenesis. Curcumin potently inhibits the nuclear factor-kappa B (NF-κB) pathway, a master regulator of inflammation, survival, and therapy resistance that is highly active in CSCs [119–121]. Moreover, it also directly regulates the Wnt/β-Catenin pathway by enhancing the degradation of β-catenin, thereby inhibiting the self-renewal of CSCs [122]. Data collected from preclinical models demonstrate that curcumin is very effective at eliminating stem-like cancer cells in breast and colon tumours. A notable finding is the ability of curcumin to work together with standard chemotherapeutic agents, thereby addressing the challenging issue of CSC resistance.

7.3.1.2 Resveratrol (from grapes/berries): modulating Notch and inducing apoptosis in glioblastoma CSCs

Resveratrol is a type of stilbene largely present in grapes, berries, and certain nuts. Crucially, resveratrol modulates the Notch signalling pathway in CSCs, which is essential for the maintenance of stemness, especially in solid tumours such as glioblastoma [123]. Via the Notch-mediated transcription pathway, resveratrol, which is an anticancer agent, not only inhibits the proliferation of CSCs but also causes apoptosis, thus lowering the tumour-initiating capacity further [124]. Due to the lipophilic nature of resveratrol, it has a special property to be permeable to the barrier of the blood–brain barrier (BBB), thereby providing a critical advantage for the treatment of the primary and metastatic brain cancers that are resistant to therapy [125]. Apart from the Notch pathway, resveratrol has been shown to activate Sirtuin 1 (SIRT1), an epigenetic regulator associated with both anti-ageing properties and the ability to disrupt the epigenetic stability required for the CSC phenotype.

7.3.1.3 Epigallocatechin gallate (from green tea): targeting CSC markers and inhibiting sphere formation

EGCG is the most bioactive compound found in green tea, offering a powerful natural approach to treat CSCs. It works by directly targeting the key surface markers of stem cells, disrupting their ability to self-renew and form tumour spheres, which serve as a gold standard for measuring stemness in the lab.

EGCG reduces the expression of CSC markers (such as CD44, CD133, and ALDH1) and interferes with the interaction between CSCs and their niche components. Significantly, EGCG is an effective inhibitor of DNA methyltransferases (DNMTs) and histone deacetylases (HDACs) [126,127], which reprogrammes the CSC epigenome to a more differentiated, less aggressive state by reversing DNA hypermethylation and reactivating silenced tumour suppressors [128].

These compounds demonstrate a remarkable breadth of action, spanning curcumin's suppression of NF-κB/Wnt, resveratrol's modulation of Notch/SIRT1, and the epigenetic impact of EGCG. Such multi-pathway efficacy highlights the potential of polyphenols to serve as the backbone of network-based, CSC-directed combination strategies.

7.3.2 Terpenoids and saponins

The pharmacological actions of artemisinin and ursolic acid serve as a testament to the power of natural compounds in drug discovery. While artemisinin provides an acute, toxic strike against CSCs by capitalizing on iron accumulation, ursolic acid acts through a sophisticated dismantling of the CSC phenotype, specifically by reversing EMT.

It is precisely this combination of strategies – one a metabolic "sledgehammer" and the other a phenotypic "architect" – that makes terpenoids so valuable. These compounds will be discussed in detail in the following sections.

7.3.2.1 *Artemisinin derivatives (from* Artemisia annua*): mechanism of action against CSCs through iron-dependent cytotoxicity*

Terpenoids and saponins form another chemically diverse and pharmacologically rich class of natural products exhibiting significant anti-CSC potential. Terpenoids, which are derived from isoprene units, range from small monoterpenes to complex triterpenoids. The clinical urgency to target CSCs has brought renewed attention to agents like the artemisinin derivatives (such as artesunate and artemether), isolated from the herb Artemisia annua. While initially developed as antimalarials, their efficacy against cancer is predicated on a unique, iron-dependent mechanism of action [129,130]. CSCs often display elevated levels of labile intracellular iron, which is necessary to fuel their rapid proliferation and protect against oxidative stress. Artemisinin, upon reacting with this high concentration of iron, generates highly cytotoxic free radicals, known as ROS. This targeted ferroptosis-like damage exploits the metabolic vulnerability of CSCs, selectively leading to cellular necrosis or apoptosis while sparing quiescent, healthy cells [129,130].

7.3.2.2 *Ursolic acid and triterpenoids: inhibition of CSC viability and epithelial-mesenchymal transition (EMT)*

Ursolic acid, a complex pentacyclic triterpenoid found in apples, rosemary, and holy basil, offers a unique, non-iron-dependent mode of CSC inhibition.

Its structural analogues display broad-spectrum anticancer activity, mainly by suppressing CSC characteristics and inhibiting their viability through the modulation of multiple pro-survival signalling pathways, including PI3K/AKT and Wnt/β-catenin [131,132]. These triterpenoids are potent inhibitors of the EMT in CSCs. Ursolic acid downregulates key EMT-inducing transcription factors and restores epithelial markers. Thus, it reverses the plasticity responsible for metastasis and chemoresistance in cancers, such as breast and ovarian carcinomas [133]. This ability to reverse mesenchymal (stem-like) cells back into a differentiated, epithelial state is a key principle of an effective CSC-targeting strategy, as it converts these aggressive cells into manageable populations.

7.3.3 Alkaloids

7.3.3.1 *Berberine (from* Coptis chinensis*): suppression of CSC properties via AMPK and Wnt pathways*

Alkaloids are a diverse group of nitrogen-containing phytochemicals that show some of the most potent CSC-targeting activities. They usually act by interfering with the fundamental cellular and metabolic processes. These heterocyclic compounds, which are principally derived from botanical sources, exhibit high selectivity against fast-dividing cells and metabolically hyperactive cells, including CSCs.

Berberine (BBR) is an isoquinoline alkaloid extracted from plants like *Coptis chinensis*. It is a highly regarded example whose anticancer properties emerge from its ability to modulate metabolic and self-renewal pathways. BBR's core mechanism involves the activation of the AMP-activated protein kinase (AMPK) pathway [134], an energy sensor that acts as a tumour suppressor by inhibiting downstream targets like mTOR and suppressing FA synthesis [135]. This metabolic stress effectively limits CSCs and disrupts the flexible metabolic state that is critical for their survival under nutrient-poor or hypoxic conditions.

BBR's anti-CSC efficacy extends beyond metabolic regulation to critical developmental signalling cascades. It has been shown to suppress stemness and tumour heterogeneityby modulating the Wnt/beta-catenin pathway. Furthermore, BBR acts as a suppressor of other essential CSC pathways, including Hedgehog signalling, thereby inhibiting the malignant phenotypes of colorectal CSCs [136]. The combined disruption of energy homeostasis (via AMPK) and key stemness pathways (Wnt, Hedgehog) makes BBR a promising lead for the eradication of the dormant or inactive fraction of CSCs.

7.3.3.2 *Sanguinarine and related compounds: inducing CSC differentiation and reducing chemoresistance*

Given their established role in disrupting metabolic regulation (BBR) and inducing differentiation/reversing drug resistance (sanguinarine), these alkaloids represent essential components in the evolving strategy of

CSC-directed precision oncology. They aim to achieve durable clinical control by eliminating the root cause of tumour recurrence.

7.3.4 Other phytochemicals

Other natural compounds that are not categorized under the classification of terpenoids, alkaloids, polyphenols, saponins, or flavonoids have shown promising effects in CSC targeting. Some of them are discussed in the following section:

7.3.4.1 *Sulforaphane (from broccoli): activation of Nrf2 and suppression of ALDH1 activity*

Beyond classical polyphenols, several other phytochemical classes demonstrate powerful anti-CSC activity through unique mechanisms of action. Among these, Sulforaphane (SFN), an isothiocyanate derived from cruciferous vegetables like broccoli and broccoli sprouts, is highly regarded for its chemopreventive properties and its efficacy against CSCs. SFN exerts its primary anti-CSC effect by activating the nuclear factor erythroid 2-related factor 2 (Nrf2) pathway, a master regulator of cellular defence against oxidative stress. While Nrf2 activation is typically cytoprotective, in many CSCs, SFN effectively modulates this pathway to restore redox homeostasis and eliminate self-renewal capacity, making it a powerful chemotherapeutic adjunct [137]. The therapeutic potential of SFN lies in its ability to breach CSC defences by targeting the intersection of detoxification and stemness.

In solid tumours, CSCs often ensure their survival through the overexpression of aldehyde dehydrogenase 1 (ALDH1), an enzyme that neutralizes cytotoxic agents and oxidative stress. SFN directly addresses this vulnerability. Studies show that SFN effectively suppresses ALDH1 activity, leading to the functional elimination of the CSC population identified by this marker in various cancers, including oral squamous cell carcinoma. Furthermore, SFN's activity extends to inhibiting critical CSC signalling pathways, such as Wnt/β-catenin, Notch, and Hedgehog, providing a multi-pronged attack on the hierarchical stem cell population [138]. Consequently, this multi-modal efficacy positions SFN as a leading candidate in clinical trials for prostate and breast cancer therapies.

7.3.4.2 *Genistein (from soy): modulation of microRNA expression in CSCs*

Genistein is the main isoflavone found in soybeans and is recognized for its phytoestrogenic and potent anticancer effects. Genistein's relevance in CSC targeting lies in its capacity to act as a nutri-epigenetic modulator, specifically by targeting microRNA (miRNA) expression. miRNAs are small

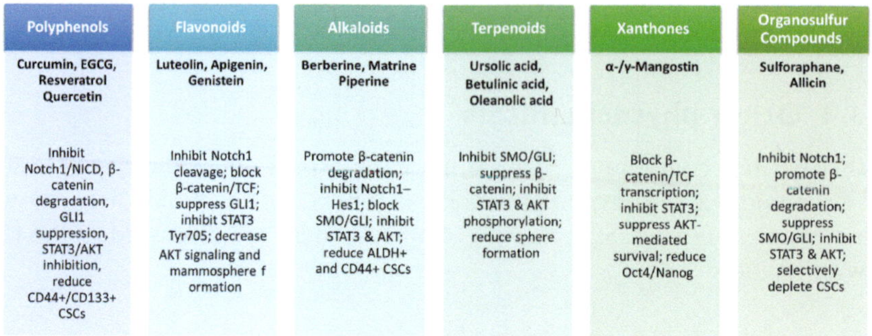

Figure 7.4 Phytochemical groups targeting CSC pathways.

non-coding RNAs that play crucial roles in CSC maintenance by regulating entire oncogenic or tumour-suppressive networks. Genistein is documented to downregulate oncogenic miRNAs, such as miR-155, which is often implicated in breast cancer progression and therapy resistance [139]. By suppressing such oncogenic miRNAs, genistein restores the expression of critical tumour suppressors and sensitizes resistant CSCs to standard-of-care agents like tamoxifen, demonstrating its utility in overcoming drug resistance in oestrogen receptor-positive (ER+) breast cancer models.

The modulation of the non-coding RNA landscape by genistein is highly attractive for precision oncology, as it allows for the functional reprogramming of the CSC phenotype without necessitating high, potentially toxic doses. In cancers like pancreatic carcinoma, genistein has been shown to downregulate specific onco-miRs like miR-223, subsequently restoring tumour-suppressive pathways and inducing apoptosis [140]. This ability to fine-tune the genetic regulatory network, coupled with its known modulation of signalling cascades like PI3K/AKT and NF-κB, underscores genistein's role as a potent agent capable of integrating dietary chemoprevention with targeted CSC intervention. Phytochemicals from different classes can regulate major pathways affected in CSCs. Their examples and effects are illustrated in Figure 7.4.

7.4 MECHANISMS OF ACTION: INDUCING DIFFERENTIATION AND OVERCOMING RESISTANCE

Natural compounds have shown substantial promise for promoting cellular differentiation and overcoming drug resistance, especially in cancer treatment and regenerative medicine. Various phytochemicals, such as curcumin and resveratrol, can modulate epigenetic mechanisms, including DNA methylation and histone modification, to activate genes crucial for differentiation [120,141]. Natural compounds can potentially change critical pathways. For instance, retinoic acid (RA), a derivative of vitamin A, activates RA receptors, thereby promoting cellular differentiation [142].

Natural compounds like quercetin and berberine can inhibit efflux pumps, such as P-glycoprotein. This makes cells more susceptible to traditional chemotherapy treatments and increases drug accumulation inside them [143,144]. These natural compounds could help overcome drug resistance, which is a good thing. Studies show that certain chemicals, like sulforaphane and EGCG, can block CSC pathways like Wnt/β-catenin and Notch [145,146]. This makes the cells less able to renew themselves, which reduces treatment efficacy. In addition to established pharmaceuticals, natural compounds may enhance differentiation and circumvent resistance mechanisms. These numerous activities illustrate the potential efficacy of natural compounds in this context.

7.4.1 Natural compound-mediated CSC differentiation

Natural compounds promote the differentiation of CSCs by blocking vital self-renewal pathways like Wnt/β-catenin, Hedgehog, Notch, and PI3K/AKT/mTOR. Stemness markers, including CD133, CD44, and ALDH1, are reduced, which inhibits the cells' ability to cause tumours and promotes their differentiation into non-tumorigenic cells. By blocking ABC transporters and drug efflux mechanisms in CSCs linked to breast, colon, lung, and pancreatic cancers, these highly non-toxic phytochemicals increase the effectiveness of chemotherapy. Turmeric-derived curcumin suppresses the Wnt/β-catenin, SHH, and JAK/STAT3 signalling pathways. In breast and colorectal cancer, this makes it more difficult for CSCs to form spheres and makes them more vulnerable to doxorubicin and mitomycin [147–149]. Multicellular organisms require specific intercellular communication networks to maintain the physiological functions of each organ throughout life and to coordinate their complex body plan throughout development [150]. RAs are signalling substances that generate genetic communication networks that are essential for embryonic development. They act through their nuclear retinoid (RARα, β, and γ) and rexinoid (RXRα, β, and γ) receptors. They also have important physiological roles, especially in the brain and reproductive system, influencing organogenesis, organ homeostasis, and cell proliferation, differentiation, and death [151,152]. As transcription factors, retinoic acid receptors (RARs) and retinoid X receptors (RXRs)primarily operate as RAR–RXR heterodimers. These heterodimers have two different roles: first, they affect how often target gene transcription starts after binding to RA response elements (RAREs) in their promoters; second, they affect how well other signalling pathways work (also known as "cross-talk") through mechanisms that have not yet been found. This relationship implies that different pathways may affect retinoid receptors. A well-known example of this is how RARs and the AP1 (c-JUN–c-FOS) transcription factor stop each other's transcriptional activity [153]. Retinoids can be both friends and enemies. The beneficial outcomes of retinoid-based "differentiation therapy" have been convincingly evidenced in the context of acute promyelocytic leukaemia (APL). The pleiotropic activities and their capacity to induce cellular differentiation and apoptosis have considerable potential for cancer treatment. But much like the sorcerer's apprentice, RAs are also very strong teratogens. Before we can use these strong compounds efficiently, we need to learn more about what they do.

Many factors can contribute to tumorigenesis, including inherited and acquired genetic changes, chromosomal rearrangements, epigenetic phenomena, and chemical carcinogenesis. RAs can interfere with these events at several levels, their principal known actions being the induction of differentiation and/or apoptosis of tumour cells, and the inhibition of tumour promotion in chemically induced cancers. In keeping with their ability to regulate growth and induce differentiation throughout life, retinoids affect the growth of many tumour cell lines in culture.

APL is a variant of acute myeloid leukaemia (AML) that is distinguished by a significantly diminished response to conventional cytotoxic therapy. The therapy of APL has developed with the discovery of all-trans retinoic acid (ATRA), which promotes leukemic blast differentiation and can induce temporary remissions when administered alone. Patients with APL frequently experience enduring remission when ATRA is used in conjunction with arsenic trioxide [154]. Researchers are now working on differentiation-based treatments for other types of AML because ATRA works for APL. These medications have inhibitors that only work on FLT3, DHODH, and the epigenetic regulator LSD1 [155]. Targeted therapy aimed at targeting mutant IDH1 (ivosidenib) and mutant IDH2 (enasidenib) provides differentiation and clinical responses in around 40% of patients with IDH-mutant AML [156].

Recent studies have greatly enhanced our understanding of differentiation therapy utilizing a transplantable p53 wild-type APL mouse model. Some studies have shown that ATRA and its isomer, 13-cis RA, induce APL cell differentiation in living organisms. After transplantation, partially differentiated APL cells retained their ability to generate leukaemia, but more fully differentiated cells lost this ability [157]. These results demonstrate the existence of a critical differentiation threshold that irreversibly diminishes leukaemogenic capacity and illustrates the inherent plasticity of APL blasts.

According to some studies, abnormal RARβ/TET2 signalling may cause aggressive breast cancers to develop in a fraction of breast CSCs resistant to ATRA therapy. Interventions such as direct PKCβ inhibition or miR-200c therapy can improve the differentiation of resistant CSCs, which may reduce the number of CSCs by enhancing differentiation and reducing stem cell characteristics. PKCβ and other members of the atypical protein kinase C (aPKC) family have a variety of roles in different forms of cancer, either promoting or suppressing tumour growth [158,159]. Recent experimental findings are among the mounting evidence that PKCβ increases the aggressiveness and growth of breast cancer tumours. It has been demonstrated that ATRA therapy particularly lowers PKCβ levels while having no effect on the related isoform PKCα. When PKCβ levels are decreased, phosphorylated NUMB (p-NUMB) is always decreased, whether this is accomplished via gene editing, gene silencing, or pharmaceutical intervention. This effect is not produced when PKβ is inhibited and studies have shown, through both in vitro and in vivo investigations, that PKCβ is the primary PKC isoform that sustains the breast CSC cell pool, hence promoting tumour development. The data identify PKCβ as a potential target for therapeutic treatments designed to diminish CSCs and impede the progression of breast cancer [154].

7.4.2 Reversing epithelial-mesenchymal transition (EMT)

EMT is a well-controlled physiological process that is important for tissue healing and embryonic development [160]. During the EMT, epithelial cells undergo a sequence of morphological, metabolic, and genetic alterations that allow them to take on mesenchymal traits [89,161,162]. The early signs of the process are the loss of polarity between the cell's apical and basal surfaces, the breakdown of intercellular connections, and the degeneration of microvilli. This helps with movement by separating the front and back parts of the body. Cells in the later stages of EMT are more able to penetrate the fibrotic compact zone and break down the basement membrane. Mesenchymal-morphotype cells demonstrate superior capabilities in synthesizing ECM components and secreting various pro-inflammatory, fibrotic, and angiogenic factors, including vascular endothelial growth factor (VEGF), which promotes angiogenesis [163–165]. Natural plant compounds (NPCs) have historically functioned as therapeutic agents and fundamental components for drug development owing to their distinctive chemical structures that enable the discovery of novel treatments [166,167]. These NPCs are bioactive substances with anti-inflammatory, anti-fibrotic, or antioxidant effects that may impact the EMT [168]. This research examines the mechanisms by which bioactive natural compounds impede the EMT.

EMT, a crucial component of cancer metastasis and treatment resistance, may be inhibited by natural compounds, according to an increasing number of studies. To stop TGF-β from causing EMT in lung cancer cells, arctigenin suppresses Smad2/3 activity and increases E-cadherin expression. It also suppresses β-catenin signalling and the phosphorylation of ERK [169].

By suppressing Slug and the NF-κB pathway, baicalin and baicalein prevent TGF-β1-induced EMT in breast and pancreatic cancer models [170]. Berberine fights cancer in several ways. E-cadherin is increased, mesenchymal markers including vimentin, fibronectin, and N-cadherin are downregulated, and transcription factors like Snail, Slug, and ZEB1 are inhibited. Additionally, it alters the PI3K/Akt and RAR signalling pathways [143,171]. By modifying EMT markers through the TGF-β pathway, betanin from beetroot reduces kidney fibrosis in diabetic nephropathy [172]. By reducing fibronectin and vimentin through the Focal Adhesion Kinase (FAK) pathway, increasing E-cadherin levels, and changing cell shape to an epithelial phenotype, black rice anthocyanins (BRACs) exhibit anti-metastatic properties [173]. By altering the levels of E-cadherin, vimentin, and Twist, Brusatol reverses EMT and increases the susceptibility of pancreatic cancer cells to chemotherapy [174]. Rosemary carnosol helps reactivate p53, suppresses EMT, and specifically targets the phenotypes of CSCs [169]. Cairicoside E suppresses the Smad2/3 and AQP5 signalling pathways to prevent EMT in colorectal cancer [175,176]. However, in lung fibrosis models, celastrol reduces the expression of Snail/Slug and mesenchymal markers. It does this, in part, by changing the levels of inflammatory cytokines and blocking HSP90. By inhibiting the VEGF, MAPK, and Akt pathways and increasing E-cadherin expression, Celastrus orbiculatus Thunb extracts stop EMT and metastasis in gastric cancer [177].

Cinnamomum cassia extracts decrease TGF-β1-induced motility and EMT by inhibiting matrix metalloproteinases (MMPs), vimentin, and fibronectin, while simultaneously increasing the expression of E-cadherin [176]. Cancer cell motility is inhibited by codonolactone and cordycepin because they prevent the synthesis of mesenchymal markers and transcription factors that support EMT [177]. By blocking Smad2 phosphorylation, cryptotanshinone reduces renal fibrosis [178].

On the other hand, curcumin affects several signalling pathways linked to EMT, including the TGF-β/SMAD and PI3K/Akt/NF-κB pathways. It has demonstrated promise in the treatment of pancreatic cancer and hepatocytes with minimal toxicity [119,122]. Green tea, jatrophone, and duchesnea extracts all contain EGCG, which inhibits EMT, reduces cellular motility, and increases the expression of epithelial markers in a variety of cancer types [177].

It has also been demonstrated that eupatolide, gallic acid, gambogic acid, gedunin, genistein, and gentiopicroside suppress the EMT by altering mesenchymal and epithelial markers or by disrupting the TGF-β/Smad, Notch, and NF-κB signalling pathways [177]. By blocking Wnt/β-catenin and EMT signalling, ginsenosides increase the sensitivity of CSCs [162]. Additionally, key transcription factors and signalling pathways connected to the EMT and metastasis are blocked by luteolin, honokiol, isoviolanthin, and ligustrazine. By downregulating Notch, N-cadherin, and vimentin, and by influencing the FAK and Hedgehog pathways, nimbolide and nitidine chloride prevent cancer cell migration and invasion [177].

Other natural inhibitors that interfere with transcriptional regulation and EMT signalling include oxymatrine, resveratrol, parthenolide, plectranthoic acid, phoyunnanin-E, piperlongumine, plumbagin, polyphyllin I, and osthole. All of these compounds show effectiveness in lowering fibrosis and metastasis in preclinical models [177,179,180]. Natural compounds that can affect EMT through several signalling pathways, such as Smad, NF-κB, MAPK, and Wnt/β-catenin, include salvianolic acid A, *Sedum sarmentosum* extract, sulforaphane, tannic acid, withaferin A, and alpha-solanine. Together, these results show that plant-derived chemicals may be used therapeutically to fight metastasis, reverse EMT, and possibly even overcome cancer therapy resistance [177].

7.4.3 Natural compounds as chemosensitizers

Chemotherapy is the principal treatment for cancer; nevertheless, its efficacy is limited, especially in cases of metastasis or drug resistance. This resistance significantly complicates the treatment of numerous cancer types. To address this issue, researchers are exploring innovative strategies such as combination therapy, which integrates traditional chemotherapeutic agents with natural compounds found in foods and plants. Researchers have examined various natural substances to assess their potential in cancer therapy, including resveratrol from grapes, quercetin from apples and onions, and curcumin from turmeric [181]. These natural ingredients can be employed to commence chemotherapy or to enhance the efficacy of other drugs. For example:

- Curcumin disrupts the routes that cancer cells utilize to evade treatment and synergizes effectively with conventional chemotherapy.
- Chemotherapy-resistant cancer cells may exhibit increased susceptibility to resveratrol. Consequently, chemotherapeutic drugs operate more efficiently and eliminate a greater number of cancer cells.
- Quercetin enhances the capacity of cancer cells to uptake medication while simultaneously mitigating the side effects of chemotherapy. When administered alongside conventional chemotherapy agents, it effectively targets breast CSCs, leading to improved therapeutic results.

A major barrier to obtaining successful treatment results in cancer is multidrug resistance (MDR), which is defined by the ineffectiveness of different chemotherapeutic drugs [182]. Understanding MDR's mechanisms – such as how it enhances drug efflux, inhibits drug influx, repairs damage caused by drugs, changes drug targets, and evades apoptosis – is crucial to fighting it [183]. Drug efficacy is compromised by increased drug efflux, which is mostly caused by overexpressed ABC transporters and significantly reduces intracellular drug concentrations [184].

P-glycoprotein (P-gp), lung resistance-related protein (LRP), breast cancer resistance protein (BCRP), and MDR-related proteins (MRPs) are some of the most significant proteins associated with MDR. The ABCB1 gene generates P-gp, which protects cancer cells from cytotoxic chemicals by transporting drugs out of them via ATP hydrolysis [185]. Different cancer types have different degrees of P-gp expression, with colon, pancreatic, hepatic, adrenal, and renal cancers having much higher baseline levels, and lung, breast, oesophageal, and ovarian cancers having elevated levels when chemotherapy is used [186]. Because P-gp overexpression can lower the number of medicines in cells, it is closely associated with MDR [187].

P-gp inhibitors have been developed over multiple generations to treat MDR and restore chemosensitivity in malignancies that are unresponsive to chemotherapy [188]. Resistance may be circumvented by combining cytotoxic medications with MDR chemosensitizers or P-gp modulators. Novel anticancer agents that can alter drug resistance mechanisms have been found in natural sources. Although preclinical and clinical research have shown the effectiveness of several natural substances, their routine use in treating MDR is still limited. To determine whether they can be used in conjunction with standard treatments, further research is required.

7.4.4 Targeting the CSC epigenome

Key epigenetic alterations encompass DNA methylation, chromatin remodelling, post-translational modifications of histones, and gene silencing mediated by non-coding RNAs [189]. Histone acetylation and DNA methylation are essential epigenetic mechanisms that preserve genomic integrity. Changes in these epigenetic patterns are among the first molecular events identified at the onset of the tumorigenesis. The development of HDAC inhibitors is a promising approach to cancer treatment since HDACs are essential for the growth and metastasis of cancer [190]. A variety of diseases, including cancer, neurological disorders, parasitic infections, and obesity, have been associated with disturbances in these epigenetic mechanisms [191,192]. Cancer is closely associated with abnormal epigenetic alterations that can silence tumour suppressor genes. These changes promote tumour growth and dissemination [193,194].

Natural products derived from diverse sources, such as plants, microorganisms, and marine sponges, have been shown to induce epigenetic modifications [195,196]. Epigenetics refers to heritable alterations in chromatin structure and gene expression that occur without modifications to the underlying DNA sequence [197]. Various critical enzymes, including DNMTs, HATs and histone deacetylases (HDTs), histone lysine and arginine methyltransferases, and histone demethylases, are responsible for these modifications [196].

After the preclinical and clinical validation of HDAC inhibitors' (HDACi) efficacy, the Food and Drug Administration (FDA) has approved drugs like vorinostat (SAHA), romidepsin (FK288), panobinostat (LBH589), and belinostat (PXD101) for use in oncological therapy [198,199]. Furthermore, several additional HDAC inhibitors are now being assessed in Phase II and III clinical trials [200].

Although industry and academia are working hard to create new HDAC inhibitors in the lab, natural products continue to be a great source of bioactive compounds with unique structures [201,202]. Natural products can produce powerful medicines, as evidenced by the long history of drugs synthesized from them [203]. To find improved lead compounds for next-generation HDAC inhibitors, it is therefore promising to test recently identified natural products for HDAC inhibitory activity. According to recent studies, the effective use of HDAC inhibitors in conjunction with traditional cytostatic drugs may improve the effectiveness of cancer treatments [200].

7.5 FORMULATION AND DELIVERY SYSTEMS FOR PRECISION TARGETING

7.5.1 Challenges of natural compound drug ability

There are several challenges associated with natural compounds. The section below discusses these challenges.

7.5.1.1 Low solubility, poor bioavailability, and rapid systemic clearance

Natural products being used in drug discovery face a big challenge where solubility, bioavailability, and rapid systemic clearance can hinder their clinical translation. Phytochemicals such as curcumin, quercetin, resveratrol, berberine, or EGCG are some examples facing this challenge. They are highly lipophilic and can form hydrogen bonds, which makes them insoluble in water, thus severely limiting their absorption in the gastrointestinal tract [204]. Even if absorbed, their swift phase I/II metabolism and P-glycoprotein's ability to pump them out of the body further reduce the amount of these substances that can reach the systemic circulation. Curcumin's oral bioavailability is less than 1% because it is rapidly broken down by glucuronidation, and resveratrol gets thoroughly sulphated, which gives rise to plasma levels that are not sufficient to produce any therapeutic effects [205]. Similarly, berberine is subjected to fast breakdown in the liver and active efflux, which restricts its therapeutic application. Similarly, berberine is subjected to high first-pass metabolism and transporter-mediated efflux, restricting its therapeutic application despite its potent antimicrobial and anticancer effects [206]. These undesired pharmacokinetic profiles, such as quick metabolism, rapid clearance, short half-lives, and high inter-individual variability, reduce the chances of clinical translation for many promising natural agents.

7.5.1.2 Tissue-specific targeting limitations

Another critical challenge involves tissue-specific targeting limitations, which restrict the ability of natural compounds to accumulate at pathological sites. Most unmodified phytochemicals distribute broadly and non-selectively, resulting in sub-therapeutic concentrations within tumours, inflamed tissues, or microbial niches. For anticancer compounds such as curcumin, genistein, and withaferin A, biological barriers, including the blood–brain barrier, dense tumour stroma, an acidic microenvironment, and heterogeneous vasculature, impede effective tumour penetration [207,208]. Even compounds with strong in vitro selectivity often fail to demonstrate robust in vivo targeting due to metabolic instability and rapid systemic dilution. Additionally, many natural molecules are substrates for efflux pumps (e.g., BCRP, MRP1), further limiting their intracellular accumulation in cancer cells [209]. Antibiotic natural products such as allicin and thymol face similar issues, achieving limited localization at infection sites due to volatility, instability, or rapid systemic elimination [210]. Together, these limitations reduce potency, necessitate higher dosing, and increase the risk of systemic toxicity.

Overall, the poor druggability of many natural compounds demonstrates the need for advanced delivery strategies, including nanocarriers, lipid-based systems, polymeric nanoparticles, prodrugs, and structural optimization, to enhance solubility, improve pharmacokinetics, and enable tissue-specific delivery [211,212]. Overcoming these challenges is essential to fully harness the therapeutic potential of natural product-derived agents.

7.5.2 Nanotechnology-based delivery systems

7.5.2.1 Liposomes and micelles: encapsulation for enhanced solubility and circulation time

Nanotechnology-based delivery platforms such as liposomes and polymeric micelles have emerged as powerful strategies to overcome the solubility, stability, and pharmacokinetic limitations of natural compounds. Liposomes are composed of phospholipid bilayers that encapsulate hydrophilic or lipophilic drugs, enhancing their aqueous solubility while protecting the labile phytochemicals from enzymatic degradation and rapid clearance.

Their surface can be PEGylated to reduce opsonization, enabling extended systemic circulation and improved tumour accumulation through enhanced permeability and retention (EPR) effect [213]. Several natural compounds such as curcumin, resveratrol, quercetin, and EGCG have demonstrated significant stability and bioavailability when formulated in liposomes, leading to enhanced anticancer efficacy in preclinical CSC models [214].

Polymeric micelles are formed via the self-assembly of amphiphilic block copolymers. In these micelles, a hydrophobic core is formed that holds and solubilizes the hydrophobic phytochemicals. Meanwhile, a hydrophilic corona around the hydrophobic core helps in sterically stabilizing the nano-formulation while it circulates in plasma. These nanosystems increase the solubility of paclitaxel, curcumin, berberine, and other poorly water-soluble compounds and allow controlled release [188,215]. Nanoparticles improve tumour penetration, and targeted ligand incorporation – which may include folate or CD44-binding hyaluronate – further, helps enhance CSC-directed delivery [216].

Both liposomes and micelles alleviate efflux-mediated drug loss by increasing intracellular retention. This is beneficial for natural compounds that are substrates of ABC transporters such as P-gp, BCRP, and MRP1. Encapsulation protects these agents from direct pump recognition, contributing to an improved therapeutic index and CSC-specific cytotoxicity [217,218]. Overall, these nanocarriers present a clinically significant and practical approach to address the fundamental druggability barriers of natural products, promoting their translation into effective anticancer therapies.

7.5.2.2 Polymeric nanoparticles: controlled release and improved tumour penetration

Polymeric nanoparticles (PNPs) have emerged as versatile nanocarriers capable of enhancing therapeutic performance through the improvement of solubility, stability, and sustained release profiles of natural compounds. Constructed from biocompatible polymers like poly(lactic-co-glycolic acid (PLGA), polylactic acid (PLA), polycaprolactone (PCL), and polyethylene glycol (PEG), and chitosan, these nanoparticles enable precise control over drug loading and release kinetics, thus maintaining therapeutic concentrations for extended timeframes [219]. This controlled release minimizes one of the major barriers to natural compound druggability – rapid systemic clearance – and reduces off-target toxicity while improving tumour-specific delivery.

PNPs also show better tumour penetration, facilitated by their nanoscale size (50–200 nm), which allows for EPR-mediated accumulation in tumour tissue. Surface modification with PEG (PEGylation) or targeting ligands for

CSC-specific receptors, such as CD44, EpCAM, or folate receptors, further enhances tumour and CSC specificity [220]. For example, curcumin-loaded PLGA nanoparticles exhibit improved intracellular uptake, reduced efflux via ABC transporters, and increased apoptosis in CSC-enriched populations when compared to free curcumin [221]. Resveratrol-, quercetin-, and berberine-loaded polymeric nanoparticles exhibit enhanced bioavailability and stronger anticancer effects because of improved penetration into hypoxic tumour cores and CSC niches [222].

Importantly, PNPs allow co-encapsulation of multiple natural compounds or natural compound–chemotherapy combinations, enabling synergistic inhibition of CSC pathways such as Wnt/β-catenin, Hedgehog, and PI3K/AKT. Moreover, their capability to evade efflux-mediated MDR – thanks to enhanced internalization and endosomal escape – results in higher intracellular accumulation of phytochemical substrates of P-gp, BCRP, or MRP1 transporters [188,223,224]. Collectively, polymeric nanoparticles provide a robust, clinically translatable platform for controlled release, deep tumour penetration, and CSC-targeted delivery of natural products, substantially improving the therapeutic index of agents with otherwise poor druggability.

7.5.2.3 Metallic nanoparticles: gold and silver nanoparticles complexed with natural agents

Metallic nanoparticles such as gold (AuNPs) and silver (AgNPs) have emerged as versatile carriers for delivering natural products because of their size, tunability, high surface area, and easy functionalization. Conjugation of various phytochemicals – including curcumin, quercetin, gallic acid, EGCG and berberine – onto AuNPs or AgNPs increases solubility, stability, and intracellular delivery, overcoming the significant druggability challenges of natural compounds [214].

AuNPs are especially appealing due to their biocompatibility and potential for targeted or stimuli-triggered release. For instance, curcumin–AuNP conjugates demonstrate enhanced uptake, ROS generation, and suppression of CSC-related pathways such as Wnt/β-catenin and Notch, showing compelling anticancer and anti-CSC effects when compared to the free form of curcumin [225,226]. Furthermore, their photothermal characteristics could promote combined chemo–photothermal approaches.

AuNPs as well as AgNPs enhance intracellular drug retention and diminish efflux by ABC transporters like P-gp and BCRP, which together improve the bioavailability and CSC-specific cytotoxicity [218]. Overall, metallic nanoparticles provide a versatile and scalable approach for enhanced delivery and therapeutic efficacy of natural compounds in the context of cancer and CSC-targeted therapy.

7.5.3 Active targeting strategies

7.5.3.1 Surface modification with CSC-specific ligands

Active targeting remarkably increases the specificity of nanocarrier–natural compound systems through the selective recognition of CSC markers. Currently, the highly expressed CSC surface receptor CD44 can be specifically targeted using natural ligands with high CD44 affinity, such as hyaluronic acid (HA). HA-functionalized nanoparticles – for example, HA-PLGA, HA-liposomes, and HA-gold nanoparticles – have shown increased cellular uptake and penetration in CSC niches, as well as continuous intracellular retention of natural drugs like curcumin, quercetin, and resveratrol [227,228].

This approach not only enhances the targeting of $CD44^+$ CSCs but also decreases off-target toxicity. HA-coated nanocarriers demonstrated enhanced inhibition of CSC pathways (Wnt/β-catenin, Hedgehog) and reduced sphere-forming ability in different tumour models. Likewise, other ligands – such as EpCAM antibodies, folate, and CD133 aptamers, and transferrin – have been used to functionalize natural products for CSC-targeted delivery [227,229].

7.5.3.2 Receptor-mediated endocytosis for targeted CSC delivery

The strategy of ligand-based targeting is well established. For example, CD44 is widely targeted using its natural ligand, HA. HA-decorated nanocarriers, such as liposomes, polymeric NPs, and micelles, have repeatedly demonstrated enhanced uptake in $CD44^+$ cells via HA–CD44 RME [230].

A classic example is HA-modified albumin nanoparticles encapsulating a differentiating agent, such as all-trans retinoic acid (ATRA, a natural derivative), which have shown high affinity for CD44-rich cells, efficient internalization, a reduced "side population" (CSC-enriched), and inhibited lung-metastatic tumour growth in mice [190].

Beyond CD44/HA, other CSC-relevant receptors have been exploited. Antibody- or aptamer-based targeting of CD133 or EpCAM on CSCs, followed by RME, has been reported, allowing preferential payload delivery to CSC subpopulations [231].

Nanocarriers come in many forms – such as PNPs, liposomes, micelles, and inorganic nanoparticles – enabling the encapsulation of diverse molecules including natural products, differentiation agents, nucleic acids, or chemotherapeutics used to eliminate CSCs or re-sensitize them to therapy [232].

7.5.4 Combination therapy formulation

7.5.4.1 *Co-delivery of a natural compound and a conventional drug in a single nanocarrier*

Co-delivery of a natural compound and conventional chemotherapeutics within a single nanocarrier provides harmonized pharmacokinetics, improved intracellular accumulation, and synergistic anticancer activity, which is especially important for CSC targeting. Nanocarriers such as polymeric nanoparticles, solid lipid nanoparticles (SLNs), and niosomes (non-ionic vesicles) allow co-encapsulation, controlled release, and deeper tumour penetration, thereby enhancing the overall therapeutic performance.

A combination of paclitaxel (PTX) and curcumin (CUR) encapsulated into polymeric nanoparticles is a well-studied example. PTX–CUR nanoparticles (~28 nm) displayed sustained dual-drug release and significantly higher apoptosis in MCF-7 cells (~64.3%) compared with free PTX and CUR (~34.2%) [233]. In vivo, the co-delivery of PTX and CUR led to stronger tumour inhibition, reduced Ki-67 proliferation rates, increased TUNEL-positive apoptotic cells, and lower systemic toxicity. Another formulation includes PTX and CUR in SLNs. These SLNs (≈190 nm) show over 94% encapsulation efficiency for both PTX and CUR, controlled release, and higher cytotoxicity in A549 lung cancer cells as compared with single-drug SLNs [234]. The dual-loaded SLNs caused a 12-fold reduction in tumour volume as compared to the control while maintaining stable body weight, indicating reduced toxicity. Co-delivery of doxorubicin (DOX) and CUR in pH-responsive niosomes provides another validated combination. Encapsulation efficiencies reached ~77%–79%, and an acidic pH triggered enhanced release. In MCF-7 cells, DOX–CUR niosomes led to ~79% cell death, which is higher than that observed with single-drug formulations (~39% –43%) [235].

In addition to chemotherapeutics, co-delivery platforms that combine natural compounds and siRNA have been tested for targeting CSC signalling. For example, both CUR and Bcl-2 siRNA co-loaded nanoparticles) to synergistically close the self-re-establishment pathways (by controlling distinctive gatherings of qualities like Nanog, Bmi1, and Wnt family) also turned around chemoresistance [236–238].

Overall, co-delivery nanocarriers integrate natural bioactive molecules with established chemotherapy agents to achieve synergistic pathway regulation, enhancing tumour and CSC elimination. The reduction in efflux-mediated resistance and superior therapeutic outcomes also demonstrate the strong translational potential of these natural products.

7.6 FUTURE DIRECTIONS: NATURAL COMPOUNDS IN PERSONALIZED AND PRECISION ONCOLOGY

7.6.1 CSC-targeting natural compounds in clinical trials

The severe side effects associated with conventional chemotherapy and radiotherapy have created an urgent need for efficient, cost-effective alternatives. Consequently, the scientific community has intensified its investigation into the therapeutic potential of naturally occurring compounds. While these natural compounds have traditionally been used to provide supportive or palliative care, clinical trials are increasingly focusing on CSC-targeting treatments. Although numerous natural compounds – including curcumin, EGCG, resveratrol, and apigenin – exhibit considerable anticancer efficacy in preclinical models, their therapeutic potential in controlled clinical trials remains a persistent and significant challenge. Further research focusing on the pharmacokinetics and pharmacodynamics of the natural products is mandatory to establish their efficacy and human safety profiles. Current research frequently focuses on combinatorial approaches that use phytochemicals like curcumin or quercetin to enhance the efficacy of conventional treatments in tumours characterized by a high CSC load, such as colorectal and pancreatic tumours [8]. For instance, studies are evaluating how natural product derivatives can suppress the self-renewal pathways (Wnt, Notch) of highly aggressive cancers, thereby preventing post-treatment relapse [239]. The most challenging aspect of these studies is translating successful anti-tumour responses into specific, on-target CSC elimination, which typically requires biomarker-driven patient stratification based on stemness features.

7.6.2 Integrating genomic and proteomic data

One of the main challenges to be faced in the research of CSCs is their inherent heterogeneity and adaptability, which, in turn, necessitates the availability of high-resolution genomic and proteomic data. The former is essential for the identification of CSC-specific biomarkers such as (e.g., highly expressed ALDH1A1 isoforms or unique cell surface markers) whose expression is positively correlated with sensitivity to a specific natural compound [240]. Compound selection and dosage are determined through this personalized strategy. Additionally, translational oncology has become increasingly dependent on the use of advanced screening instruments, such as patient-derived organoids (PDOs) and tumour spheroids – 3D models that closely correspond to the genomic complexity and CSC phenotype of the original tumour. These are currently being used for high-throughput screening (HTS) of novel natural product libraries [148]. This advanced and *ex vivo* system rapidly determines the most effective CSC-targeting natural agent or combination strategy for an individual patient.

7.6.3 Computational and *in silico* approaches

Optimization of multi-targeting agents requires advanced computational and *in silico* approaches. For instance, network pharmacology is increasingly utilized to determine how a natural extract works at the level of the entire biological network, rather than through single-target findings that only provide single-pathway insights. Thus, network pharmacology allows for rational drug design through the mapping of overlapping, simultaneous Wnt, Notch, and NF-κB modulation networks [239]. In addition, molecular docking and structure-activity relationship (SAR) studies guide chemists to prepare interested derivatives with increased potency for increased action with increased drug-like properties. Ultimately, the final derivative must be highly active towards the interested CSC and possess sufficient drug-like properties for better distribution to the site of action, a process that can only be achieved through systems biology and theoretical chemistry working together to micro-direct from specific crude extracts [241].

7.6.4 Regulatory and commercialization hurdles

However, despite these exciting preclinical and biotechnological advances, significant regulatory and commercial barriers prevent these natural products from achieving clinical translational success. These challenges include poor aqueous solubility, physical stability, and a lack of standardization for industrially scalable manufacturing [242]. Regulatory bodies, such as the FDA, maintain stringent CMC requirements to prevent herb-drug adverse reactions (e.g., the inhibition or stimulation of cytochrome P450 enzymes) by ensuring strict batch-to-batch consistency [243]. Thus, as the research community establishes standardized extraction procedures and Quality Control (QC) protocols utilizing marker compounds and advanced formulation methods (e.g., nanocarriers), the clinical translation of these natural products into FDA-approved, CSC-targeting therapeutics ultimately depends on industry efforts to bridge decades of ethnopharmacological applications with modern regulatory requirements.

7.7 CONCLUSION AND FUTURE PERSPECTIVES

The persistence of CSCs represents the most significant obstacle in modern oncology due to their characteristics such as drug resistance, metastasis, and relapse. The studies presented in this chapter suggest that natural compounds or phytochemicals can offer a multi-target approach that can regulate the complex survival networks of CSCs.

Bioactive compounds like curcumin, resveratrol, and sulforaphane distinguish themselves by their ability to target multiple signalling pathways simultaneously, including Wnt, Notch, and NF-κB. By modulating these networks and disabling resistance mechanisms like efflux pumps, these phytochemicals can effectively "re-sensitize" refractory cells, making them susceptible to conventional therapy once again.

However, advanced holistic approaches are required to fully understand their potential. In this view, nanotechnology can play a key role in evolving natural products from simple supplements into potent combinatorial therapeutics. Similarly, by utilizing targeted liposomes and polymeric micelles, these compounds can be delivered directly into the challenging microenvironment of the hypoxic CSC niche. Furthermore, combining these natural agents with novel cell death mechanisms can create a synergistic effect that modulates the challenging resistance of CSCs.

To upgrade this translational route, advanced computational approaches to structural and pharmacological optimization must be adopted. The integration of computational and *in silico* allows for the rational design of enhanced drug carriers and the prediction of biomarkers for patient profiling. Simultaneously, the combination of systems biology, structural modification, and predictive computation can enable the creation of potent derivatives that overcome the structural limitations of their parent molecules.

Nevertheless, significant translational hurdles remain. Certain issues regarding solubility, stability, and manufacturing standardization – required to move these compounds from the source to the target – must be addressed. Strict regulatory adherence and robust QC must be established to ensure safety and efficacy. Finally, the successful utilization of natural compounds against CSCs leverages their inherent multi-target advantages via precision delivery and combination therapy, which is critical for shifting cancer care from temporary control to long-term elimination.

REFERENCES

1. Chu X, Tian W, Ning J, Xiao G, Zhou Y, Wang Z, et al. Cancer stem cells: advances in knowledge and implications for cancer therapy. *Signal Transduct Target Ther* 2024;9:170. https://doi.org/10.1038/s41392-024-01851-y.
2. Lee H, Kim B, Park J, Park S, Yoo G, Yum S, et al. Cancer stem cells: landscape, challenges and emerging therapeutic innovations. *Signal Transduct Target Ther* 2025;10:248. https://doi.org/10.1038/s41392-025-02360-2.
3. Chin VL, Lim CL. Epithelial-mesenchymal plasticity: engaging stemness in an interplay of phenotypes. *Stem Cell Investig* 2019;6:25–25. https://doi.org/10.21037/sci.2019.08.08.

4. Kaur J, Bhattacharyya S. Cancer stem cells: metabolic characterization for targeted cancer therapy. *Front Oncol* 2021;11:756888. https://doi.org/10.3389/fonc.2021.756888.
5. Irshad R, Husain M. Natural products in the reprogramming of cancer epigenetics. *Toxicol Appl Pharmacol* 2021;417:115467. https://doi.org/10.1016/j.taap.2021.115467.
6. Liu K, Guo L, Miao L, Bao W, Yang J, Li X, et al. Ursolic acid inhibits epithelial–mesenchymal transition by suppressing the expression of astrocyte-elevated gene-1 in human nonsmall cell lung cancer A549 cells. *Anticancer Drugs* 2013;24:494–503. https://doi.org/10.1097/CAD.0b013e328360093b.
7. Kyakulaga AH, Aqil F, Munagala R, Gupta RC. Withaferin A inhibits epithelial to mesenchymal transition in non-small cell lung cancer cells. *Sci Rep* 2018;8:15737. https://doi.org/10.1038/s41598-018-34018-1.
8. Debnath I, Kundu M. Therapeutic potential of natural compounds in targeting cancer stem cells: a promising approach for cancer treatment. *Discov Oncol* 2025;16:1433. https://doi.org/10.1007/s12672-025-03190-y.
9. Fong D, Chan MM. Dietary Phytochemicals Target Cancer Stem Cells for Cancer Chemoprevention. In: Chandra D, editor. *Mitochondria Targets Phytochem.* Cancer Prev. Ther., New York, NY: Springer New York; 2013, p. 85–125.
10. Liskova A, Kubatka P, Samec M, Zubor P, Mlyncek M, Bielik T, et al. Dietary phytochemicals targeting cancer stem cells. *Molecules* 2019;24:899. https://doi.org/10.3390/molecules24050899.
11. Bozzuto G, Calcabrini A, Colone M, Condello M, Dupuis ML, Pellegrini E, et al. Phytocompounds and nanoformulations for anticancer therapy: a review. *Molecules* 2024;29:3784. https://doi.org/10.3390/molecules29163784.
12. Liu Y, Wang H. Biomarkers and targeted therapy for cancer stem cells. *Trends Pharmacol Sci* 2024;45:56–66. https://doi.org/10.1016/j.tips.2023.11.006.
13. Mostafa MAH, Khojah HMJ. Nanoparticle-based delivery systems for phytochemicals in cancer therapy: molecular mechanisms, clinical evidence, and emerging trends. *Drug Dev Ind Pharm* 2025;51:1105–1121. https://doi.org/10.1080/03639045.2025.2483425.
14. Balekundri A, Mannur V. Quality control of the traditional herbs and herbal products: a review. *Future J Pharm Sci* 2020;6:67. https://doi.org/10.1186/s43094-020-00091-5.
15. Govindaraghavan S, Sucher NJ. Quality assessment of medicinal herbs and their extracts: criteria and prerequisites for consistent safety and efficacy of herbal medicines. *Epilepsy Behav* 2015;52:363–371. https://doi.org/10.1016/j.yebeh.2015.03.004.
16. Chaffer CL, Marjanovic ND, Lee T, Bell G, Kleer CG, Reinhardt F, et al. Poised chromatin at the ZEB1 promoter enables breast cancer cell plasticity and enhances tumorigenicity. *Cell* 2013;154:61–74. https://doi.org/10.1016/j.cell.2013.06.005.
17. Gupta PB, Fillmore CM, Jiang G, Shapira SD, Tao K, Kuperwasser C, et al. Stochastic state transitions give rise to phenotypic equilibrium in populations of cancer cells. *Cell* 2011;146:633–644. https://doi.org/10.1016/j.cell.2011.07.026.
18. Chen W, Dong J, Haiech J, Kilhoffer M-C, Zeniou M. Cancer stem cell quiescence and plasticity as major challenges in cancer therapy. *Stem Cells Int* 2016;2016:1740936. https://doi.org/10.1155/2016/1740936.

19. Zhang H, Deng Y, Liu X, Sun J, Ma L, Ding Y, et al. Glass fiber-supported hybrid monolithic spin tip for enrichment of phosphopeptides from urinary extracellular vesicles. *Anal Chem* 2020;92:14790–14797. https://doi.org/10.1021/acs.analchem.0c03557.
20. Wei J, Zheng X, Li W, Li X, Fu Z. Sestrin2 reduces cancer stemness via Wnt/β-catenin signaling in colorectal cancer. *Cancer Cell Int* 2022;22:75. https://doi.org/10.1186/s12935-022-02498-x.
21. Zhang Z, Westover D, Tang Z, Liu Y, Sun J, Sun Y, et al. Wnt/β-catenin signaling in the development and therapeutic resistance of non-small cell lung cancer. *J Transl Med* 2024;22:565. https://doi.org/10.1186/s12967-024-05380-8.
22. Pak S, Park S, Kim Y, Park J-H, Park C-H, Lee K-J, et al. The small molecule WNT/β-catenin inhibitor CWP232291 blocks the growth of castration-resistant prostate cancer by activating the endoplasmic reticulum stress pathway. *J Exp Clin Cancer Res* 2019;38:342. https://doi.org/10.1186/s13046-019-1342-5.
23. Zhang Y, Wang X. Targeting the Wnt/β-catenin signaling pathway in cancer. *J Hematol Oncol* 2020;13:165. https://doi.org/10.1186/s13045-020-00990-3.
24. Rizzo P, Osipo C, Foreman K, Golde T, Osborne B, Miele L. Rational targeting of Notch signaling in cancer. *Oncogene* 2008;27:5124–5131. https://doi.org/10.1038/onc.2008.226.
25. Takebe N, Miele L, Harris PJ, Jeong W, Bando H, Kahn M, et al. Targeting Notch, Hedgehog, and Wnt pathways in cancer stem cells: clinical update. *Nat Rev Clin Oncol* 2015;12:445–464. https://doi.org/10.1038/nrclinonc.2015.61.
26. Ingham PW, McMahon AP. Hedgehog signaling in animal development: paradigms and principles. *Genes Dev* 2001;15:3059–3087. https://doi.org/10.1101/gad.938601.
27. Wang F, Ma L, Zhang Z, Liu X, Gao H, Zhuang Y, et al. Hedgehog signaling regulates epithelial-mesenchymal transition in pancreatic cancer stem-like cells. *J Cancer* 2016;7:408–417. https://doi.org/10.7150/jca.13305.
28. Das S, Samant RS, Shevde LA. Nonclassical activation of Hedgehog signa ling enhances multidrug resistance and makes cancer cells refractory to smoothened-targeting Hedgehog inhibition. *J Biol Chem* 2013;288:11824–11833. https://doi.org/10.1074/jbc.M112.432302.
29. Nguyen NM, Cho J. Hedgehog pathway inhibitors as targeted cancer therapy and strategies to overcome drug resistance. *Int J Mol Sci* 2022;23:1733. https://doi.org/10.3390/ijms23031733.
30. Yu H, Lee H, Herrmann A, Buettner R, Jove R. Revisiting STAT3 signalling in cancer: new and unexpected biological functions. *Nat Rev Cancer* 2014;14:736–46. https://doi.org/10.1038/nrc3818.
31. Karin M. Nuclear factor-κB in cancer development and progression. *Nature* 2006;441:431–436. https://doi.org/10.1038/nature04870.
32. Karin M. NF-B as a critical link between inflammation and cancer. *Cold Spring Harb Perspect Biol* 2009;1:a000141. https://doi.org/10.1101/cshperspect.a000141.
33. Lee H, Herrmann A, Deng J-H, Kujawski M, Niu G, Li Z, et al. Persistently activated stat3 maintains constitutive NF-κB activity in tumors. *Cancer Cell* 2009;15:283–293. https://doi.org/10.1016/j.ccr.2009.02.015.

34. Taniguchi K, Karin M. NF-κB, inflammation, immunity and cancer: coming of age. *Nat Rev Immunol* 2018;18:309–324. https://doi.org/10.1038/nri.2017.142.
35. Kryczek I, Liu S, Roh M, Vatan L, Szeliga W, Wei S, et al. Expression of aldehyde dehydrogenase and CD133 defines ovarian cancer stem cells. *Int J Cancer* 2012;130:29–39. https://doi.org/10.1002/ijc.25967.
36. Dzobo K, Sinkala M. Cancer stem cell marker CD44 plays multiple key roles in human cancers: immune suppression/evasion, drug resistance, epithelial–mesenchymal transition, and metastasis. *OMICS J Integr Biol* 2021;25:313–332. https://doi.org/10.1089/omi.2021.0025.
37. Yaghobi Z, Movassaghpour A, Talebi M, Abdoli Shadbad M, Hajiasgharzadeh K, Pourvahdani S, et al. The role of CD44 in cancer chemoresistance: a concise review. *Eur J Pharmacol* 2021;903:174147. https://doi.org/10.1016/j.ejphar.2021.174147.
38. Luker GD, Yang J, Richmond A, Scala S, Festuccia C, Schottelius M, et al. At the bench: pre-clinical evidence for multiple functions of CXCR4 in cancer. *J Leukoc Biol* 2021;109:969–989. https://doi.org/10.1002/JLB.2BT1018-715RR.
39. Rueda A, Serna N, Mangues R, Villaverde A, Unzueta U. Targeting the chemokine receptor CXCR4 for cancer therapies. *Biomark Res* 2025;13:68. https://doi.org/10.1186/s40364-025-00778-y.
40. Avnet S, Di Pompo G, Chano T, Errani C, Ibrahim-Hashim A, Gillies RJ, et al. Cancer-associated mesenchymal stroma fosters the stemness of osteosarcoma cells in response to intratumoral acidosis via NF-κB activation: tumor acidic microenvironment fosters osteosarcoma stemness via mesenchymal stroma. *Int J Cancer* 2017;140:1331–1345. https://doi.org/10.1002/ijc.30540.
41. Le Grand M, Mukha A, Püschel J, Valli E, Kamili A, Vittorio O, et al. Interplay between MycN and c-Myc regulates radioresistance and cancer stem cell phenotype in neuroblastoma upon glutamine deprivation. *Theranostics* 2020;10:6411–6429. https://doi.org/10.7150/thno.42602.
42. Ma D-Q, Zhang Y-H, Ding D-P, Li J, Chen L-L, Tian Y-Y, et al. Effect of Bmi-1-mediated NF-κB signaling pathway on the stem-like properties of CD133+ human liver cancer cells. *Cancer Biomark* 2018;22:575–585. https://doi.org/10.3233/CBM-181329.
43. Tatetsu H, Kong NR, Chong G, Amabile G, Tenen DG, Chai L. SALL4, the missing link between stem cells, development and cancer. *Gene* 2016;584:111–119. https://doi.org/10.1016/j.gene.2016.02.019.
44. Vaziri N, Shariati L, Zarrabi A, Farazmand A, Haghjooy Javanmard S. Cancer-associated fibroblasts regulate the plasticity of breast cancer stemness through the production of leukemia inhibitory factor. *Life* 2021;11:1298. https://doi.org/10.3390/life11121298.
45. Pastrana E, Silva-Vargas V, Doetsch F. Eyes wide open: a critical review of sphere-formation as an assay for stem cells. *Cell Stem Cell* 2011;8:486–498. https://doi.org/10.1016/j.stem.2011.04.007.
46. Behbod F, Vivanco MDM. Side Population. In: Vivanco MDM, editor. *Mammary Stem Cells*, vol. 1293, New York, NY: Springer New York; 2015, p. 73–81.

47. Jaggupilli A, Elkord E. Significance of CD44 and CD24 as cancer stem cell markers: an enduring ambiguity. *Clin Dev Immunol* 2012;2012:1–11. https://doi.org/10.1155/2012/708036.
48. Sheridan C, Kishimoto H, Fuchs RK, Mehrotra S, Bhat-Nakshatri P, Turner CH, et al. CD44+/CD24-breast cancer cells exhibit enhanced invasive properties: an early step necessary for metastasis. *Breast Cancer Res* 2006;8:R59. https://doi.org/10.1186/bcr1610.
49. Tiezzi DG, Sicchieri RD, Antonio HM, Mouro LR, Da Silva JS, De Andrade JM, et al. Cancer stem cell markers in locally advanced breast cancer. *J Clin Oncol* 2012;30:1075–1075. https://doi.org/10.1200/jco.2012.30.15_suppl.1075.
50. Bao Q, Zhao Y, Renner A, Niess H, Seeliger H, Jauch K-W, et al. Cancer stem cells in pancreatic cancer. *Cancers* 2010;2:1629–1641. https://doi.org/10.3390/cancers2031629.
51. Ferrara B, Dugnani E, Sordi V, Pasquale V, Pellegrini S, Reni M, et al. A comprehensive characterization of stemness in cell lines and primary cells of pancreatic ductal adenocarcinoma. *Int J Mol Sci* 2022;23:10663. https://doi.org/10.3390/ijms231810663.
52. Kure S, Matsuda Y, Hagio M, Ueda J, Naito Z, Ishiwata T. Expression of cancer stem cell markers in pancreatic intraepithelial neoplasias and pancreatic ductal adenocarcinomas. *Int J Oncol* 2012;41:1314–1324. https://doi.org/10.3892/ijo.2012.1565.
53. Ohara Y, Oda T, Sugano M, Hashimoto S, Enomoto T, Yamada K, et al. Histological and prognostic importance of $CD44^{+}/CD24^{+}/EpCAM^{+}$ expression in clinical pancreatic cancer. *Cancer Sci* 2013;104:1127–1134. https://doi.org/10.1111/cas.12198.
54. Ungureanu BS, Gheonea DI, Turcu-Stiolica A, Schenker M, Pirici D, Vere C-C, et al. Clinicopathologic relevance of EpCAM and CD44 in pancreatic cancer: insights from a meta-analysis. *Stem Cell Res Ther* 2025;16:463. https://doi.org/10.1186/s13287-025-04601-1.
55. Chen J, Xia Q, Jiang B, Chang W, Yuan W, Ma Z, et al. Prognostic value of cancer stem cell marker ALDH1 expression in colorectal cancer: a systematic review and meta-analysis. *PLoS One* 2015;10:e0145164. https://doi.org/10.1371/journal.pone.0145164.
56. O'Brien CA, Pollett A, Gallinger S, Dick JE. A human colon cancer cell capable of initiating tumour growth in immunodeficient mice. *Nature* 2007;445:106–110. https://doi.org/10.1038/nature05372.
57. Ricci-Vitiani L, Lombardi DG, Pilozzi E, Biffoni M, Todaro M, Peschle C, et al. Identification and expansion of human colon-cancer-initiating cells. *Nature* 2007;445:111–115. https://doi.org/10.1038/nature05384.
58. Schmitt M, Metzger M, Gradl D, Davidson G, Orian-Rousseau V. CD44 functions in Wnt signaling by regulating LRP6 localization and activation. *Cell Death Differ* 2015;22:677–689. https://doi.org/10.1038/cdd.2014.156.
59. Wang K, Yuan Y, Cho J-H, McClarty S, Baxter D, Galas DJ. Comparing the MicroRNA spectrum between serum and plasma. *PLoS One* 2012;7:e41561. https://doi.org/10.1371/journal.pone.0041561.
60. Zhao H, Yan C, Hu Y, Mu L, Huang K, Li Q, et al. Sphere-forming assay vs. organoid culture: Determining long-term stemness and the chemoresistant capacity of primary colorectal cancer cells. *Int J Oncol* 2019. https://doi.org/10.3892/ijo.2019.4683.

61. Gangemi RMR, Griffero F, Marubbi D, Perera M, Capra MC, Malatesta P, et al. SOX2 Silencing in glioblastoma tumor-initiating cells causes stop of proliferation and loss of tumorigenicity. *Stem Cells* 2009;27:40–48. https://doi.org/10.1634/stemcells.2008-0493.
62. Hatanpaa KJ, Hu T, Vemireddy V, Foong C, Raisanen JM, Oliver D, et al. High expression of the stem cell marker nestin is an adverse prognostic factor in WHO grade II–III astrocytomas and oligoastrocytomas. *J Neurooncol* 2014;117:183–189. https://doi.org/10.1007/s11060-014-1376-7.
63. Lathia JD, Mack SC, Mulkearns-Hubert EE, Valentim CLL, Rich JN. Cancer stem cells in glioblastoma. *Genes Dev* 2015;29:1203–1217. https://doi.org/10.1101/gad.261982.115.
64. Singh SK, Hawkins C, Clarke ID, Squire JA, Bayani J, Hide T, et al. Identification of human brain tumour initiating cells. *Nature* 2004;432:396–401. https://doi.org/10.1038/nature03128.
65. Dreyzin A, Holtzman NG, Bonifant CL. CD123-targeting immunotherapeutic approaches in acute myeloid leukaemia. *Br J Haematol* 2025;207:1178–1191. https://doi.org/10.1111/bjh.70019.
66. Mendoza M, Ishida T, Le Q, Kraskouskas K, Hylkema T, Dozono S, et al. 3128-Identification and characterization of $CD96^{+}CD70^{+}$ leukemic stem cells and niche-mediated resistance in CBFA2T3-GLIS2 AMKL. *Exp Hematol* 2024;137:104449. https://doi.org/10.1016/j.exphem.2024.104449.
67. Mohamed Soliman AS, Osman AA, Diaa El-Dine DA. Evaluation of CD96 expression in acute myeloid leukemia patients: relation to prognosis and induction chemotherapy. *QJM Int J Med* 2024;117:hcae175.180. https://doi.org/10.1093/qjmed/hcae175.180.
68. Stelmach P, Trumpp A. Leukemic stem cells and therapy resistance in acute myeloid leukemia. *Haematologica* 2023;108:353–366. https://doi.org/10.3324/haematol.2022.280800.
69. Zhai X, Jiang X. Properties of leukemic stem cells in regulating drug resistance in acute and chronic myeloid leukemias. *Biomedicines* 2022;10:1841. https://doi.org/10.3390/biomedicines10081841.
70. Daver NG, Montesinos P, DeAngelo DJ, Wang ES, Papadantonakis N, Todisco E, et al. Pivekimab sunirine (IMGN632), a novel CD123-targeting antibody–drug conjugate, in relapsed or refractory acute myeloid leukaemia: a phase 1/2 study. *Lancet Oncol* 2024;25:388–399. https://doi.org/10.1016/S1470-2045(23)00674-5.
71. Acikgoz E, Soner BC, Ozdil B, Guven M. $CD133^{+}/CD44^{+}$ prostate cancer stem cells exhibit embryo-like behavior patterns. *Acta Histochem* 2021;123:151743. https://doi.org/10.1016/j.acthis.2021.151743.
72. Al Salhi Y, Sequi MB, Valenzi FM, Fuschi A, Martoccia A, Suraci PP, et al. Cancer stem cells and prostate cancer: a narrative review. *Int J Mol Sci* 2023;24:7746. https://doi.org/10.3390/ijms24097746.
73. Kanwal R, Shukla S, Walker E, Gupta S. Acquisition of tumorigenic potential and therapeutic resistance in $CD133^{+}$ subpopulation of prostate cancer cells exhibiting stem-cell like characteristics. *Cancer Lett* 2018;430:25–33. https://doi.org/10.1016/j.canlet.2018.05.014.
74. Tu S-M, Lin S-H. Prostate cancer stem cells. *Clin Genitourin Cancer* 2012;10:69–76. https://doi.org/10.1016/j.clgc.2012.01.002.

75. Zhang K, Zhou S, Wang L, Wang J, Zou Q, Zhao W, et al. Current stem cell biomarkers and their functional mechanisms in prostate cancer. *Int J Mol Sci* 2016;17:1163. https://doi.org/10.3390/ijms17071163.
76. Zhou K, Lu H, Zhang J, Shen Q, Liu P, Xu Q, et al. Prostate cancer stem cells: an updated mini-review. *J Cancer* 2024;15:6570–6576. https://doi.org/10.7150/jca.100604.
77. Frąszczak K, Barczyński B. The role of cancer stem cell markers in ovarian cancer. *Cancers* 2023;16:40. https://doi.org/10.3390/cancers16010040.
78. Królewska-Daszczyńska P, Wendlocha D, Smycz-Kubańska M, Stępień S, Mielczarek-Palacz A. Cancer stem cells markers in ovarian cancer: clinical and therapeutic significance (review). *Oncol Lett* 2022;24:465. https://doi.org/10.3892/ol.2022.13585.
79. Takahashi A, Hong L, Chefetz I. How to win the ovarian cancer stem cell battle: destroying the roots. *Cancer Drug Resist* 2021. https://doi.org/10.20517/cdr.2020.93.
80. Tomao F, Papa A, Rossi L, Strudel M, Vici P, Lo Russo G, et al. Emerging role of cancer stem cells in the biology and treatment of ovarian cancer: basic knowledge and therapeutic possibilities for an innovative approach. *J Exp Clin Cancer Res* 2013;32:48. https://doi.org/10.1186/1756-9966-32-48.
81. Zong X, Nephew KP. Ovarian cancer stem cells: role in metastasis and opportunity for therapeutic targeting. *Cancers* 2019;11:934. https://doi.org/10.3390/cancers11070934.
82. Alamgeer M, Peacock CD, Matsui W, Ganju V, Watkins DN. Cancer stem cells in lung cancer: evidence and controversies. *Respirology* 2013;18:757–764. https://doi.org/10.1111/resp.12094.
83. Satar N, Fakiruddin K, Lim M, Mok P, Zakaria N, Fakharuzi N, et al. Novel triple-positive markers identified in human non-small cell lung cancer cell line with chemotherapy-resistant and putative cancer stem cell characteristics. *Oncol Rep* 2018. https://doi.org/10.3892/or.2018.6461.
84. Zakaria N, Satar NA, Abu Halim NH, Ngalim SH, Yusoff NM, Lin J, et al. Targeting lung cancer stem cells: research and clinical impacts. *Front Oncol* 2017;7:80. https://doi.org/10.3389/fonc.2017.00080.
85. Zhang H, Wang Y, Yuan X, Zou Y, Xiong H. Research progress on lung cancer stem cells in epidermal growth factor receptor–tyrosine kinase inhibitor targeted therapy resistance in lung adenocarcinoma. *Oncol Transl Med* 2024;10:42–46. https://doi.org/10.1097/ot9.0000000000000028.
86. Zhao W, Luo Y, Li B, Zhang T. Tumorigenic lung tumorospheres exhibit stem-like features with significantly increased expression of CD133 and ABCG2. *Mol Med Rep* 2016;14:2598–2606. https://doi.org/10.3892/mmr.2016.5524.
87. Keith B, Simon MC. Hypoxia-inducible factors, stem cells, and cancer. *Cell* 2007;129:465–472. https://doi.org/10.1016/j.cell.2007.04.019.
88. Heddleston JM, Li Z, McLendon RE, Hjelmeland AB, Rich JN. The hypoxic microenvironment maintains glioblastoma stem cells and promotes reprogramming towards a cancer stem cell phenotype. *Cell Cycle* 2009;8:3274–3284. https://doi.org/10.4161/cc.8.20.9701.
89. Kalluri R. The biology and function of fibroblasts in cancer. *Nat Rev Cancer* 2016;16:582–598. https://doi.org/10.1038/nrc.2016.73.

90. Mariathasan S, Turley SJ, Nickles D, Castiglioni A, Yuen K, Wang Y, et al. TGFβ attenuates tumour response to PD-L1 blockade by contributing to exclusion of T cells. *Nature* 2018;554:544–548. https://doi.org/10.1038/nature25501.
91. Binnewies M, Roberts EW, Kersten K, Chan V, Fearon DF, Merad M, et al. Understanding the tumor immune microenvironment (TIME) for effective therapy. *Nat Med* 2018;24:541–550. https://doi.org/10.1038/s41591-018-0014-x.
92. Rezayatmand H, Razmkhah M, Razeghian-Jahromi I. Drug resistance in cancer therapy: the Pandora's Box of cancer stem cells. *Stem Cell Res Ther* 2022;13:181. https://doi.org/10.1186/s13287-022-02856-6.
93. Sancho P, Barneda D, Heeschen C. Hallmarks of cancer stem cell metabolism. *Br J Cancer* 2016;114:1305–1312. https://doi.org/10.1038/bjc.2016.152.
94. Viale A, Pettazzoni P, Lyssiotis CA, Ying H, Sánchez N, Marchesini M, et al. Oncogene ablation-resistant pancreatic cancer cells depend on mitochondrial function. *Nature* 2014;514:628–632. https://doi.org/10.1038/nature13611.
95. Zheng J. Energy metabolism of cancer: glycolysis versus oxidative phosphorylation (review). *Oncol Lett* 2012;4:1151–1157. https://doi.org/10.3892/ol.2012.928.
96. Rasool Hassan BA. Medicinal plants (importance and uses). *Pharm Anal Acta* 2012;03. https://doi.org/10.4172/2153-2435.1000e139.
97. Thallaj N, Agha MIH, Nattouf AH, Khatib C, Karaali A, Moustapha A, et al. Evaluation of antimicrobial activities and bioactive compounds of different extracts related to Syrian traditional products of Damask Rose (Rosa damascena). *OALib* 2020;07:1–21. https://doi.org/10.4236/oalib.1106302.
98. Chopra A, Doiphode VV. Ayurvedic medicine: core concept, therapeutic principles, and current relevance. *Med Clin North Am* 2002;86:75–89. https://doi.org/10.1016/S0025-7125(03)00073-7.
99. Maurya R, Singh G, Yadav PP. Antiosteoporotic Agents From Natural Sources. In: *Stud Nat Prod Chem*, vol. 35, Elsevier; 2008, p. 517–548. https://doi.org/10.1016/S1572-5995(08)80013-0
100. Akula R, Ravishankar GA. Influence of abiotic stress signals on secondary metabolites in plants. *Plant Signal Behav* 2011;6:1720–1731. https://doi.org/10.4161/psb.6.11.17613.
101. Thirumurugan D, Cholarajan A, Raja SSS, Vijayakumar R. An Introductory Chapter: Secondary Metabolites. In: Secondary Metabolites- Sources and applications. Vijayakumar R, Raja SSS, editors. InTech; 2018. DOI: 10.5772/intechopen.79766.
102. Zhao Y, Liu G, Yang F, Liang Y, Gao Q, Xiang C, et al. Multilayered regulation of secondary metabolism in medicinal plants. *Mol Hortic* 2023;3:11. https://doi.org/10.1186/s43897-023-00059-y.
103. Verpoorte R. Exploration of nature's chemodiversity: the role of secondary metabolites as leads in drug development. *Drug Discov Today* 1998;3: 232–238. https://doi.org/10.1016/S1359-6446(97)01167-7.
104. Savithramma N, Rao M, Ankanna S. Screening of medicinal plants for secondary metabolites. *Middle-East J Sci Res* 2011;8: 643–647.
105. Yadav M, Kumar V, Sehrawat N, editors. *Preface. Ind. Biotechnol.*, De Gruyter; 2019, p. V–VI.

106. Jamwal K, Bhattacharya S, Puri S. Plant growth regulator mediated consequences of secondary metabolites in medicinal plants. *J Appl Res Med Aromat Plants* 2018;9:26–38. https://doi.org/10.1016/j.jarmap.2017.12.003.
107. Sanchita SA. Gene Expression Analysis in Medicinal Plants under Abiotic Stress Conditions. In: *Plant Metabolites and Regulation Under Environmental Stress*, Elsevier; 2018, p. 407–414.
108. Croteau R, Kutchan TM, Lewis NG. Natural products (secondary metabolites). *Biochem Mol Biol Plants* 2000;24, 1250–1319. - References - Scientific Research Publishing n.d. https://www.scirp.org/reference/referencespapers?referenceid=1770695 (accessed January 14, 2026).
109. Pagare S, Bhatia M, Tripathi N, Bansal YK. Secondary metabolites of plants and their role: overview. *Curr Trends Biotechnol Pharm* 2015;9:293–304.
110. Mimche PN, Taramelli D, Vivas L. The plant-based immunomodulator curcumin as a potential candidate for the development of an adjunctive therapy for cerebral malaria. *Malar J* 2011;10:S10. https://doi.org/10.1186/1475-2875-10-S1-S10.
111. Alhazmi HA, Najmi A, Javed SA, Sultana S, Al Bratty M, Makeen HA, et al. Medicinal plants and isolated molecules demonstrating immunomodulation activity as potential alternative therapies for viral diseases including COVID-19. *Front Immunol* 2021;12:637553. https://doi.org/10.3389/fimmu.2021.637553.
112. Norahmad NA, Mohd Abd Razak MR, Mohmad Misnan N, Md Jelas NH, Sastu UR, Muhammad A, et al. Effect of freeze-dried Carica papaya leaf juice on inflammatory cytokines production during dengue virus infection in AG129 mice. *BMC Complement Altern Med* 2019;19:44. https://doi.org/10.1186/s12906-019-2438-3.
113. Varsha K, Sharma A, Kaur A, Madan J, Pandey RS, Jain UK, et al. Natural Plant-Derived Anticancer Drugs Nanotherapeutics: A Review on Preclinical to Clinical Success. In:Micro and Nano Technologies, *Nanostructures for Cancer Therapy*, Ficai A, Grumezescu AM, editors Elsevier; 2017, p. 775–809. https://doi.org/10.1016/B978-0-323-46144-3.00028-3.
114. Bariwal J, Kumar V, Chen H, Bhattarai RS, Peng Y, Li W, et al. Nanoparticulate delivery of potent microtubule inhibitor for metastatic melanoma treatment. *J Controlled Release* 2019;309:231–243. https://doi.org/10.1016/j.jconrel.2019.07.025.
115. Tungmunnithum D, Thongboonyou A, Pholboon A, Yangsabai A. Flavonoids and other phenolic compounds from medicinal plants for pharmaceutical and medical aspects: an overview. *Medicines* 2018;5:93. https://doi.org/10.3390/medicines5030093.
116. Li L, Jiang S, Li X, Liu Y, Su J, Chen J. Recent advances in trimethoxyphenyl (TMP) based tubulin inhibitors targeting the colchicine binding site. *Eur J Med Chem* 2018;151:482–494. https://doi.org/10.1016/j.ejmech.2018.04.011.
117. Dubey KK, Behera BK. Statistical optimization of process variables for the production of an anticancer drug (colchicine derivatives) through fermentation: at scale-up level. *New Biotechnol* 2011;28:79–85. https://doi.org/10.1016/j.nbt.2010.07.008.
118. Kozaka T, Nakagawa-Goto K, Shi Q, Lai C-Y, Hamel E, Bastow KF, et al. Antitumor agents 273. Design and synthesis of N-alkyl-thiocolchicinoids as potential antitumor agents. *Bioorg Med Chem Lett* 2010;20:4091–4094. https://doi.org/10.1016/j.bmcl.2010.05.081.

119. Hewlings S, Kalman D. Curcumin: a review of its effects on human health. *Foods* 2017;6:92. https://doi.org/10.3390/foods6100092.
120. Kunnumakkara AB, Bordoloi D, Padmavathi G, Monisha J, Roy NK, Prasad S, et al. Curcumin, the golden nutraceutical: multitargeting for multiple chronic diseases. *Br J Pharmacol* 2017;174:1325–1348. https://doi.org/10.1111/bph.13621.
121. Li W, Jiang Z, Xiao X, Wang Z, Wu Z, Ma Q, et al. Curcumin inhibits superoxide dismutase-induced epithelial-to-mesenchymal transition via the PI3K/Akt/NF-κB pathway in pancreatic cancer cells. *Int J Oncol* 2018. https://doi.org/10.3892/ijo.2018.4295.
122. Wang J-Y, Wang X, Wang X-J, Zheng B-Z, Wang Y, Wang X, et al. Curcumin inhibits the growth via Wnt/β-catenin pathway in non-small-cell lung cancer cells. *Eur Rev Med Pharmacol Sci* 2018;22:7492–7499. https://doi.org/10.26355/eurrev_201811_16290.
123. Asemi R, Omidi Najafabadi E, Baradaran A, Sharifi M, Mansournia MA, Asemi Z. Resveratrol-mediated suppression of glioblastoma and glioma: targeting apoptosis, differentiation, and stem cell dynamics. *Cancer Cell Int* 2025;25:411. https://doi.org/10.1186/s12935-025-04048-7.
124. Giordano F, Montalto FI, Panno ML, Andò S, De Amicis F. A Notch inhibitor plus resveratrol induced blockade of autophagy drives glioblastoma cell death by promoting a switch to apoptosis. *Am J Cancer Res* 2021;11:5933–5950.
125. Nowacka A, Śniegocka M, Smuczyński W, Liss S, Ziółkowska E, Bożiłow D, et al. The potential application of resveratrol and its derivatives in central nervous system tumors. *Int J Mol Sci* 2024;25:13338. https://doi.org/10.3390/ijms252413338.
126. Agarwal A, Kansal V, Farooqi H, Prasad R, Singh VK. Epigallocatechin gallate (EGCG), an active phenolic compound of green tea, inhibits tumor growth of head and neck cancer cells by targeting DNA hypermethylation. *Biomedicines* 2023;11:789. https://doi.org/10.3390/biomedicines11030789.
127. Negri A, Naponelli V, Rizzi F, Bettuzzi S. Molecular targets of epigallocatechin: gallate (EGCG): a special focus on signal transduction and cancer. *Nutrients* 2018;10:1936. https://doi.org/10.3390/nu10121936.
128. Ciesielski O, Biesiekierska M, Balcerczyk A. Epigallocatechin-3-gallate (EGCG) alters histone acetylation and methylation and impacts chromatin architecture profile in human endothelial cells. *Molecules* 2020;25:2326. https://doi.org/10.3390/molecules25102326.
129. Wang Y, Yuan X, Ren M, Wang Z. Ferroptosis: a new research direction of artemisinin and its derivatives in anti-cancer treatment. *Am J Chin Med* 2024;52:161–181. https://doi.org/10.1142/S0192415X24500071.
130. Xia Y, Shi C, Lu J, Zhu Z, Li M, Pan Y, et al. Artemisinin and its derivatives from molecular mechanisms to clinical applications: new horizons beyond antimalarials. *Int J Mol Sci* 2025;26:8409. https://doi.org/10.3390/ijms26178409.
131. Limami Y, Pinon A, Wahnou H, Oudghiri M, Liagre B, Simon A, et al. Ursolic acid's alluring journey: one triterpenoid vs. cancer hallmarks. *Molecules* 2023;28:7897. https://doi.org/10.3390/molecules28237897.
132. Zhang J, Wang W, Qian L, Zhang Q, Lai D, Qi C. Ursolic acid inhibits the proliferation of human ovarian cancer stem-like cells through epithelial-mesenchymal transition. *Oncol Rep* 2015;34:2375–2384. https://doi.org/10.3892/or.2015.4213.

133. Liao W-L, Liu Y-F, Ying T-H, Shieh J-C, Hung Y-T, Lee H-J, et al. Inhibitory effects of ursolic acid on the stemness and progression of human breast cancer cells by modulating argonaute-2. *Int J Mol Sci* 2022;24:366. https://doi.org/10.3390/ijms24010366.
134. Li W, Hua B, Saud SM, Lin H, Hou W, Matter MS, et al. Berberine regulates AMP-activated protein kinase signaling pathways and inhibits colon tumorigenesis in mice. *Mol Carcinog* 2015;54:1096–1109. https://doi.org/10.1002/mc.22179.
135. Gu S, Song X, Xie R, Ouyang C, Xie L, Li Q, et al. Berberine inhibits cancer cells growth by suppressing fatty acid synthesis and biogenesis of extracellular vesicles. *Life Sci* 2020;257:118122. https://doi.org/10.1016/j.lfs.2020.118122.
136. Sun Q, Tao Q, Ming T, Tang S, Zhao H, Liu M, et al. Berberine is a suppressor of Hedgehog signaling cascade in colorectal cancer. *Phytomedicine* 2023;114:154792. https://doi.org/10.1016/j.phymed.2023.154792.
137. Su X, Jiang X, Meng L, Dong X, Shen Y, Xin Y. Anticancer activity of sulforaphane: the epigenetic mechanisms and the Nrf2 signaling pathway. *Oxid Med Cell Longev* 2018;2018:5438179. https://doi.org/10.1155/2018/5438179.
138. Coutinho LDL, Junior TCT, Rangel MC. Sulforaphane: an emergent anti-cancer stem cell agent. *Front Oncol* 2023;13:1089115. https://doi.org/10.3389/fonc.2023.1089115.
139. Shpigel J, Luciano EF, Ukandu B, Sauane M, De La Parra C. Soy isoflavone genistein enhances tamoxifen sensitivity in breast cancer via microRNA and glucose metabolism modulation. *Int J Mol Sci* 2025;26:733. https://doi.org/10.3390/ijms26020733.
140. Ma J, Cheng L, Liu H, Zhang J, Shi Y, Zeng F, et al. Genistein down-regulates miR-223 expression in pancreatic cancer cells. *Curr Drug Targets* 2013;14:1150–1156. https://doi.org/10.2174/13894501113149990187.
141. Haolong C, Du N, Hongchao T, Yang Y, Wei Z, Hua Z, et al. Enterovirus 71 VP1 activates calmodulin-dependent protein kinase II and results in the rearrangement of vimentin in human astrocyte cells. *PLoS One* 2013;8:e73900. https://doi.org/10.1371/journal.pone.0073900.
142. Altucci L, Gronemeyer H. The promise of retinoids to fight against cancer. *Nat Rev Cancer* 2001;1:181–193. https://doi.org/10.1038/35106036.
143. Tillhon M, Guamán Ortiz LM, Lombardi P, Scovassi AI. Berberine: new perspectives for old remedies. *Biochem Pharmacol* 2012;84:1260–1267. https://doi.org/10.1016/j.bcp.2012.07.018.
144. Zou H, Ye H, Kamaraj R, Zhang T, Zhang J, Pavek P. A review on pharmacological activities and synergistic effect of quercetin with small molecule agents. *Phytomedicine* 2021;92:153736. https://doi.org/10.1016/j.phymed.2021.153736.
145. Sabaawy HE, Broggini M, Gupta SK. Editorial: Patient-derived tumor models for drug development. *Front Oncol* 2023;13:1243534. https://doi.org/10.3389/fonc.2023.1243534.
146. Yang L, Shi P, Zhao G, Xu J, Peng W, Zhang J, et al. Targeting cancer stem cell pathways for cancer therapy. *Signal Transduct Target Ther* 2020;5:8. https://doi.org/10.1038/s41392-020-0110-5.
147. Mishra A, Pathak Y, Mishra SK, Prakash H, Tripathi V. Natural compounds as a potential modifier of stem cells renewal: comparative analysis. *Eur J Pharmacol* 2023;938:175412. https://doi.org/10.1016/j.ejphar.2022.175412.

148. Opara JK, Sanchez S, Thomas SM, Anant S. Natural products targeting cancer stem cells: a promising therapeutic approach. *Explor Drug Sci* 2025;3:1008106. https://doi.org/10.37349/eds.2025.1008106.
149. Taylor WF, Jabbarzadeh E. The use of natural products to target cancer stem cells. *Am J Cancer Res* 2017;7:1588–1605.
150. Niederreither K, Subbarayan V, Dollé P, Chambon P. Embryonic retinoic acid synthesis is essential for early mouse post-implantation development. *Nat Genet* 1999;21:444–448. https://doi.org/10.1038/7788.
151. Dupé V, Ghyselinck NB, Thomazy V, Nagy L, Davies PJA, Chambon P, et al. Essential roles of retinoic acid signaling in interdigital apoptosis and control of BMP-7 expression in mouse autopods. *Dev Biol* 1999;208:30–43. https://doi.org/10.1006/dbio.1998.9176.
152. Garattini E, Gianni' M, Terao M. Cytodifferentiation by Retinoids, a Novel Therapeutic Option in Oncology: Rational Combinations with Other Therapeutic Agents. In: *Vitam. Horm.*, vol. 75, Elsevier; 2007, p. 301–354.
153. Kastner P, Mark M, Ghyselinck N, Krezel W, Dupé V, Grondona JM, et al. Genetic evidence that the retinoid signal is transduced by heterodimeric RXR/RAR functional units during mouse development. *Development* 1997;124:313–326. https://doi.org/10.1242/dev.124.2.313.
154. Benson R, Kachhwaha A, Satadeve P, Gupta A, Kumar K, Singh S, et al. Treatment outcomes in acute promyelocytic leukemia patients using a uniform ATRA+ATO consolidation therapy regimen across all risk groups: a single centre study from India. *Blood* 2024;144:1491–1491. https://doi.org/10.1182/blood-2024-210206.
155. Zhao JC, Agarwal S, Ahmad H, Amin K, Bewersdorf JP, Zeidan AM. A review of FLT3 inhibitors in acute myeloid leukemia. *Blood Rev* 2022;52:100905. https://doi.org/10.1016/j.blre.2021.100905.
156. Cerchione C, Romano A, Daver N, DiNardo C, Jabbour EJ, Konopleva M, et al. IDH1/IDH2 inhibition in acute myeloid leukemia. *Front Oncol* 2021;11:639387. https://doi.org/10.3389/fonc.2021.639387.
157. Liang C, Qiao G, Liu Y, Tian L, Hui N, Li J, et al. Overview of all-trans-retinoic acid (ATRA) and its analogues: structures, activities, and mechanisms in acute promyelocytic leukaemia. *Eur J Med Chem* 2021;220:113451. https://doi.org/10.1016/j.ejmech.2021.113451.
158. Antal CE, Hudson AM, Kang E, Zanca C, Wirth C, Stephenson NL, et al. Cancer-associated protein kinase C mutations reveal kinase's role as tumor suppressor. *Cell* 2015;160:489–502. https://doi.org/10.1016/j.cell.2015.01.001.
159. Blais JD, Filipenko V, Bi M, Harding HP, Ron D, Koumenis C, et al. Activating transcription factor 4 is translationally regulated by hypoxic stress. *Mol Cell Biol* 2004;24:7469–82. https://doi.org/10.1128/MCB.24.17.7469-7482.2004.
160. Thiery JP, Acloque H, Huang RYJ, Nieto MA. Epithelial-mesenchymal transitions in development and disease. *Cell* 2009;139:871–890. https://doi.org/10.1016/j.cell.2009.11.007.
161. Savagner P. The epithelial–mesenchymal transition (EMT) phenomenon. *Ann Oncol* 2010;21:vii89–92. https://doi.org/10.1093/annonc/mdq292.
162. Deng S, Wong CKC, Lai H-C, Wong AST. Ginsenoside-Rb1 targets chemotherapy-resistant ovarian cancer stem cells via simultaneous inhibition of Wnt/β-catenin signaling and epithelial-to-mesenchymal transition. *Oncotarget* 2017;8:25897–25914. https://doi.org/10.18632/oncotarget.13071.

163. Boutet A, De Frutos CA, Maxwell PH, Mayol MJ, Romero J, Nieto MA. Snail activation disrupts tissue homeostasis and induces fibrosis in the adult kidney. *EMBO J* 2006;25:5603–5613. https://doi.org/10.1038/sj.emboj.7601421.
164. Lovisa S, LeBleu VS, Tampe B, Sugimoto H, Vadnagara K, Carstens JL, et al. Epithelial-to-mesenchymal transition induces cell cycle arrest and parenchymal damage in renal fibrosis. *Nat Med* 2015;21:998–1009. https://doi.org/10.1038/nm.3902.
165. Cho HS, Kim JH, Jang HN, Lee TW, Jung MH, Kim TH, et al. Alpha-lipoic acid ameliorates the epithelial mesenchymal transition induced by unilateral ureteral obstruction in mice. *Sci Rep* 2017;7:46065. https://doi.org/10.1038/srep46065.
166. Newman DJ, Cragg GM. Natural products as sources of new drugs over the last 25 years. *J Nat Prod* 2007;70:461–477. https://doi.org/10.1021/np068054v.
167. Khan T, Gurav P. Phytonanotechnology: enhancing delivery of plant based anti-cancer drugs. *Front Pharmacol* 2018;8:1002. https://doi.org/10.3389/fphar.2017.01002.
168. Boldbaatar A, Lee S, Han S, Jeong A, Ka H, Buyanravjikh S, et al. Eupatolide inhibits the TGF-β1-induced migration of breast cancer cells via downregulation of SMAD3 phosphorylation and transcriptional repression of ALK5. *Oncol Lett* 2017. https://doi.org/10.3892/ol.2017.6957.
169. Xu Y, Lou Z, Lee S-H. Arctigenin represses TGF-β-induced epithelial mesenchymal transition in human lung cancer cells. *Biochem Biophys Res Commun* 2017;493:934–939. https://doi.org/10.1016/j.bbrc.2017.09.117.
170. Chung H, Choi HS, Seo E-K, Kang D-H, Oh E-S. Baicalin and baicalein inhibit transforming growth factor-β1-mediated epithelial-mesenchymal transition in human breast epithelial cells. *Biochem Biophys Res Commun* 2015;458:707–713. https://doi.org/10.1016/j.bbrc.2015.02.032.
171. Kou Y, Li L, Li H, Tan Y, Li B, Wang K, et al. Berberine suppressed epithelial mesenchymal transition through cross-talk regulation of PI3K/AKT and RARα/RARβ in melanoma cells. *Biochem Biophys Res Commun* 2016;479:290–296. https://doi.org/10.1016/j.bbrc.2016.09.061.
172. Sutariya B, Saraf M. Betanin, isolated from fruits of Opuntia elatior Mill attenuates renal fibrosis in diabetic rats through regulating oxidative stress and TGF-β pathway. *J Ethnopharmacol* 2017;198:432–443. https://doi.org/10.1016/j.jep.2016.12.048.
173. Zhou J, Zhu Y-F, Chen X-Y, Han B, Li F, Chen J-Y, et al. Black rice-derived anthocyanins inhibit HER-2-positive breast cancer epithelial-mesenchymal transition-mediated metastasis in vitro by suppressing FAK signaling. *Int J Mol Med* 2017. https://doi.org/10.3892/ijmm.2017.3183.
174. Vergara D, Simeone P, Bettini S, Tinelli A, Valli L, Storelli C, et al. Antitumor activity of the dietary diterpene carnosol against a panel of human cancer cell lines. *Food Funct* 2014;5:1261. https://doi.org/10.1039/c4fo00023d.
175. Chen C, Ma T, Zhang C, Zhang H, Bai L, Kong L, et al. Down-regulation of aquaporin 5-mediated epithelial-mesenchymal transition and anti-metastatic effect by natural product Cairicoside E in colorectal cancer. *Mol Carcinog* 2017;56:2692–2705. https://doi.org/10.1002/mc.22712.

176. Lin C-Y, Hsieh Y-H, Yang S-F, Chu S-C, Chen P-N, Hsieh Y-S. Cinnamomum cassia extracts reverses TGF-β1-induced epithelial-mesenchymal transition in human lung adenocarcinoma cells and suppresses tumor growth in vivo. *Environ Toxicol* 2017;32:1878–1887. https://doi.org/10.1002/tox.22410.
177. Avila-Carrasco L, Majano P, Sánchez-Toméro JA, Selgas R, López-Cabrera M, Aguilera A, et al. Natural plants compounds as modulators of epithelial-to-mesenchymal transition. *Front Pharmacol* 2019;10:715. https://doi.org/10.3389/fphar.2019.00715.
178. Wang W, Zhou P-H, Hu W, Xu C-G, Zhou X-J, Liang C-Z, et al. Cryptotanshinone hinders renal fibrosis and epithelial transdifferentiation in obstructive nephropathy by inhibiting TGF-β1/Smad3/integrin β1 signal. *Oncotarget* 2018;9:26625–26637. https://doi.org/10.18632/oncotarget.23803.
179. Akhtar N, Syed DN, Lall RK, Mirza B, Mukhtar H. Targeting epithelial to mesenchymal transition in prostate cancer by a novel compound, plectranthoic acid, isolated from Ficus microcarpa. *Mol Carcinog* 2018;57:653–663. https://doi.org/10.1002/mc.22790.
180. Petpiroon N, Sritularak B, Chanvorachote P. Phoyunnanin E inhibits migration of non-small cell lung cancer cells via suppression of epithelial-to-mesenchymal transition and integrin αv and integrin β3. *BMC Complement Altern Med* 2017;17:553. https://doi.org/10.1186/s12906-017-2059-7.
181. Colobatiu L, Gavrilas L, Mocan A. Natural Compounds as Chemosensitizers: A Lesson from Plants. In: *In Cancer Sensitizing Agents for Chemotherapy, pH-Interfering Agents as Chemosensitizers in Cancer Therapy*, Supuran CT, Carradori S, editors, Elsevier; 2021, p. 147–165. https://doi.org/10.1016/B978-0-12-820701-7.00009-9
182. Duan C, Yu M, Xu J, Li B-Y, Zhao Y, Kankala RK. Overcoming cancer multi-drug resistance (MDR): reasons, mechanisms, nanotherapeutic solutions, and challenges. *Biomed Pharmacother* 2023;162:114643. https://doi.org/10.1016/j.biopha.2023.114643.
183. Emran TB, Shahriar A, Mahmud AR, Rahman T, Abir MH, Siddiquee Mohd F-R, et al. Multidrug resistance in cancer: understanding molecular mechanisms, immunoprevention and therapeutic approaches. *Front Oncol* 2022;12:891652. https://doi.org/10.3389/fonc.2022.891652.
184. Choi Y, Yu A-M. ABC transporters in multidrug resistance and pharmacokinetics, and strategies for drug development. *Curr Pharm Des* 2014;20:793–807. https://doi.org/10.2174/138161282005140214165212.
185. Tian Y, Lei Y, Wang Y, Lai J, Wang J, Xia F. Mechanism of multidrug resistance to chemotherapy mediated by P-glycoprotein (review). *Int J Oncol* 2023;63:119. https://doi.org/10.3892/ijo.2023.5567.
186. Robinson K, Tiriveedhi V. Perplexing role of P-glycoprotein in tumor microenvironment. *Front Oncol* 2020;10:265. https://doi.org/10.3389/fonc.2020.00265.
187. Tia ST, Luo M, Fan W. Mapping the role of P-gp in multidrug resistance: insights from recent structural studies. *Int J Mol Sci* 2025;26:4179. https://doi.org/10.3390/ijms26094179.
188. Yang X, Li Z, Wang N, Li L, Song L, He T, et al. Curcumin-encapsulated polymeric micelles suppress the development of colon cancer in vitro and in vivo. *Sci Rep* 2015;5:10322. https://doi.org/10.1038/srep10322.

189. Allis CD, Jenuwein T. The molecular hallmarks of epigenetic control. *Nat Rev Genet* 2016;17:487–500. https://doi.org/10.1038/nrg.2016.59.
190. Li Y, Seto E. HDACs and HDAC inhibitors in cancer development and therapy. *Cold Spring Harb Perspect Med* 2016;6:a026831. https://doi.org/10.1101/cshperspect.a026831.
191. Duraisingh MT, Skillman KM. Epigenetic variation and regulation in malaria parasites. *Annu Rev Microbiol* 2018;72:355–375. https://doi.org/10.1146/annurev-micro-090817-062722.
192. Ferioli M, Zauli G, Maiorano P, Milani D, Mirandola P, Neri LM. Role of physical exercise in the regulation of epigenetic mechanisms in inflammation, cancer, neurodegenerative diseases, and aging process. *J Cell Physiol* 2019;234:14852–14864. https://doi.org/10.1002/jcp.28304.
193. Esteller M. Epigenetics in cancer. *N Engl J Med* 2008;358:1148–1159. https://doi.org/10.1056/NEJMra072067.
194. Jones PA, Baylin SB. The epigenomics of cancer. *Cell* 2007;128:683–692. https://doi.org/10.1016/j.cell.2007.01.029.
195. Lascano S, Lopez M, Arimondo PB. Natural products and chemical biology tools: alternatives to target epigenetic mechanisms in cancers. *Chem Rec* 2018;18:1854–1876. https://doi.org/10.1002/tcr.201800133.
196. Deng X, Qiu Q, He K, Cao X. The seekers: how epigenetic modifying enzymes find their hidden genomic targets in Arabidopsis. *Curr Opin Plant Biol* 2018;45:75–81. https://doi.org/10.1016/j.pbi.2018.05.006.
197. Dawson MA. The cancer epigenome: concepts, challenges, and therapeutic opportunities. *Science* 2017;355:1147–1152. https://doi.org/10.1126/science.aam7304.
198. Mehndiratta S, Lin M-H, Wu Y-W, Chen C-H, Wu T-Y, Chuang K-H, et al. N-alkyl-hydroxybenzoyl anilide hydroxamates as dual inhibitors of HDAC and HSP90, downregulating IFN-γ induced PD-L1 expression. *Eur J Med Chem* 2020;185:111725. https://doi.org/10.1016/j.ejmech.2019.111725.
199. Nguyen DD, Wu C-H, Moree WJ, Lamsa A, Medema MH, Zhao X, et al. MS/MS networking guided analysis of molecule and gene cluster families. *Proc Natl Acad Sci* 2013;110. https://doi.org/10.1073/pnas.1303471110.
200. Suraweera A, O'Byrne KJ, Richard DJ. Combination therapy with histone deacetylase inhibitors (HDACi) for the treatment of cancer: achieving the full therapeutic potential of HDACi. *Front Oncol* 2018;8:92. https://doi.org/10.3389/fonc.2018.00092.
201. Herrmann J, Fayad AA, Müller R. Natural products from myxobacteria: novel metabolites and bioactivities. *Nat Prod Rep* 2017;34:135–160. https://doi.org/10.1039/C6NP00106H.
202. Chen Y, Garcia De Lomana M, Friedrich N-O, Kirchmair J. Characterization of the chemical space of known and readily obtainable natural products. *J Chem Inf Model* 2018;58:1518–1532. https://doi.org/10.1021/acs.jcim.8b00302.
203. Dias DA, Urban S, Roessner U. A historical overview of natural products in drug discovery. *Metabolites* 2012;2:303–336. https://doi.org/10.3390/metabo2020303.
204. Anand P, Kunnumakkara AB, Newman RA, Aggarwal BB. Bioavailability of curcumin: problems and promises. *Mol Pharm* 2007;4:807–818. https://doi.org/10.1021/mp700113r.

205. Williams HD, Trevaskis NL, Charman SA, Shanker RM, Charman WN, Pouton CW, et al. Strategies to address low drug solubility in discovery and development. *Pharmacol Rev* 2013;65:315–499. https://doi.org/10.1124/pr.112.005660.
206. Zhang C, Sheng J, Li G, Zhao L, Wang Y, Yang W, et al. Effects of berberine and its derivatives on cancer: a systems pharmacology review. *Front Pharmacol* 2020;10:1461. https://doi.org/10.3389/fphar.2019.01461.
207. Blanco E, Shen H, Ferrari M. Principles of nanoparticle design for overcoming biological barriers to drug delivery. *Nat Biotechnol* 2015;33:941–951. https://doi.org/10.1038/nbt.3330.
208. Sindhwani S, Syed AM, Ngai J, Kingston BR, Maiorino L, Rothschild J, et al. The entry of nanoparticles into solid tumours. *Nat Mater* 2020;19:566–575. https://doi.org/10.1038/s41563-019-0566-2.
209. Chen X-Y, Wang J-Q, Yang Y, Li J, Chen Z-S. Natural product as substrates of ABC transporters: a review. *Recent Patents Anticancer Drug Discov* 2021;16:222–238. https://doi.org/10.2174/1574892816666210218220943.
210. Salehi B, Sharifi-Rad J, Cappellini F, Reiner Ž, Zorzan D, Imran M, et al. The therapeutic potential of anthocyanins: current approaches based on their molecular mechanism of action. *Front Pharmacol* 2020;11:1300. https://doi.org/10.3389/fphar.2020.01300.
211. Atanasov AG, Zotchev SB, Dirsch VM, the International Natural Product Sciences Taskforce, Orhan IE, Banach M, et al. Natural products in drug discovery: advances and opportunities. *Nat Rev Drug Discov* 2021;20:200–216. https://doi.org/10.1038/s41573-020-00114-z.
212. Cragg GM, Newman DJ. Natural products: a continuing source of novel drug leads. *Biochim Biophys Acta BBA Gen Subj* 2013;1830:3670–3695. https://doi.org/10.1016/j.bbagen.2013.02.008.
213. Khan YY, Suvarna V. Liposomes containing phytochemicals for cancer treatment: an update. *Int J Curr Pharm Res* 2016;9:20. https://doi.org/10.22159/ijcpr.2017v9i1.16629.
214. Parvin N, Aslam M, Joo SW, Mandal TK. Nano-phytomedicine: harnessing plant-derived phytochemicals in nanocarriers for targeted human health applications. *Molecules* 2025;30:3177. https://doi.org/10.3390/molecules30153177.
215. Huh KM, Lee SC, Cho YW, Lee J, Jeong JH, Park K. Hydrotropic polymer micelle system for delivery of paclitaxel. *J Controlled Release* 2005;101:59–68. https://doi.org/10.1016/j.jconrel.2004.07.003.
216. Li Y, Shi S, Ming Y, Wang L, Li C, Luo M, et al. Specific cancer stem cell-therapy by albumin nanoparticles functionalized with CD44-mediated targeting. *J Nanobiotechnol* 2018;16:99. https://doi.org/10.1186/s12951-018-0424-4.
217. Monteiro CJ, Monteiro PF, Travanut A, Gulfam M, Heery DM, Grabowska A, et al. Redox-responsive micellar-like nanoparticles can overcome intrinsic multi-drug resistance in tumour spheroids of triple negative breast cancer. *RSC Pharm* 2025;2:644–656. https://doi.org/10.1039/D4PM00336E.
218. Yadav P, Ambudkar SV, Rajendra Prasad N. Emerging nanotechnology-based therapeutics to combat multidrug-resistant cancer. *J Nanobiotechnol* 2022;20:423. https://doi.org/10.1186/s12951-022-01626-z.

219. Bai X, Smith Z, Wang Y, Butterworth S, Tirella A. Sustained drug release from smart nanoparticles in cancer therapy: a comprehensive review. *Micromachines* 2022;13:1623. https://doi.org/10.3390/mi13101623.
220. Beach MA, Nayanathara U, Gao Y, Zhang C, Xiong Y, Wang Y, et al. Polymeric nanoparticles for drug delivery. *Chem Rev* 2024;124:5505–5616. https://doi.org/10.1021/acs.chemrev.3c00705.
221. Shukla S, Zaher H, Hartz A, Bauer B, Ware JA, Ambudkar SV. Curcumin inhibits the activity of ABCG2/BCRP1, a multidrug resistance-linked ABC drug transporter in mice. *Pharm Res* 2009;26:480–487. https://doi.org/10.1007/s11095-008-9735-8.
222. Annaji M, Poudel I, Boddu SHS, Arnold RD, Tiwari AK, Babu RJ. Resveratrol-loaded nanomedicines for cancer applications. *Cancer Rep* 2021;4:e1353. https://doi.org/10.1002/cnr2.1353.
223. Hedayati N, Safari MH, Milasi YE, Kahkesh S, Farahani N, Khoshnazar SM, et al. Modulation of the PI3K/Akt signaling pathway by resveratrol in cancer: molecular mechanisms and therapeutic opportunity. *Discov Oncol* 2025;16:669. https://doi.org/10.1007/s12672-025-02471-w.
224. Zhou Y, Gong J, Deng X, Shen L, Wu S, Fan H, et al. Curcumin and nanodelivery systems: New directions for targeted therapy and diagnosis of breast cancer. *Biomed Pharmacother* 2024;180:117404. https://doi.org/10.1016/j.biopha.2024.117404.
225. Aborig M, Alsefaou M, Osei E, Wettig S. Engineered dual-functional gold nanoparticles enhance radiosensitization in prostate cancer cells: synergistic action of curcumin and gold. *Nanoscale Adv* 2025;7:5964–5977. https://doi.org/10.1039/D5NA00376H.
226. Wahnou H, El Kebbaj R, Liagre B, Sol V, Limami Y, Duval RE. Curcumin-based nanoparticles: advancements and challenges in tumor therapy. *Pharmaceutics* 2025;17:114. https://doi.org/10.3390/pharmaceutics17010114.
227. Al Jayoush AR, Haider M, Khan SA, Hussain Z. Hyaluronic acid-functionalized nanomedicines for CD44-receptors-mediated targeted cancer therapy: a review of selective targetability and biodistribution to tumor microenvironment. *Int J Biol Macromol* 2025;308:142486. https://doi.org/10.1016/j.ijbiomac.2025.142486.
228. Hong L, Li W, Li Y, Yin S. Nanoparticle-based drug delivery systems targeting cancer cell surfaces. *RSC Adv* 2023;13:21365–21382. https://doi.org/10.1039/D3RA02969G.
229. Sun S, Yang Q, Jiang D, Zhang Y. Nanobiotechnology augmented cancer stem cell guided management of cancer: liquid-biopsy, imaging, and treatment. *J Nanobiotechnol* 2024;22:176. https://doi.org/10.1186/s12951-024-02432-5.
230. Wickens JM, Alsaab HO, Kesharwani P, Bhise K, Amin MCIM, Tekade RK, et al. Recent advances in hyaluronic acid-decorated nanocarriers for targeted cancer therapy. *Drug Discov Today* 2017;22:665–680. https://doi.org/10.1016/j.drudis.2016.12.009.
231. Mitranovici M-I, Caravia LG, Moraru L, Puşcaşiu L. Targeting cancer stemness using nanotechnology in a holistic approach: a narrative review. *Pharmaceutics* 2025;17:277. https://doi.org/10.3390/pharmaceutics17030277.
232. Lu B, Huang X, Mo J, Zhao W. Drug delivery using nanoparticles for cancer stem-like cell targeting. *Front Pharmacol* 2016;7. https://doi.org/10.3389/fphar.2016.00084.

233. Xiong K, Zhang Y, Wen Q, Luo J, Lu Y, Wu Z, et al. Co-delivery of paclitaxel and curcumin by biodegradable polymeric nanoparticles for breast cancer chemotherapy. *Int J Pharm* 2020;589:119875. https://doi.org/10.1016/j.ijpharm.2020.119875.
234. Li M, Fang G, Zahid F, Saleem R, Ishrat G, Ali Z, et al. RETRACTED: Co-delivery of paclitaxel and curcumin loaded solid lipid nanoparticles for improved targeting of lung cancer: in vitro and in vivo investigation. *Heliyon* 2024;10:e30290. https://doi.org/10.1016/j.heliyon.2024.e30290.
235. Saharkhiz S, Zarepour A, Nasri N, Cordani M, Zarrabi A. A comparison study between doxorubicin and curcumin co-administration and co-loading in a smart niosomal formulation for MCF-7 breast cancer therapy. *Eur J Pharm Sci* 2023;191:106600. https://doi.org/10.1016/j.ejps.2023.106600.
236. Ashrafizadeh M, Zarrabi A, Hushmandi K, Hashemi F, Rahmani Moghadam E, Raei M, et al. Progress in natural compounds/siRNA co-delivery employing nanovehicles for cancer therapy. *ACS Comb Sci* 2020;22:669–700. https://doi.org/10.1021/acscombsci.0c00099.
237. Ghaffari M, Dehghan G, Baradaran B, Zarebkohan A, Mansoori B, Soleymani J, et al. Co-delivery of curcumin and Bcl-2 siRNA by PAMAM dendrimers for enhancement of the therapeutic efficacy in HeLa cancer cells. *Colloids Surf B Biointerfaces* 2020;188:110762. https://doi.org/10.1016/j.colsurfb.2019.110762.
238. Xing Z-H, Wei J-H, Cheang T-Y, Wang Z-R, Zhou X, Wang S-S, et al. Bifunctional pH-sensitive Zn(II)–curcumin nanoparticles/siRNA effectively inhibit growth of human bladder cancer cells in vitro and in vivo. *J Mater Chem B* 2014;2:2714. https://doi.org/10.1039/c3tb21625j.
239. Ghosh R, Biswas S, Samanta P, Sarkar R, Pakhira S, Mondal M, et al. Targeting cancer stem cells by natural compounds in combination with conventional chemotherapy: an ideal strategy for combating cancer recurrence. *J Curr Oncol Trends* 2024;1:14–21. https://doi.org/10.4103/JCOT.JCOT_9_23.
240. Obisi JN, Abimbola ANJ, Babaleye OA, Atidoglo PK, Usin SG, Nwanaforo EO, et al. Unveiling the future of cancer stem cell therapy: a narrative exploration of emerging innovations. *Discov Oncol* 2025;16:373. https://doi.org/10.1007/s12672-025-02102-4.
241. Chen Z, Liu M. Natural compounds in cancer therapy: revealing the role of flavonoids in renal cell carcinoma treatment. *Biomolecules* 2025;15:620. https://doi.org/10.3390/biom15050620.
242. Yang L-Y, Lei S-Z, Xu W-J, Lai Y-X, Zhang Y-Y, Wang Y, et al. Rising above: exploring the therapeutic potential of natural product-based compounds in human cancer treatment. *Tradit Med Res* 2025;10:18. https://doi.org/10.53388/TMR20240618001.
243. Andreani T, Cheng R, Elbadri K, Ferro C, Menezes T, Dos Santos MR, et al. Natural compounds-based nanomedicines for cancer treatment: future directions and challenges. *Drug Deliv Transl Res* 2024;14:2845–2916. https://doi.org/10.1007/s13346-024-01649-z.

Chapter 8

Computational approaches in medicinal compound discovery

From *in silico* screening to ADMET prediction

Anoop Kumar[1], *Roma Pandey*[2], *Praveen Rai*[3], *and Gauri Misra*[1]

[1]Molecular Diagnostic Laboratory, National Institute of Biologicals, Noida, Uttar Pradesh, India

[2]Department of Biotechnology, IILM University, Greater Noida, India

[3]Department of Infectious Diseases and Microbial Genomics, Nitte University Centre for Science Education and Research (NUCSER), Paneer Campus, Deralakatte, Mangaluru, Karnataka, India

8.1 INTRODUCTION

Drug discovery is a rigorous scientific process that identifies, develops, and manufactures new drugs to cure, mitigate, or prevent diseases and health problems. One term used to describe drug development is a "pipeline." According to Wagner et al., this implies that basic and clinical research leads directly from hit/lead to a candidate and, eventually, to a drug [1]. The integration of computational methods into traditional drug development procedures has resulted in a significant shift in medicinal chemistry. Traditionally, drug discovery took a long time and cost a lot of money. It often relied on luck, extensive trial and error, and the high-throughput screening (HTS) of large chemical libraries, which usually meant that many drugs failed to make it to the market [2,3]. At present, computational methodologies have revolutionized early-stage discovery by enabling the virtual screening of millions of compounds, thereby allowing the prioritization of promising candidates based on projected efficacy and safety prior to any wet-lab synthesis or testing [4,5].

In silico methods like molecular docking, virtual screening, and quantitative structure–activity relationship (QSAR) modeling are essential to these advances [6,7]. Structure-based virtual screening (SBVS) employs resolved three-dimensional protein structures to predict the binding affinity and

DOI: 10.1201/9781003733010-8

orientation of potential ligands, thereby narrowing the chemical universe to a manageable array of viable candidates [8,9]. The identification and optimization of lead compounds without prior knowledge of target structures are made possible by ligand-based methodologies, such as QSAR, which use statistical and machine learning (ML) techniques to elucidate correlations between molecular descriptors and biological activity [4].

However, unfavorable pharmacokinetic or toxicity profiles (ADMET: absorption, distribution, metabolism, excretion, and toxicity) may cause even highly effective and selective medications found through virtual screening to fail preclinical or clinical trials [10,11]. Predicting ADMET characteristics early on is crucial to reducing the risk of drug attrition later [12]. To quickly and accurately evaluate drug-likeness, oral bioavailability, metabolic stability, and toxicity in silico, contemporary computational platforms such as SwissADME, pkCSM, and admetSAR use ML and large bioactivity datasets [10,11,13].

The various computational techniques used in drug development, including docking, virtual screening, molecular simulation, QSAR modeling, and ADMET prediction, are explained in this chapter. It demonstrates how these techniques can be combined to expedite the search for new drugs.

8.2 TRADITIONAL VS. COMPUTATIONAL APPROACHES FOR DRUG DISCOVERY

Traditionally, the majority of drug development relied on trial-and-error methodologies to assess compounds for their therapeutic efficacy [14]. These traditional methods are vital, but they have a high attrition rate and take a long time to scale. Many candidates fail during preclinical or clinical screening, and it might take years and significant financial and technological investment to identify the first chemical and get it licensed for use as a medicine [15]. Furthermore, traditional techniques result in a limited understanding of complex biological networks and disease etiology, thereby obstructing the development of coherent and targeted therapy strategies [16].

In silico screening approaches have transformed early-stage drug discovery by utilizing computational methods to search through vast chemical libraries and identify bioactive molecules targeting specific therapeutic indications. SBVS, ligand-based virtual screening (LBVS), molecular docking, and pharmacophore modeling are all significant techniques used to identify and improve hits. SBVS leverages 3D protein structures from crystallography or prediction models like AlphaFold2, as well as software like AutoDock, GOLD, and Glide, to identify optimal binders. Recent advancements in molecular dynamics facilitate the resolution of protein flexibility issues. LBVS is used when structural data is unavailable. It evaluates structural similarity and shared pharmacophoric qualities among known active substances. Molecular docking combines both rigid and flexible docking methods to predict ligands-receptor binding modes and binding affinities. Pharmacophore modeling identifies key steric and electronic features required for molecular recognition. It also combines data from both ligand- and structure-based sources to screen multiple targets. Generally, these computer programs speed up, lower the cost of, and make the drug discovery process more efficient than traditional experimental methods [9,17–19]. Figure 8.1 shows the dates, steps, and analytical parts of traditional drug development next to modern computational methods.

Deep learning (DL) techniques and artificial intelligence (AI) have greatly benefited this field. We can find complex, nonlinear correlations in large chemical and biological datasets by using graph-based methods and deep neural networks. This enables the optimization of leads and the generation of novel compounds across multiple parameters [20,21]. Computing is improving due to the expansion of public chemical databases such as ChEMBL, ZINC, and PubChem [22–24]. These databases include the biological and chemical information required for model testing and training. Despite these improvements, there are challenges. The quality and variety of the data used determine how accurate computer forecasts are. On the other hand, the full in silico simulation of pharmacological activity is difficult due to the complexity of human biology [25]. We must continue to advance computer power, algorithms, and data handling to maximize the benefits of computer-aided drug development.

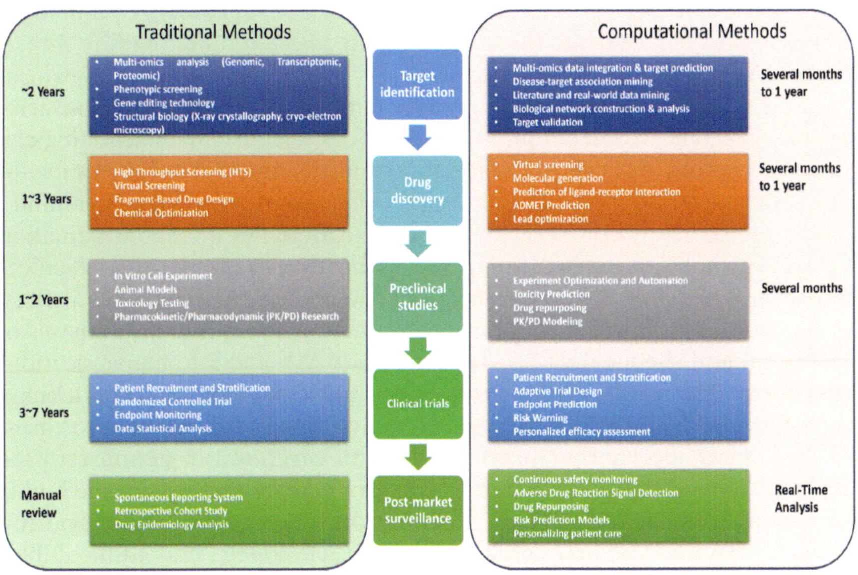

Figure 8.1 Comparison of traditional and computational approaches in the drug development pipeline.

Through the ability to quickly analyze vast chemical libraries, HTS accelerated the process of identifying hits and optimizing leads. This changed the early stages of the drug discovery process [26]. Our knowledge of biological pathways, disease variability, and molecular interactions has improved as a result of the concurrent development of genomics, proteomics, metabolomics, and other omics disciplines [27]. In addition to these advancements, computer technologies like molecular modeling, structure-based design, and in silico simulations have proved highly useful for improving chemical scaffolds, predicting the behavior of compounds, and simplifying investigations [28].

8.3 IMPORTANCE OF COMPUTATIONAL DRUG DISCOVERY

A thorough understanding of chemistry, biology, pharmacology, computer science, and clinical research is necessary for the creation, discovery, and development of pharmaceuticals. People note how expensive and time-consuming this process is. As of 2015, it can cost over $58.8 billion and take 10–15 years to bring a single medication to market [29,30]. Typically, only 200–250 of the first 10,000 chemical compounds are evaluated in humans. Only around ten of these move on to more advanced in vivo research in humans. According to an important study by the Tufts Center for the Study of Drug Development (1995–2007), only 11.83% of pharmaceutical substances that undergo Phase I clinical trials get approved by regulators. These figures demonstrate the dangers, high costs, and high failure rate of the traditional drug-making process.

These challenges have prompted scientists to search for innovative approaches. One excellent option that provides scientists with tools to assist them in locating the best candidates more quickly and reduce the number of development failures is computer-aided drug design, or CADD. Computer modeling, structural biology, and predictive analytics are all combined in CADD to enable scientists to more strategically explore the vast chemical universe. This expedites, concentrates, and reduces the expense of drug research. By enabling the automatic examination of enormous databases of biological and chemical data, AI, particularly ML and DL, has revolutionized the pharmaceutical industry. Millions of compounds can be quickly sorted using AI-powered virtual screening to identify the ones that are most likely to meet specific therapeutic goals. This can result in significant time and cost savings when compared to traditional testing [31]. ADME/T profiles, which include absorption, distribution, metabolism, excretion, and toxicity, may now be precisely predicted by AI models. According to Lim et al. [32], this enables researchers to select candidates with favorable pharmacokinetic and pharmacodynamic properties.

8.4 IN SILICO SCREENING TECHNIQUES

8.4.1 Structure-based virtual screening (SBVS)

Finding the most effective manner for ligands to attach to a molecular target in order to form a complex is the goal of SBVS, or target-based virtual screening (TBVS). Thus, the ligands are ranked according to the strength of their predicted binding affinity, with the stronger ones at the top. For in silico prediction of interactions between the target and each chemical molecule, SBVS approaches require prior knowledge of the three-dimensional structure of the target protein [33]. Using this method, molecules are selected from a database and arranged according to how well they fit into the receptor site.

SBVS relies heavily on molecular docking. It has a reputation for being reasonably priced and producing favorable results [34]. During the 1980s, Kuntz et al. [35] developed and evaluated a number of methods for examining the geometrical alignments that could exist between a ligand and its target. Although it first seemed like a great idea, it didn't gain traction until the 1990s, when advancements in computing power, algorithms, and the ability to obtain structural information on target molecules became available. In an SBVS implementation, the chemicals that are evaluated are arranged according to their binding affinity within the receptor site. These ligands are more likely to interact pharmacologically with the molecular target. To determine a binding site's viability, scientists employ scoring functions. These functions show the degree to which the ligand and target are compatible. According to Leelananda and Lindert [36], the docking strategy necessitates an efficient grading system.

Although SBVS has many advantages, it also has several significant limitations. When examining millions of small molecules, one of its best features is that it saves a significant amount of time and money. Compared to other methods of conducting research, this makes it far more efficient. SBVS also makes it more straightforward to examine substances that have not yet been synthesized. This makes it possible for computers to examine them before they are chemically produced. Modern software tools and algorithms greatly increase its usefulness for drug discovery. SBVS, however, still faces notable limitations. According to Lionta et al. [37], many strategies are less useful in larger or more complex systems since they are only effective in specific contexts. Furthermore, it is challenging to distinguish between active and inactive molecules and to predict the actual binding configuration due to the complexity of ligand-receptor interactions. Furthermore SBVS can produce false positives and false negatives, which complicates subsequent testing.

8.4.2 Ligand-based virtual screening (LBVS)

A ligand-centric computational paradigm enables ligand-oriented pharmaceutical development. The idea of chemical similarity suggests that structurally similar molecules will bind to the same targets, have similar binding affinities, and elicit comparable biological responses [38]. The notion of similarity has been widely employed in the identification and augmentation of leads, especially owing to its incorporation of the polypharmacological attributes of medications [39]. Additionally, it is crucial to quickly analyze both primary and secondary targets while ensuring that the target families are carefully selected. This strategy relies on the interaction between particular protein targets and ligands that possess similar chemical structures, properties, and pharmacophoric characteristics to facilitate the prediction of biological targets. To do this, chemogenomic databases containing biologically characterized compounds with known targets are combined with the structures of phenotypic chemical modulators, such as natural products or orphan compounds [40]. In this context, cheminformatics and bioinformatics approaches employ mapping models—including topology-based, Bayesian classification models, and atom pair models—that apply ML and statistical methods to bioactivity data [41]. These models are used to map compounds into the bioactivity database or chemogenomic space. When generating target predictions, they use chemical fingerprints and 2D or 3D molecular descriptors to evaluate structural similarity. Chemical fingerprints can be used to develop models that predict interactions between known proteins and compounds with orphan molecules. This is predicated on the idea that chemically related compounds exhibit similar affinity-based chemical fingerprints [40]. Molecular descriptors are numerical representations of a compound's molecular properties, while chemical fingerprints are high-dimensional vectors that indicate the presence of sub-structural components. Furthermore, ligand-based methods help elucidate the relationships between physically distinct yet functionally similar proteins based on ligand similarity. This helps generate a hypothesis that we can test with numbers. The ligand-based technique gives pharmacology, medicinal chemistry, and biochemistry more usable information, similar to biospectra [41,42]. Researchers can use the Minkowski distance metrics, the Tanimoto similarity coefficient, and the Soergel distance to see how alike two structures are. The Tanimoto similarity coefficient is also applicable to 3D structures; nevertheless, it is influenced by molecular size as it disregards minor properties of larger molecules, resulting in a contradicting size dependency. Scientists have developed tools like the Polypharmacology Browser [43], SpiDER [44], Swiss Target Prediction (STP) [45], HitPick [46], Similarity Ensemble Approach (SEA) [41], SuperPred [47], Prediction of Activity Spectra for Biologically Active Substances (PASS) [48], the Candidate Ligand Identification Program (CLIP) [49], the MOst-Similar ligand-based Target inference approach (MOST), and the Chemical Similarity Network Analysis Pulldown (CSNAP) [50] by studying how compounds act. The ligand-based target prediction method does not work for predicting targets that have few or no bioactive ligands, or for ligands that have activity cliffs, which are defined as two ligands that have similar structures but different activity levels.

8.4.3 Molecular docking

Molecular docking is a fundamental computational technique that is employed in drug development today. It helps scientists determine how ligands will interact with targets and is crucial for virtual screening, hit detection, and lead optimization. Following the Human Genome Project, more potential therapeutic targets emerged [51]. At the same time, advances in X-ray crystallography, NMR, and high-throughput protein purification led to the creation of enormous structural datasets that aid in computer-aided drug design [52,53]. Consequently, docking has become a significant component of pharmaceutical research [5,54,55]. It evolved from the classical lock-and-key paradigm, which focused on geometric complementarity [56], to the induced-fit model, which allows for receptor flexibility [57]. Docking uses scoring algorithms that evaluate intermolecular energetics to predict the best binding poses and affinities [58]. Methods often include simulation-based and shape-complementarity approaches, along with search algorithms that can explore large conformational spaces and sometimes need as many as 10^7 trials [59]. Docking can be rigid [60], flexible [61,62], induced-fit [63], ligand-based [64,65], protein–protein based [64], or blind for unidentified binding sites [66], depending on the goals of the research. In a typical workflow, protein structures or validated homology models are carefully prepared [67], ligands are sorted based on their physicochemical and ADME-Tox properties [68], and pose generation is executed using systematic, stochastic, or hybrid algorithms [69]. To evaluate a substance, it is necessary to compare hydrogen bonding, charge complementarity, and hydrophobic interactions with established structural data [70]. AutoDock, DOCK, Glide, GEMDOCK, FlexX, SwissDock, and other programs are commonly used [59]. Docking assists scientists in discovering novel pharmaceuticals by enabling rapid screening of virtual compounds, enhancing binding affinity and specificity, predicting ADMET profiles, identifying targets, elucidating structures, assessing nutraceuticals, and simulating bioremediation [71–73]. Molecular docking is becoming increasingly dependable and useful for accelerating in silico drug discovery, thanks to ongoing advances in structural biology and computer science. There are still some challenges, though, such as inconsistent accuracy, scoring functions that lack precision, protein flexibility, and false positives and negatives. This accuracy varies from 0% to 92% [74].

Molecular docking is a useful computational method that demonstrates how small molecules bind to proteins. When experimental methods like X-ray crystallography or NMR are too challenging, take too long, or lack adequate resources, docking serves as a viable alternative. Docking models give medicinal chemists important structural information, highlight critical contact points, and facilitate systematic drug design. However, ligand binding is far more complicated than a simple lock-and-key interaction. The binding outcome depends on the flexibility of both the ligand and protein, the kinetics of the induced-fit effect, the dynamic conformational changes of side chains over time, and the complex energy environment of the binding site. Standard docking approaches often utilize simple scoring functions and rigid or semi-rigid models, which fail to capture these details [75]. The protein's microenvironment significantly influences ligand behavior. Local electrostatics can change how the charges on ligands are spread out, and the dynamics of structured water molecules along with solvation effects significantly influence the binding process. Standard docking methods frequently neglect the essential functions of water molecules and entropic factors. Docking is an important part of current drug research; however, it has problems such as sampling challenges, scoring mistakes, and overly simplistic molecular interaction models. This means that researchers need to develop more advanced methodologies. Ensemble docking, improved sampling methods, free-energy calculations (such as MM-PBSA/MM-GBSA), and ML-assisted scoring are becoming increasingly crucial for obtaining accurate and realistic insights into molecular recognition [76].

8.4.4 Pharmacophore modeling

Pharmacophore modeling is a sophisticated in silico technique utilized in drug design that defines the optimal spatial arrangement of steric and electronic characteristics required for ligand interaction with a biological target. These characteristics—hydrogen-bond donors (HBD), hydrogen-bond acceptors (HBA), hydrophobic regions (H), aromatic rings (ARs), positively/negatively ionizable groups (PI/NI), and exclusion volumes (XVOL)—are converted into geometric shapes (spheres, vectors, planes) to illustrate how ligands can associate with various scaffolds [77]. The two main methods of pharmacophore modeling are "ligand-based" and "structure-based." These techniques are distinguished primarily by their input data. The structure-based technique examines the structural characteristics of target proteins, such as receptors or enzymes, to identify potential medicinal compounds. The ligand-based method exclusively employs the physicochemical features of known ligand molecules to create 3D pharmacophore models and model quantitative structure–activity relationships (QSAR) or quantitative structure-property relationships (QSPR). Several factors must be considered when choosing the optimal method, such as data availability, quality, computing power, and the intended application of the pharmacophore models. The following sections discuss both strategies and their application in virtual screening. The goal is to help non-specialists implement these protocols by explaining the fundamental concepts behind them.

Pharmacophore-based virtual screening (VS) is an important method for finding new bioactive compounds. It swiftly selects compounds from big libraries that are most likely to bind to a target and can identify molecules that might be hazardous or exhibit adverse pharmacological or pharmacokinetic features [78]. Pharmacophore models are especially powerful for filtering chemical libraries to obtain hits that match essential interaction features and enable scaffold hopping due to their abstraction of chemical functionalities rather than fixed groups [77]. Furthermore, VS can be performed through tools used for pharmacophore generation or open-access platforms such as Pharmit, which supports user-defined queries and rapid ranking of results [79], and ZINCPharmer, which screens purchasable ZINC compounds using Pharmer technology [80].

The process begins by consulting compound databases such as the Protein Data Bank (PDB), PubChem, ChEMBL, ZINC, and DrugBank. Given the size of these repositories, datasets are commonly filtered using criteria such as Lipinski's Rule of Five, PAINS removal, and ADMET-based parameters to reduce the computational load [37,78,81], with the understanding that these guidelines (e.g., RO5) are not absolute. Tools such as the ACD/LogD Suite and the ADME Suite help predict physicochemical and ADME properties [82,83]. When structural information is available, hits are typically refined by molecular docking using programs like DOCK or AutoDock to estimate binding affinity and obtain preliminary complexes, followed by molecular dynamics simulations to assess the stability of binding poses and improve affinity predictions [84–86]. Ultimately, the VS-identified candidates must be validated experimentally to confirm their *in silico*-predicted activity.

8.5 QUANTITATIVE STRUCTURE–ACTIVITY RELATIONSHIP (QSAR) MODELING

QSAR modeling is a simple way to use math to find connections between chemical structures and their biological properties or actions. The basic premise is that the way a molecule acts in living things is directly affected by its physical, chemical, structural, electronic, and steric properties. These characteristics are quantified as molecular descriptors, encompassing conformational and electrical properties, functional groups, stereochemical features, elemental composition, or topological indices. Typically, descriptors are divided into three categories: 1D, 2D, and 3D, which provide increasingly detailed structural information. For instance, they can provide information on the electrostatic potential distributions, topological connectivity, three-dimensional structure, and molecular weights or atom counts.

To build effective QSAR models, researchers must employ techniques for feature selection and dimensionality reduction. This prevents potential overfitting and redundancy caused by an excessive number of descriptors. Conventional techniques like recursive feature elimination (RFE) and principal component analysis (PCA) are still used [86,87]. Regularization-based methods, such as LASSO, and information-theoretic ranking algorithms, have become popular recently for identifying the most relevant molecular properties and eliminating redundant ones. Improved descriptors not only increase prediction accuracy but also facilitate model interpretability. This is essential for obtaining regulatory approval and developing novel, evidence-based concepts in medicinal chemistry.

The traditional 1D–3D paradigm for descriptor dimensionality has been expanded by recent developments in computational chemistry. Instead of providing a single static structure, 4D descriptors incorporate molecular flexibility within biological contexts by using conformational ensembles. These have been particularly beneficial in enhancing QSAR workflows and ligand-based pharmacophore modeling [88,89]. Quantum chemical descriptors like HOMO–LUMO gaps, dipole moments, orbital energies, and electrostatic potential patterns increase the accuracy of QSAR models more accurate, especially when electronic factors significantly influence activity [4,90]. Consequently, software packages like DRAGON, PaDEL, and RDKit have made it straightforward to calculate 3D and quantum descriptors. The term "deep descriptors" refers to new ideas in DL that use graph neural networks (GNNs) and autoencoders to automatically learn latent molecular embeddings from SMILES strings or molecular graphs [91,92]. These representations have enabled QSAR models to generalize across chemically heterogeneous datasets and to gather hierarchical chemical information that exceeds the limitations of manually constructed descriptors.

Many QSAR methods have been used to study drug-relevant endpoints like absorption, distribution, metabolism, excretion, and toxicity (ADMET). More and more structural data for the key ADMET proteins have made it possible to use hybrid approaches, which combine standard descriptor-driven QSAR with structure-based docking and scoring. Docking on ADMET proteins that are very flexible is still hard since their binding holes are so big and changeable. However, enzymes that break down drugs have shown good results. For example, structure-based modeling that takes receptor flexibility into consideration has helped phase II metabolizing sulfotransferases a lot [93]. These improvements mean that structure-based ADMET profiling will continue to work with QSAR operations in the years to come.

QSAR methodology has seen substantial evolution, transitioning from early linear regression models utilizing chemically homogeneous datasets to modern frameworks capable of handling large, heterogeneous libraries through ML and virtual screening techniques [94,95]. QSAR models can be either linear (like MLR and PLS) or nonlinear (like SVM, random forests, and artificial neural networks) [96,97]. This depends on how complicated the link between the descriptor and the activity is. Descriptor dimensionality provides the foundation for an additional important classification:

3D-QSAR, which includes CoMFA and CoMSIA, has three-dimensional conformational features, while 2D-QSAR is based on two-dimensional structural representations [98,99]. Subsequent advancements encompassed 4D-QSAR, which employs ensemble sampling of ligand orientations and conformations [100,101]; 5D-QSAR, which considers receptor-driven conformational alterations to incorporate induced-fit effects [102,103]; and 6D-QSAR, which utilizes various solvation states and ligand-solvent interactions to enhance binding representations [89]. Molecular descriptors that can have dimensions from one to six. Table 8.1 shows how each type of descriptor gathers more and more information about chemicals, structures, and the environment. It also shows how they could be used to locate new medications, do QSAR, and do virtual screening.

Non-mechanistic QSAR, exemplified by CoMFA-based models, emphasizes prediction accuracy despite unclear mechanistic details. In contrast, mechanistic QSAR employs interpretable descriptors that elucidate the underlying biological mechanisms [104,105]. Datasets can yield local QSAR models derived from structurally analogous compounds or global QSAR models based on structurally different substances; however, the latter often exhibit diminished predictive accuracy due to chemical heterogeneity [106]. Additional methodologies include topological QSAR, quantum mechanical QSAR, and hologram QSAR (HQSAR), the latter of which uses fragment-based molecular holograms as regression variables.

QSAR can serve as a regression framework to forecast continuous variables such as permeability, oral bioavailability, or IC_{50}, or as a classification tool to differentiate between active and inactive compounds, contingent upon the endpoint [107,108]. A typical application of the method has been to predict interactions with biological targets, toxicity endpoints, and ADME characteristics. PAMPA-based models, which represented a breakthrough in 1998 [109], made it possible to use QSAR for early predictions of passive transport in permeability studies. Later research employed QSAR to predict the oral absorption of flavonoids from *Silybum marianum* [110–112]. Boik and Newman (2008) used data on oral clearance, LD_{50}, and cytotoxicity to construct classification models that evaluated more than 115,000 natural chemicals. This shows that QSAR models have made it possible to conduct large-scale toxicological screening at the organismal level.

Improvements in ML algorithms, structural modeling, and chemical descriptors have all made QSAR a powerful and adaptable framework. Modern QSAR helps with more than just virtual screening and lead optimization; it also supports toxicity profiling, mechanistic understanding, and ADMET prediction for both synthetic and natural drugs. As integrative methods continue to combine QSAR with DL and structure-based modeling, the field will become increasingly significant in the prediction of chemical biology and rational drug discovery. Table 8.2 summarizes the major

Table 8.1 Key molecular descriptor types (1D–6D) with definitions, examples, and applications in QSAR and drug discovery

Descriptor type	*Definition*	*Key features captured*	*Examples of descriptors*	*Applications*
1D Descriptors	Simple count-based or global molecular properties independent of molecular connectivity	Atomic composition, molecular formula, basic physicochemical properties	Molecular weight, atom count, number of hydrogen bond donors/acceptors, logP, pKa	Initial screening, drug-likeness assessment, ADME filters
2D Descriptors	Derived from molecular topology; based on 2D structure and atom-bond connectivity	Structural fragments, topological indices, functional groups	Topological indices (Wiener index, Kier–Hall indices), substructure flags, fragment counts, fingerprints	Classical QSAR models, similarity searching, virtual screening
3D Descriptors	Use three-dimensional molecular conformation	Shape, electrostatics, steric fields, molecular surfaces	Molecular surface area/volume, 3D pharmacophores, CoMFA and CoMSIA fields	3D-QSAR, receptor–ligand modeling, conformation-dependent predictions
4D Descriptors	Capture conformational flexibility using ensemble sampling rather than a single static structure	Multiple conformations, orientations, time-averaged molecular properties	Ensemble-averaged steric/electrostatic fields, conformer grids	Flexible ligand modeling, pharmacophore refinement, advanced 3D-QSAR
5D Descriptors	Integrate receptor-induced fit effects into ligand modeling	Dynamic conformational changes during binding	Interaction energies under varying receptor conformations	Induced-fit modeling, receptor–ligand dynamics in QSAR
6D Descriptors	Incorporate solvation effects and multiple solvation states	Solvent–ligand interactions, solvation energies, environment-dependent conformations	Solvation energy states, solvent-accessible molecular fields	Modeling binding in physiological environments, high-fidelity QSAR for aqueous/biological systems

Table 8.2 QSAR and QSPR modeling strategies for predicting drug absorption and ADMET-related permeability endpoints

QSAR model/approach	*Endpoint predicted*	*Descriptors used*	*Dataset/experimental basis*	*Strengths*	*Limitations*
Hansch-type linear QSAR	General absorption parameters (e.g., permeability, lipophilicity impact)	logP, electronic (σ), steric constants	Early drug datasets	Simple, interpretable; useful for early screening	Cannot model nonlinear relationships; limited to structurally similar compounds
PAMPA-based QSAR models	Passive permeability; oral absorption potential	logP, TPSA, H-bond donors/acceptors, MW	PAMPA permeability assays (Kansy et al. [109])	High-throughput; good for passive diffusion drugs	Poor prediction for transporter-mediated absorption
Caco-2 QSAR models	Intestinal permeability; efflux ratio	TPSA, pKa, logD, molecular volume, polar descriptors	Caco-2 monolayer assays	Widely used; good transporter/efflux sensitivity	Overestimates permeability for highly lipophilic compounds
HIA (human intestinal absorption) QSAR	% absorbed in humans after oral administration	Topological indices, MW, HBA/HBD, flexibility	Clinical HIA data + machine learning	Direct clinical relevance; good for oral drug prediction	Data heterogeneous; biological variability affects accuracy
QSPR-based absorption models	Various absorption properties (including rate constants)	Quantum descriptors, solvation energies	Calculated physicochemical datasets	Accounts for electronic and QM effects	Requires heavy computation for QM descriptors
4D-QSAR absorption models	Conformation-dependent absorption predictions	Conformational ensembles, spatial field descriptors	Computational ligand ensembles	Captures flexibility; better for diverse molecules	High computational cost; complex interpretation

(*Continued*)

Table 8.2 (Continued) QSAR and QSPR modeling strategies for predicting drug absorption and ADMET-related permeability endpoints

QSAR model/approach	*Endpoint predicted*	*Descriptors used*	*Dataset/experimental basis*	*Strengths*	*Limitations*
5D-QSAR (induced fit consideration)	Absorption influenced by dynamic conformation states	Binding mode variations, conformational response	Receptor–ligand dynamic models	Incorporates induced-fit; relevant for transporter-binding absorption	Requires structural data of transporters; limited datasets
6D-QSAR (solvation-inclusive models)	Solvent-influenced absorption, membrane interactions	Solvent parameters, solvation free energy	Multi-solvation simulations	Best models for real physiological absorption	Computationally expensive; fewer published models
Machine learning QSAR (RF, SVM, ANN)	HIA, Caco-2 Papp, PAMPA permeability, oral bioavailability	Large descriptor sets (2D–3D physicochemical, fingerprints, ML descriptors)	High-quality curated datasets (ADMETlab, DrugBank, ChEMBL)	High accuracy; handles nonlinear patterns; scalable	Requires large training data; model interpretability varies
Deep learning–based QSAR	Multi-endpoint absorption: Papp, HIA, F%	Graph-based features, molecular embeddings	Big-data ADME/t datasets	State-of-the-art predictive performance	Black box nature; needs high-quality big datasets

QSAR and QSPR modeling strategies used to predict drug absorption and related ADMET permeability endpoints, highlighting their typical descriptors, experimental data sources, strengths, and limitations.

8.6 MOLECULAR DYNAMICS AND SIMULATION

Molecular dynamics (MD) is a key simulation technique used to study the structural, dynamic, and thermodynamic behavior of biomolecular systems [113]. Simulations typically begin with a protein or enzyme structure obtained from the PDB, or from homology/comparative modeling when no experimental structure exists [114]. In atomistic MD (AAMD), both the solute and solvent are represented explicitly as interacting atoms whose motions follow Newton's laws, which are integrated using algorithms such as velocity-Verlet or leap-frog [113]. Particle forces are computed from molecular mechanics force fields that describe bonded and non-bonded interactions; widely used examples include force fields for proteins, lipids, and small molecules, which are also used in MM energy calculations [115]. Simulations are commonly run under constant temperature and pressure, often using periodic boundary conditions and Ewald-based methods to treat long-range electrostatics. Trajectories are stored at regular intervals, and simulations typically use 1–2 fs time steps for systems containing tens of thousands of atoms.

Following target validation, drug discovery advances to lead identification and subsequent optimization to generate candidate drugs. At this stage, the central objective is to design molecules that exhibit strong binding affinity and high selectivity toward the intended target. To rapidly predict the locations of optimal ligands and the strength of their binding, researchers usually employ conventional docking and scoring techniques. However, these scoring algorithms frequently suffer from significant limitations that hinder their ability to determine binding strength [116,117]. By explicitly incorporating solvent effects and capturing the conformational flexibility of both the ligand and receptor, MD successfully overcomes these limitations and enables a more accurate and realistic assessment of binding affinity [117]. MD-based methods are used by scientists to determine the strength and rate at which ligands adhere to objects. For modeling peptide ligands with several functions, hybrid docking–MD techniques are effective.

8.7 ADMET PREDICTION

Pharmacodynamics (PD) and pharmacokinetics (PK) must be optimally balanced to ensure drug efficacy and safety. This indicates that the medication should be highly potent, exhibit a high affinity and selectivity for the target

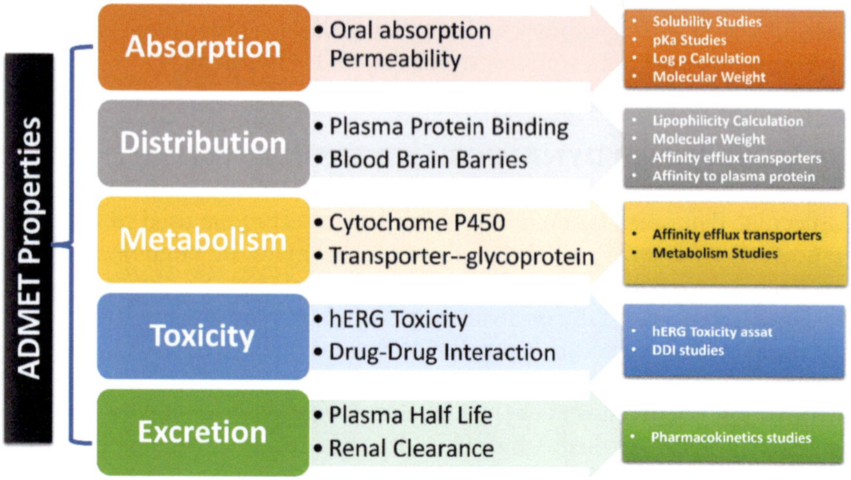

Figure 8.2 Overview of ADMET components and the primary physicochemical and biological factors assessed for drug-likeness.

molecule, and possess favorable ADMET characteristics for toxicity, metabolism, excretion, and absorption. An overview of ADMET components and the primary physicochemical and biological parameters evaluated during in silico drug-likeness assessment is provided in Figure 8.2. Coordinated optimization of these interconnected attributes continues to pose a significant challenge for pharmaceutical research and development (R&D) [118]. Numerous initiatives have been undertaken to utilize technology for the early prediction of pharmacodynamics (PD) and pharmacokinetics (PK) during the hit-to-lead and lead-optimization phases [119,120]. The acronym ADMET represents the basic physiological processes that control how a chemical is absorbed, moved to different parts of the body, broken down by enzymes, removed, and tested for toxicity. These properties are crucial for drug development and chemical safety testing because they control how drugs and xenobiotics work in living things and help regulators figure out the risks of pesticides, food additives, and environmental pollutants. The order of ADME impacts how much of the drug gets into the body, how it is absorbed, and how long it lasts. Synthetic approaches are developed in the first stages of discovery to formulate drugs with advantageous ADME characteristics and elevated potency [121]. It is now common practice to simultaneously evaluate a drug's biopharmaceutical qualities and pharmacological potency. This is because a medication must not only affect its primary target but also reach it at the appropriate dosage and for the required duration within the body. Since many clinical trials fail because of poor ADMET profiles rather than a lack of efficacy, modern pharmacological research has standardized early ADMET profiling to lower the frequency of failed clinical trials. In this regard, in silico ADMET models enable rapid and initial screening before experimental assessment, facilitating the discovery of promising candidates [122–124]. Although ADMET profiles can be estimated using a variety of commercial and open-access computational tools, these methods are typically not particularly accurate. To overcome these limitations, researchers have developed integrated systems—such as ADMET Lab—in order to overcome these issues. By standardizing data curation, these platforms increase predictive accuracy and facilitate the inclusion of comprehensive ADMET endpoints.

AI has undergone significant advancements in response to the growing demand for early-stage ADMET optimization. In the past, ADMET testing relied on expensive and time-consuming in vitro and in vivo assays, which were limited to a small number of compounds. Currently, AI techniques enable the accurate, rapid, and cost-effective prediction of ADMET properties. In the later stages, this reduces the likelihood of failures [125,126]. Large chemical datasets are used by ML and DL models to identify nonlinear trends that influence ADMET activity.

In addition to experimentally generated ADMET endpoints, traditional ML methods such as random forest, support vector machines (SVM), and gradient boosting machines (GBM) use carefully selected chemical descriptors. Because of their intrinsic graph representation, DL architectures—in particular, GNNs—operate on chemical structures. The atom-bond correlations that form the basis of biological interactions are thus better understood by models. GNN-based systems can produce accurate and scalable predictions for a range of ADMET attributes when trained on huge datasets. Selecting models and adjusting hyperparameters are automated using AutoML and other cutting-edge technologies. Predictive accuracy may be increased with the use of new quantum ML methods [127–130]. Table 8.3 summarizes widely used computational platforms for ADMET modeling and toxicity prediction, outlining their key endpoints, methodological basis, main advantages, and typical access modes (web, API, or commercial suites).

In a ground-breaking study, Kearnes et al. used 1,024-bit circular fingerprints across 22 Vertex ADMET endpoints, totaling over 280,000 data points, to compare multitask deep neural networks with common single-task models (random forest, logistic regression). According to their research, multitask learning offers modest but long-lasting advantages, particularly when dealing with small datasets. But as the datasets grow, these changes become less pronounced. It is important to understand that adding too much extra information does not always improve prediction accuracy. Their results indicate that the benefits of multitask architectures depend on the datasets used, emphasizing the necessity of customized modeling techniques for computational ADMET prediction [131].

Table 8.3 Computational platforms for ADMET modeling and toxicity prediction

Tool	*Key ADMET parameters predicted*	*Expanded description/ strengths*	*Access*
PBPK.ai	Whole-body PBPK, clearance, bioavailability	AI-based physiologically based pharmacokinetic simulator.	Web
MegaMolBART-ADMET	Multi-endpoint ADMET + de novo generation	Transformer model predicts and optimizes ADMET.	Cloud/API
ADMET-Boost	Solubility, clearance, BBB permeability	Ensemble model with high explainability & accuracy.	Web/GitHub
ADMET-AI	Solubility, clearance, permeability, toxicity	State-of-the-art AI ADMET modeling using GNN/transformers.	Web/API
DeepTox (Tox21)	12 Tox21 endpoints	Deep learning model outperforming others in Tox21 toxicity challenge.	Web/GitHub
ToxiGNN	Acute toxicity, LD50, organ toxicity	GNN-based toxicity predictor with high interpretability.	Web/GitHub
OCHEM-ADMET	ADMET physicochemical endpoints	Open-source consensus ML ADMET resource.	Web
TEST 5.1 (EPA)	LD50, mutagenicity, carcinogenicity	EPA-supported QSAR toxicity modeling.	Web/ Standalone
OpenTox/eTOX 2.0	In vivo + in vitro toxicity	Integrated toxicology platform with pharma + regulatory data.	Web
BioTransformer 3.0	Phase I–II metabolism, microbial metabolism	Most advanced metabolite prediction engine (>30 metabolic reactions).	Web
ToxCast dashboard	700+ in vitro toxicity endpoints	EPA HTS toxicity dataset for endocrine/neural/ developmental toxicity.	Web
VEGA HUB 3.2	Carcinogenicity, mutagenicity, skin sensitization	OECD-validated QSAR platform with regulatory quality.	Web
HergPred/ HergCentral	hERG cardiotoxicity	ML-based cardiotoxicity risk prediction.	Web
QikProp	Solubility, logP, permeability, metabolism	Industry gold-standard ADME predictor; strong experimental correlation.	Commercial

(*Continued*)

Table 8.3 (Continued) Computational platforms for ADMET modeling and toxicity prediction

Tool	*Key ADMET parameters predicted*	*Expanded description/ strengths*	*Access*
ChemProp-ADMET	Full-scale ADMET profiling	Neural message passing-based ADMET prediction framework.	Web/Python
ADMETlab 2.0	>200 ADMET endpoints	Deep-learning ADMET predictions with toxicity maps and analytics.	Web
admetSAR 2.0	50+ endpoints incl. carcinogenicity, mutagenicity, CYP interactions	Large curated ADMET dataset; consensus predictions enhance accuracy.	Web
ProTox-II	LD50, hepatotoxicity, immunotoxicity, cytotoxicity	Organ-specific toxicity; ML-based classification and LD50 prediction.	Web
SwissADME	Lipophilicity, solubility, GI absorption, BBB penetration, drug-likeness	BOILED-Egg model; excellent medicinal chemistry support; fast ADME profiling.	Web
FAF-Drugs4	Drug-likeness, ADME filters, PAINS alerts	Excellent at HTS library filtering; removes reactive groups and PAINS.	Web
pkCSM	Toxicity, permeability, clearance, hERG, metabolism	Highly accurate QSAR predictions; graph-based signatures; widely used for hit-to-lead triaging.	Web
DataWarrior	Toxicity alerts, drug-likeness, physicochemical properties	Free cheminformatics tool with toxicity alerting.	Free
DMPK Predictor	Clearance, PK parameters, PBPK-ready	Mechanistic PK modeling enabling IVIVE and PBPK.	Commercial
StarDrop (ADME/ Model Builder)	Solubility, clearance, permeability, toxicity	Multi-parameter optimization integrating potency + ADMET + toxicity.	Commercial
Discovery Studio – ADMET	CYP inhibition, PPB, hepatotoxicity	Comprehensive molecular modeling+ADMET optimization.	Commercial

8.8 INTEGRATION OF MULTI-OMICS DATA IN DRUG DISCOVERY

The advent of high-throughput omics technologies in recent years has made a huge difference in how new drugs are found. Multi-omics is the examination of data from more than one omics platform at the same time. Some of these systems are genomics, proteomics, metabolomics, transcriptomics, and epigenomics. Integrating these diverse forms of knowledge may enhance our understanding of biological processes, thereby facilitating the development and application of medications. Each layer of omics shows a different part of how living things work. Genomic analysis finds discrepancies in DNA sequences, transcriptomics makes sense of gene expression patterns, proteomics examines protein levels and their changes, and metabolomics describes metabolic profiles. By bringing these layers together, researchers may be able to watch how genes, proteins, metabolites, and other biomolecules interact with each other in real time [132]. We need to take an all-encompassing approach to learn more about how diseases work, locate novel therapeutic targets, and find biomarkers that can help doctors figure out which patients need certain medicines. It is hard to combine multi-omics data since it is so unique and requires modern computer technology. ML, network analysis, and systems biology modeling are some of the most common ways to combine and analyze multi-omics data [133]. For example, ML algorithms can find links and patterns that you can't see when you simply look at one type of omics. This makes it easier to choose targets and utilize drugs for different things. Network-based approaches can also highlight how genes, proteins, and metabolites operate together by showing important nodes and pathways that are linked to diseases [134]. These kinds of studies help figure out how drug candidates will act in a bigger biological context and find the most important therapeutic targets.

Several phases of drug development have seen the successful application of integrative multi-omics approaches. Finding driver mutations and gene signatures associated with the effectiveness of cancer treatments has become simpler thanks to the integration of transcriptomics and genomics [135]. Combining proteome and metabolomics data makes it simpler to find biomarkers for medication efficacy and safety. Drug companies are increasingly using multi-omics integration to improve precision medicine. This enables each patient to have customized, targeted treatment [136]. Moreover, you can formulate and evaluate novel concepts utilizing collaborative platforms like the Cancer Genome Atlas (TCGA) and publicly accessible multi-omics datasets. Despite its significant promise, challenges related to data standards, quality assurance, and computational infrastructure complicate the integration of multi-omics data. It is crucial to ensure interoperability of data from diverse platforms and the replicability of outcomes [132]. It is posited that enhanced data analytics and improved collaboration among regulators, corporations, and academia would facilitate the resolution of these difficulties and accelerate the search for innovative solutions. Integrating multi-omics data is transforming drug discovery by offering a full picture of biological system functionality. The integration of multi-omics for the identification of novel pharmacological targets, biomarkers, and personalized treatment protocols will gain significance as computational capabilities advance and data sharing proliferates.

8.9 CHALLENGES AND FUTURE PERSPECTIVES

Notwithstanding considerable progress, in silico drug design faces challenges due to the intricacies of biology and variable data. Extensive experimental data, encompassing protein structures, ligand libraries, and bioactivity profiles, are crucial for computer models. Vamathevan et al. [21] assert that these datasets often exhibit flaws, duplication, and inconsistencies that might propagate through computational workflows, hence diminishing the accuracy of predictions. The predominant structures identified to date were generated by cryo-electron microscopy or crystallography. These approaches may insufficiently represent the dynamics of protein mobility or the flexibility of binding sites in real organisms [5,137]. Biological systems are inherently complex and dynamic. Interactions across signaling pathways, protein–protein interactions, and genetic and epigenetic alterations are often implicated in the multifactorial etiologies of diseases. The attrition rate may rise when candidates identified through computational methods progress to experimental or clinical phases, as critical factors such as post-translational modifications, allosteric regulation, or the effects of the cellular microenvironment may be neglected [138].

Various applications of DL and other AI methodologies have demonstrated significant potential, including virtual screening, compound synthesis, and chemical property forecasting [137,139]. DL methodologies for protein structure prediction have demonstrated unprecedented accuracy [137]. Despite the necessity for substantial high-quality data for training these algorithms, the utilized data is likely to be biased. Utilizing DL models in regulatory contexts might be problematic due to their "black box" nature, which renders them tricky to employ and understand [21]. Quantum computing, capable of swiftly resolving quantum mechanical issues beyond the reach of traditional computers, represents a novel discipline with the potential to fundamentally revolutionize molecular simulations. A preliminary study suggests that quantum algorithms may improve small-scale molecular modeling. Nonetheless, the practical application in drug discovery is impeded by existing hardware limitations, including qubit coherence time, error rates, and scalability [140,141].

Further efforts are required to enhance data quality through the meticulous integration, standardization, and curation of multi-omics datasets to address these shortcomings. Employing systems biology methodologies and sophisticated modeling of protein conformational dynamics can enhance the precision of biological systems in reflecting their intrinsic complexity [138]. The drug development process can be enhanced in accuracy and transparency by the utilization of interpretable ML models and hybrid quantum-classical algorithms [141]. Open-source datasets and community challenges, such as the SAMPL series, are essential for testing and validating innovative computational algorithms [142].

8.10 CONCLUSION

Although in silico drug design has revolutionized the early stages of drug research, it still faces substantial obstacles because of a lack of data, the complexity of biological processes, and the development of new technology. Because in silico drug design has been so successful, several difficulties have arisen. In order to progress computational drug research, more data must be gathered, more sophisticated modeling tools must be developed, and community benchmarks must be established.

REFERENCES

1. Wagner, J., et al., A dynamic map for learning, communicating, navigating and improving therapeutic development. *Nature Reviews Drug Discovery*, 2018. **17**(2): p. 150–150.
2. Hughes, J., et al., Principles of early drug discovery. *British Journal of Pharmacology*, 2011. **162**(6): p. 1239–1249.
3. Paul, S.M., et al., How to improve R&D productivity: the pharmaceutical industry's grand challenge. *Nature Reviews Drug Discovery*, 2010. **9**(3): p. 203–214.
4. Cherkasov, A., et al., QSAR modeling: where have you been? Where are you going to? *Journal of Medicinal Chemistry*, 2014. **57**(12): p. 4977–5010.
5. Sliwoski, G., et al., Computational methods in drug discovery. *Pharmacological Reviews*, 2014. **66**(1): p. 334–395.
6. Lavecchia, A. and C. Giovanni, Virtual screening strategies in drug discovery: a critical review. *Current Medicinal Chemistry*, 2013. **20**(23): p. 2839–2860.
7. Pagadala, N.S., K. Syed, and J. Tuszynski, Software for molecular docking: a review. *Biophysical Reviews*, 2017. **9**(2): p. 91–102.
8. Kitchen, D.B., et al., Docking and scoring in virtual screening for drug discovery: methods and applications. *Nature Reviews Drug Discovery*, 2004. **3**(11): p. 935–949.
9. Li, Q. and S. Shah, Structure-Based Virtual Screening, in *Protein Bioinformatics*, C. H. Wu, C. N. Arighi, and K. E. Ross, Editors. 2017, Springer New York: New York, NY. p. 111–124.

10. Daina, A., O. Michielin, and V. Zoete, SwissADME: a free web tool to evaluate pharmacokinetics, drug-likeness and medicinal chemistry friendliness of small molecules. *Scientific Reports*, 2017. **7**(1): p. 42717.
11. Pires, D.E.V., T.L. Blundell, and D.B. Ascher, pkCSM: Predicting small-molecule pharmacokinetic and toxicity properties using graph-based signatures. *Journal of Medicinal Chemistry*, 2015. **58**(9): p. 4066–4072.
12. Van De Waterbeemd, H. and E. Gifford, ADMET in silico modelling: towards prediction paradise? *Nature Reviews Drug Discovery*, 2003. **2**(3): p. 192–204.
13. Cheng, F., et al., admetSAR: a comprehensive source and free tool for assessment of chemical ADMET properties. *Journal of Chemical Information and Modeling*, 2012. **52**(11): p. 3099–3105.
14. Harvey, A., The role of natural products in drug discovery and development in the new millennium. *IDrugs: The Investigational Drugs Journal*, 2010. **13**(2): p. 70–72.
15. Begley, C.G. and L.M. Ellis, Raise standards for preclinical cancer research. *Nature*, 2012. **483**(7391): p. 531–533.
16. Korinek, M., et al., Randialic acid B and tomentosolic acid block formyl peptide receptor 1 in human neutrophils and attenuate psoriasis-like inflammation in vivo. *Biochemical Pharmacology*, 2021. **190**: p. 114596.
17. Giordano, D., et al., Drug design by pharmacophore and virtual screening approach. *Pharmaceuticals*, 2022. **15**(5): p. 646.
18. Ouma, R.B.O., S.M. Ngari, and J.K. Kibet, A review of the current trends in computational approaches in drug design and metabolism. *Discover Public Health*, 2024. **21**(1): p. 108.
19. Wang, Q., et al., Progress of AI-driven drug–target interaction prediction and lead optimization. *International Journal of Molecular Sciences*, 2025. **26**(20): p. 10037.
20. Kadurin, A., et al., druGAN: an advanced generative adversarial autoencoder model for de novo generation of new molecules with desired molecular properties in silico. *Molecular Pharmaceutics*, 2017. **14**(9): p. 3098–3104.
21. Vamathevan, J., et al., Applications of machine learning in drug discovery and development. *Nature Reviews Drug Discovery*, 2019. **18**(6): p. 463–477.
22. Gaulton, A., et al., The ChEMBL database in 2017. *Nucleic Acids Research*, 2017. **45**(D1): p. D945–D954.
23. Irwin, J.J. and B.K. Shoichet, ZINC: a free database of commercially available compounds for virtual screening. *Journal of Chemical Information and Modeling*, 2005. **45**(1): p. 177–182.
24. Kim, S., et al., PubChem in 2021: new data content and improved web interfaces. *Nucleic Acids Research*, 2021. **49**(D1): p. D1388–D1395.
25. Lombardo, F., et al., In silico absorption, distribution, metabolism, excretion, and pharmacokinetics (ADME-PK): utility and best practices. An industry perspective from the international consortium for innovation through quality in pharmaceutical development: miniperspective. *Journal of Medicinal Chemistry*, 2017. **60**(22): p. 9097–9113.
26. Tingaud-Sequeira, A., O. Carnevali, and J. Cerdà, Cathepsin B differential expression and enzyme processing and activity during Fundulus heteroclitus embryogenesis. *Comparative Biochemistry and Physiology Part A: Molecular & Integrative Physiology*, 2011. **158**(2): p. 221–228.

27. Wingender, E., The TRANSFAC project as an example of framework technology that supports the analysis of genomic regulation. *Briefings in Bioinformatics*, 2008. **9**(4): p. 326–332.
28. Baig, M.H., et al., Computer aided drug design and its application to the development of potential drugs for neurodegenerative disorders. *Current Neuropharmacology*, 2018. **16**(6): p. 740–748.
29. Ghosh, B. and S. Choudhuri, Drug Design for Malaria with Artificial Intelligence (AI), in *Plasmodium Species and Drug Resistance*, R. K. Tyagi, Editor. 2021, IntechOpen. United Kingdom. p. 1–15.
30. Mullard, A., Biotech R&D spend jumps by more than 15%. *Nature Reviews Drug Discovery*, 2016. **15**(7): p. 447–447.
31. Urban, G., et al., Deep learning for drug discovery and cancer research: automated analysis of vascularization images. *IEEE/ACM Transactions on Computational Biology and Bioinformatics*, 2019. **16**(3): p. 1029–1035.
32. Lim, J., et al., Predicting drug–target interaction using a novel graph neural network with 3D structure-embedded graph representation. *Journal of Chemical Information and Modeling*, 2019. **59**(9): p. 3981–3988.
33. Liu, S., et al., Practical model selection for prospective virtual screening. *Journal of Chemical Information and Modeling*, 2018. **59**: p. 282–293, Biochemistry.
34. Meng, X.-Y., et al., Molecular docking: a powerful approach for structure-based drug discovery. *Current Computer Aided-Drug Design*, 2011. **7**(2): p. 146–157.
35. Kuntz, I.D., et al., A geometric approach to macromolecule-ligand interactions. *Journal of Molecular Biology*, 1982. **161**(2): p. 269–288.
36. Leelananda, S.P. and S. Lindert, Computational methods in drug discovery. *Beilstein Journal of Organic Chemistry*, 2016. **12**: p. 2694–2718.
37. Lionta, E., et al., Structure-based virtual screening for drug discovery: principles, applications and recent advances. *Current Topics in Medicinal Chemistry*, 2014. **14**(16): p. 1923–1938.
38. Byrne, R. and G. Schneider, In Silico Target Prediction for Small Molecules, in *Systems Chemical Biology*, S. Ziegler and H. Waldmann, Editors. 2019, Springer New York: New York, NY. p. 273–309.
39. Bender, A. and R.C. Glen, Molecular similarity: a key technique in molecular informatics. *Organic & Biomolecular Chemistry*, 2004. **2**(22): p. 3204.
40. Jenkins, J.L., A. Bender, and J.W. Davies, In silico target fishing: predicting biological targets from chemical structure. *Drug Discovery Today: Technologies*, 2006. **3**(4): p. 413–421.
41. Wang, Z., et al., Improving chemical similarity ensemble approach in target prediction. *Journal of Cheminformatics*, 2016. **8**(1): p. 20.
42. Hert, J., et al., Quantifying the relationships among drug classes. *Journal of Chemical Information and Modeling*, 2008. **48**(4): p. 755–765.
43. Awale, M. and J.-L. Reymond, The polypharmacology browser: a web-based multi-fingerprint target prediction tool using ChEMBL bioactivity data. *Journal of Cheminformatics*, 2017. **9**(1): p. 11.
44. Reker, D., et al., Identifying the macromolecular targets of de novo-designed chemical entities through self-organizing map consensus. *Proceedings of the National Academy of Sciences*, 2014. **111**(11): p. 4067–4072.

45. Gfeller, D., et al., SwissTargetPrediction: a web server for target prediction of bioactive small molecules. *Nucleic Acids Research*, 2014. **42**(W1): p. W32–W38.
46. Liu, X., et al., HitPick: a web server for hit identification and target prediction of chemical screenings. *Bioinformatics*, 2013. **29**(15): p. 1910–1912.
47. Dunkel, M., et al., SuperPred: drug classification and target prediction. *Nucleic Acids Research*, 2008. **36**(Web Server): p. W55–W59.
48. Lagunin, A., et al., PASS: prediction of activity spectra for biologically active substances. *Bioinformatics*, 2000. **16**(8): p. 747–748.
49. Rhodes, N., et al., CLIP: similarity searching of 3D databases using clique detection. *Journal of Chemical Information and Computer Sciences*, 2003. **43**(2): p. 443–448.
50. Lo, Y.-C., et al., Large-scale chemical similarity networks for target profiling of compounds identified in cell-based chemical screens. *PLOS Computational Biology*, 2015. **11**(3): p. e1004153.
51. Turner, P. and W. Denny, The genome as a drug target sequence specific minor groove binding ligands. *Current Drug Targets*, 2000. **1**(1): p. 1–14.
52. Berman, H.M., The protein data bank. *Nucleic Acids Research*, 2000. **28**(1): p. 235–242.
53. Jorgensen, W.L., The many roles of computation in drug discovery. *Science*, 2004. **303**(5665): p. 1813–1818.
54. Pei, J., et al., Systems biology brings new dimensions for structure-based drug design. *Journal of the American Chemical Society*, 2014. **136**(33): p. 11556–11565.
55. Warren, G.L., et al., Essential considerations for using protein–ligand structures in drug discovery. *Drug Discovery Today*, 2012. **17**(23–24): p. 1270–1281.
56. Morrison, J.L., et al., A lock-and-key model for protein–protein interactions. *Bioinformatics*, 2006. **22**(16): p. 2012–2019.
57. Koshland, D.E., The key–lock theory and the induced fit theory. *Angewandte Chemie International Edition in English*, 1995. **33**(23–24): p. 2375–2378.
58. Audie, J. and S. Scarlata, A novel empirical free energy function that explains and predicts protein–protein binding affinities. *Biophysical Chemistry*, 2007. **129**(2–3): p. 198–211.
59. Trott, O. and A.J. Olson, AutoDock Vina: improving the speed and accuracy of docking with a new scoring function, efficient optimization, and multi-threading. *Journal of Computational Chemistry*, 2010. **31**(2): p. 455–461.
60. Brooijmans, N. and I.D. Kuntz, Molecular recognition and docking algorithms. *Annual Review of Biophysics and Biomolecular Structure*, 2003. **32**(1): p. 335–373.
61. Liang, L., et al., Characteristics of umami peptides identified from porcine bone soup and molecular docking to the taste receptor T1R1/T1R3. *Food Chemistry*, 2022. **387**: p. 132870.
62. Lin, R., et al., Developments in molecular docking technologies for application of polysaccharide-based materials: a review. *Critical Reviews in Food Science and Nutrition*, 2024. **64**(24): p. 8540–8552.
63. Jin, Z. and Z. Wei, Molecular simulation for food protein–ligand interactions: a comprehensive review on principles, current applications, and emerging trends. *Comprehensive Reviews in Food Science and Food Safety*, 2024. **23**(1): p. e13280.

64. Yang, Y., et al., Ligand-based approach for predicting drug targets and for virtual screening against COVID-19. *Briefings in Bioinformatics*, 2021. **22**(2): p. 1053–1064.
65. Yu, Y., et al., Application of molecular simulation methods in food science: status and prospects. *Journal of Agricultural and Food Chemistry*, 2023. **71**(6): p. 2684–2703.
66. Vyas, V.K., et al., Structure- and ligand-based drug design methods for the modeling of antimalarial agents: a review of updates from 2012 onwards. *Journal of Biomolecular Structure and Dynamics*, 2022. **40**(20): p. 10481–10506.
67. Andreani, J., C. Quignot, and R. Guerois, Structural prediction of protein interactions and docking using conservation and coevolution. *WIREs Computational Molecular Science*, 2020. **10**(6): p. e1470.
68. Evers, A. and T. Klabunde, Structure-based drug discovery using GPCR homology modeling: successful virtual screening for antagonists of the Alpha1A adrenergic receptor. *Journal of Medicinal Chemistry*, 2005. **48**(4): p. 1088–1097.
69. Gastreich, M., et al., Ultrafast de novo docking combining pharmacophores and combinatorics. *Journal of Computer-Aided Molecular Design*, 2007. **20**(12): p. 717–734.
70. Chen, R. and Z. Weng, Docking unbound proteins using shape complementarity, desolvation, and electrostatics. *Proteins: Structure, Function, and Bioinformatics*, 2002. **47**(3): p. 281–294.
71. Asiamah, I., et al., Applications of molecular docking in natural products-based drug discovery. *Scientific African*, 2023. **20**: p. e01593.
72. Singh, S., Q. Bani Baker, and D.B. Singh, Molecular Docking and Molecular Dynamics Simulation, in *Bioinformatics, Dev Bukhsh Singh, Rajesh Kumar Pathak, Editors*. 2022, Elsevier. Academic Press, United States. p. 291–304.
73. Vidal-Limon, A., J.E. Aguilar-Toalá, and A.M. Liceaga, Integration of molecular docking analysis and molecular dynamics simulations for studying food proteins and bioactive peptides. *Journal of Agricultural and Food Chemistry*, 2022. **70**(4): p. 934–943.
74. Friesner, R.A., et al., Glide: a new approach for rapid, accurate docking and scoring. 1. Method and assessment of docking accuracy. *Journal of Medicinal Chemistry*, 2004. **47**(7): p. 1739–1749.
75. Li, Y., et al., Decoding the limits of deep learning in molecular docking for drug discovery. *Chemical Science*, 2025. **16**(37): p. 17374–17390.
76. Totrov, M. and R. Abagyan, Flexible ligand docking to multiple receptor conformations: a practical alternative. *Current Opinion in Structural Biology*, 2008. **18**(2): p. 178–184.
77. Vuorinen, A. and D. Schuster, Methods for generating and applying pharmacophore models as virtual screening filters and for bioactivity profiling. *Methods*, 2015. **71**: p. 113–134.
78. Maia, E.H.B., et al., Structure-based virtual screening: from classical to artificial intelligence. *Frontiers in Chemistry*, 2020. **8**: p. 343.
79. Sunseri, J. and D.R. Koes, Pharmit: interactive exploration of chemical space. *Nucleic Acids Research*, 2016. **44**(W1): p. W442–W448.

80. Koes, D.R. and C.J. Camacho, ZINCPharmer: pharmacophore search of the ZINC database. *Nucleic Acids Research*, 2012. **40**(W1): p. W409–W414.
81. Carpenter, K.A. and X. Huang, Machine learning-based virtual screening and its applications to Alzheimer's drug discovery: a review. *Current Pharmaceutical Design*, 2018. **24**(28): p. 3347–3358.
82. Song, C.M., S.J. Lim, and J.C. Tong, Recent advances in computer-aided drug design. *Briefings in Bioinformatics*, 2009. **10**(5): p. 579–591.
83. Wang, K., et al., Recent advances from computer-aided drug design to artificial intelligence drug design. *RSC Medicinal Chemistry*, 2024. **15**(12): p. 3978–4000.
84. Allen, W.J., et al., DOCK 6: Impact of new features and current docking performance. *Journal of Computational Chemistry*, 2015. **36**(15): p. 1132–1156.
85. Morris, G.M., R. Huey, and A.J. Olson, Using autodock for ligand-receptor docking. *Current Protocols in Bioinformatics*, 2008. **24**(1). https://doi.org/10.1002/0471250953.bi0814s24
86. Pal, S., et al., Ligand-based pharmacophore modeling, virtual screening and molecular docking studies for discovery of potential topoisomerase I inhibitors. *Computational and Structural Biotechnology Journal*, 2019. **17**: p. 291–310.
87. De, P., et al., Prediction reliability of QSAR models: an overview of various validation tools. *Archives of Toxicology*, 2022. **96**(5): p. 1279–1295.
88. Das, J., et al., 2-Aminothiazole as a novel kinase inhibitor template. Structure: activity relationship studies toward the discoveryof N-(2-chloro-6-methylphenyl)-2-[[6-[4-(2-hydroxyethyl)-1-piperazinyl)]-2-methyl-4-pyrimidinyl]amino)]-1,3-thiazole-5-carboxamide(Dasatinib,BMS-354825) as a potent pan-Src kinase inhibitor. *Journal of Medicinal Chemistry*, 2006. **49**(23): p. 6819–6832.
89. Vedani, A., M. Dobler, and M.A. Lill, Combining protein modeling and 6D-QSAR. Simulating the binding of structurally diverse ligands to the estrogen receptor. *Journal of Medicinal Chemistry*, 2005. **48**(11): p. 3700–3703.
90. Tropsha, A., et al., Integrating QSAR modelling and deep learning in drug discovery: the emergence of deep QSAR. *Nature Reviews Drug Discovery*, 2024. **23**(2): p. 141–155.
91. Hung, C. and G. Gini, QSAR modeling without descriptors using graph convolutional neural networks: the case of mutagenicity prediction. *Molecular Diversity*, 2021. **25**(3): p. 1283–1299.
92. Yang, K., et al., Analyzing learned molecular representations for property prediction. *Journal of Chemical Information and Modeling*, 2019. **59**(8): p. 3370–3388.
93. Moroy, G., et al., Toward in silico structure-based ADMET prediction in drug discovery. *Drug Discovery Today*, 2012. **17**(1–2): p. 44–55.
94. Muratov, E.N., et al., QSAR without borders. *Chemical Society Reviews*, 2020. **49**(11): p. 3525–3564.
95. Neves, B.J., et al., QSAR-based virtual screening: advances and applications in drug discovery. *Frontiers in Pharmacology*, 2018. **9**: p. 1275.
96. Ramandi, M., et al., Molecular docking, linear and nonlinear QSAR studies on factor Xa inhibitors. *Structural Chemistry*, 2020. **31**(5): p. 2023–2040.

97. Zhang, S., et al., MOA-based linear and nonlinear QSAR models for predicting the toxicity of organic chemicals to Vibrio fischeri. *Environmental Science and Pollution Research*, 2020. **27**(9): p. 9114–9125.
98. Olotu, F.A., et al., The recent application of 3D-QSAR and docking studies to novel HIV-protease inhibitor drug discovery. *Expert Opinion on Drug Discovery*, 2020. **15**(9): p. 1095–1109.
99. Zhao, Y., et al., QSAR in natural non-peptidic food-related compounds: current status and future perspective. *Trends in Food Science & Technology*, 2023. **140**: p. 104165.
100. Hopfinger, A.J., et al., Construction of 3D-QSAR models using the 4D-QSAR analysis formalism. *Journal of the American Chemical Society*, 1997. **119**(43): p. 10509–10524.
101. Wang, T., et al., The advancement of multidimensional QSAR for novel drug discovery: where are we headed? *Expert Opinion on Drug Discovery*, 2017: p. 1–16.
102. Suay-Garcia, B., et al., Quantitative structure–activity relationship methods in the discovery and development of antibacterials. *WIREs Computational Molecular Science*, 2020. **10**(6): p. e1472.
103. Vedani, A. and M. Dobler, 5D-QSAR: the key for simulating induced fit? *Journal of Medicinal Chemistry*, 2002. **45**(11): p. 2139–2149.
104. Benigni, R., et al., Mechanistic QSAR of aromatic amines: new models for discriminating between homocyclic mutagens and nonmutagens, and validation of models for carcinogens. *Environmental and Molecular Mutagenesis*, 2007. **48**(9): p. 754–771.
105. Leo, A.J. and C. Hansch, Role of hydrophobic effects in mechanistic QSAR. *Perspectives in Drug Discovery and Design*, 1999. **17**(1): p. 1–25.
106. Chaudhry, Q., et al., Global QSAR models of skin sensitisers for regulatory purposes. *Chemistry Central Journal*, 2010. **4**(S1): p. S5.
107. Sahlin, U., M. Filipsson, and T. Öberg, A risk assessment perspective of current practice in characterizing uncertainties in QSAR regression predictions. *Molecular Informatics*, 2011. **30**(6–7): p. 551–564.
108. Su, B.-H., et al., In silico binary classification QSAR models based on 4D-fingerprints and MOE descriptors for prediction of hERG blockage. *Journal of Chemical Information and Modeling*, 2010. **50**(7): p. 1304–1318.
109. Kansy, M., F. Senner, and K. Gubernator, Physicochemical high throughput screening: parallel artificial membrane permeation assay in the description of passive absorption processes. *Journal of Medicinal Chemistry*, 1998. **41**(7): p. 1007–1010.
110. Akamatsu, M., et al., In silico prediction of human oral absorption based on QSAR analyses of PAMPA permeability. *Chemistry & Biodiversity*, 2009. **6**(11): p. 1845–1866.
111. Diukendjieva, A., et al., ADME/Tox properties and biochemical interactions of Silybin congeners: in silico study. *Natural Product Communications*, 2017. **12**(2): p. 1934578X1701200208.
112. Boik, J.C., R.A. Newman, Structure-activity models of oral clearance, cytotoxicity, and LD50: a screen for promising anticancer compounds. *BMC Pharmacology*, 2008 Jun 13;8:12. doi: 10.1186/1471-2210-8-12. PMID: 18554402; PMCID: PMC2442056.

113. Van Gunsteren, W.F. and H.J.C. Berendsen, Computer simulation of molecular dynamics: methodology, applications, and perspectives in chemistry. *Angewandte Chemie International Edition in English*, 1990. **29**(9): p. 992–1023.
114. Kuhlman, B. and P. Bradley, Advances in protein structure prediction and design. *Nature Reviews Molecular Cell Biology*, 2019. **20**(11): p. 681–697.
115. Lin, F.-Y. and A.D. MacKerell, Force Fields for Small Molecules, in *Biomolecular Simulations*, M. Bonomi and C. Camilloni, Editors. 2019, Springer New York: New York, NY. p. 21–54.
116. Li, J., A. Fu, and L. Zhang, An overview of scoring functions used for protein–ligand interactions in molecular docking. *Interdisciplinary Sciences: Computational Life Sciences*, 2019. **11**(2): p. 320–328.
117. Pantsar, T. and A. Poso, Binding affinity via docking: fact and fiction. *Molecules*, 2018. **23**(8): p. 1899.
118. Segall, M., Advances in multiparameter optimization methods for de novo drug design. *Expert Opinion on Drug Discovery*, 2014. **9**(7): p. 803–817.
119. Cheng, F., et al., In silico ADMET prediction: recent advances, current challenges and future trends. *Current Topics in Medicinal Chemistry*, 2013. **13**(11): p. 1273–1289.
120. González-Medina, M., et al., Open chemoinformatic resources to explore the structure, properties and chemical space of molecules. *RSC Advances*, 2017. **7**(85): p. 54153–54163.
121. Sohlenius-Sternbeck, A.-K. and Y. Terelius, Evaluation of ADMET predictor in early discovery drug metabolism and pharmacokinetics project work. *Drug Metabolism and Disposition*, 2022. **50**(2): p. 95–104.
122. Hou, T., Editorial. *Advanced Drug Delivery Reviews*, 2015. **86**: p. 1.
123. Tao, L., et al., Recent progresses in the exploration of machine learning methods as in-silico ADME prediction tools. *Advanced Drug Delivery Reviews*, 2015. **86**: p. 83–100.
124. Wang, N.-N., et al., Predicting human intestinal absorption with modified random forest approach: a comprehensive evaluation of molecular representation, unbalanced data, and applicability domain issues. *RSC Advances*, 2017. **7**(31): p. 19007–19018.
125. Swanson, K., et al., ADMET-AI: a machine learning ADMET platform for evaluation of large-scale chemical libraries. *Bioinformatics*, 2024. **40**(7): p. btae416.
126. Yi, J., et al., OptADMET: a web-based tool for substructure modifications to improve ADMET properties of lead compounds. *Nature Protocols*, 2024. **19**(4): p. 1105–1121.
127. Abdelwahab, A.A., M.A. Elattar, and S.A. Fawzi, Advancing ADMET prediction for major CYP450 isoforms: graph-based models, limitations, and future directions. *BioMedical Engineering OnLine*, 2025. **24**(1): p. 93.
128. Dong, Z., et al., Research on the optimization model of anti-breast cancer candidate drugs based on machine learning. *Frontiers in Genetics*, 2025. **16**: p. 1523015.
129. Göller, A.H., et al., Machine Learning Applied to the Modeling of Pharmacological and ADMET Endpoints, in *Artificial Intelligence in Drug Design*, A. Heifetz, Editor. 2022, Springer US: New York, NY. p. 61–101.

130. Raju, B., et al., Multiple machine learning, molecular docking, and ADMET screening approach for identification of selective inhibitors of CYP1B1. *Journal of Biomolecular Structure and Dynamics*, 2022. **40**(17): p. 7975–7990.
131. Kearnes, S., B. Goldman, and V. Pande, Modeling Industrial ADMET Data with Multitask Networks. 2016, arXiv.
132. Hasin, Y., M. Seldin, and A. Lusis, Multi-omics approaches to disease. *Genome Biology*, 2017. **18**(1): p. 83.
133. Picard, M., et al., Integration strategies of multi-omics data for machine learning analysis. *Computational and Structural Biotechnology Journal*, 2021. **19**: p. 3735–3746.
134. Žitnik, M. and B. Zupan, Gene network inference by fusing data from diverse distributions. *Bioinformatics*, 2015. **31**(12): p. i230–i239.
135. Heo, Y.J., et al., Integrative multi-omics approaches in cancer research: from biological networks to clinical subtypes. *Molecules and Cells*, 2021. **44**(7): p. 433–443.
136. Karczewski, K.J. and M.P. Snyder, Integrative omics for health and disease. *Nature Reviews Genetics*, 2018. **19**(5): p. 299–310.
137. Jumper, J., et al., Highly accurate protein structure prediction with AlphaFold. *Nature*, 2021. **596**(7873): p. 583–589.
138. Jiménez-Luna, J., F. Grisoni, and G. Schneider, Drug discovery with explainable artificial intelligence. *Nature Machine Intelligence*, 2020. **2**(10): p. 573–584.
139. Zhavoronkov, A., et al., Deep learning enables rapid identification of potent DDR1 kinase inhibitors. *Nature Biotechnology*, 2019. **37**(9): p. 1038–1040.
140. Cao, Y., et al., Quantum chemistry in the age of quantum computing. *Chemical Reviews*, 2019. **119**(19): p. 10856–10915.
141. Zinner, M., et al., Quantum computing's potential for drug discovery: early stage industry dynamics. *Drug Discovery Today*, 2021. **26**(7): p. 1680–1688.
142. Rizzi, A., et al., Overview of the SAMPL6 host–guest binding affinity prediction challenge. *Journal of Computer-Aided Molecular Design*, 2018. **32**(10): p. 937–963.

Chapter 9

Theranostics in cancer management

Medicinal compounds for dual imaging and therapy

Mohini Agarwal, Manish Kumar, Srinjoy Ruj, Chanchal Meena, and Kumud Bala
Therapeutics & Molecular Diagnostic Lab,
Centre for Medical Biotechnology, Amity University
Uttar Pradesh, Noida, Uttar Pradesh, India

9.1 INTRODUCTION

Cancer remains one of the leading causes of morbidity and mortality worldwide, posing a major public health challenge across both developed and developing nations. According to the latest GLOBOCAN 2022 data, an estimated 20 million new cancer cases and 9.7 million cancer-related deaths occurred globally, and these numbers are projected to rise significantly in the coming decades [1]. This escalating burden highlights the critical need for improved strategies for early detection, accurate diagnosis, and effective therapeutic intervention [2].

Traditionally, cancer management relies on conventional diagnostic approaches such as histopathological examination, radiological imaging (X-ray, computed tomography [CT], magnetic resonance imaging [MRI]), ultrasonography, and biomarker-based tests. While these methods play an essential role in tumor detection, staging, and treatment planning, they often face limitations in terms of sensitivity, specificity, real-time monitoring capability, and the ability to assess molecular characteristics of tumors. Similarly, conventional treatment modalities including surgery, chemotherapy, radiotherapy, hormone therapy, and immunotherapy have saved countless lives but often lack precise tumor selectivity. These treatments may lead to systemic toxicity, therapy resistance, heterogeneous clinical response, and inability to personalize treatment for individual tumor biology.

These limitations have driven the need for a paradigm shift toward more precise, targeted, and personalized cancer management strategies. In this context, theranostics (a fusion of therapy and diagnostics) has emerged as a transformative approach (Figure 9.1) [3]. Theranostics integrates molecular

DOI: 10.1201/9781003733010-9

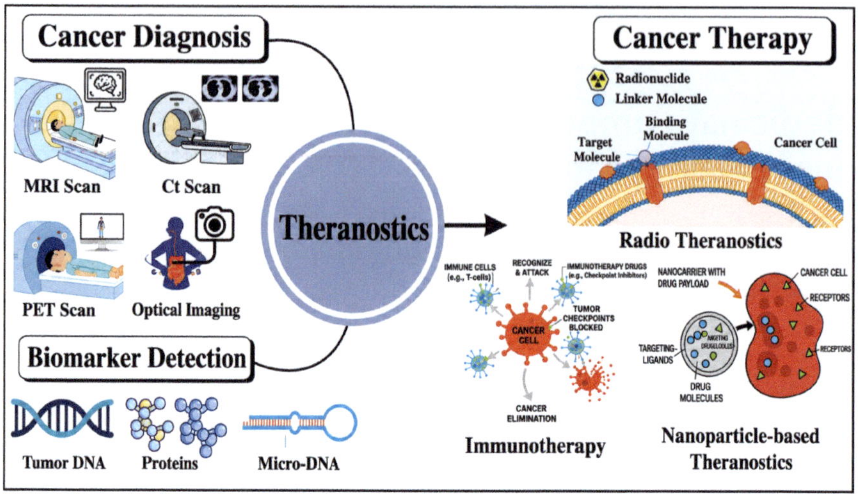

Figure 9.1 Schematic representation of theranostics illustrating the integration of cancer diagnosis, real-time molecular imaging, and targeted therapeutic interventions to enable personalized and precision-based cancer management.

imaging and targeted therapy within a single platform, enabling clinicians to visualize tumor-specific biomarkers, assess disease burden, guide patient selection, deliver targeted therapeutic agents, and monitor treatment response in real time [4]. By using the same or highly similar molecular compound for both imaging and therapy, theranostics ensures high target specificity, minimizes off-target toxicity, and facilitates personalized treatment decisions [5].

A central component of theranostic applications involves the development of medicinal compounds capable of dual imaging and therapy. These agents may include radiolabeled peptides, small molecules, antibodies, nanoparticles, or multifunctional nanocarriers engineered to deliver diagnostic isotopes, therapeutic radionuclides, chemotherapeutic drugs, or photothermal/photodynamic agents [6]. Such dual-functional compounds allow simultaneous tumor visualization and targeted treatment, thereby improving therapeutic precision, enhancing biodistribution, enabling controlled drug release, and reducing systemic side effects [7].

With continuous advancements in molecular biology, nanotechnology, radiopharmacy, and imaging sciences, theranostics is rapidly evolving into a cornerstone of precision oncology. The integration of dual imaging and therapeutic capabilities not only overcomes many limitations of conventional cancer diagnosis and treatment but also opens avenues for more efficient, personalized, and outcome-driven cancer care [8]. This chapter provides an in-depth overview of theranostics in cancer management, with a particular emphasis on medicinal compounds designed for dual imaging and therapy. The mechanisms, classes of compounds, clinical applications, challenges, and future directions of this rapidly evolving field are discussed.

9.2 HISTORY OF THERANOSTICS IN CANCER MANAGEMENT

Theranostics originated in the 1940s when Saul Hertz demonstrated that radioactive iodine could both image and treat thyroid disease, establishing ^{131}I as the first dual diagnostic–therapeutic radiopharmaceutical and laying the foundation for targeted radionuclide therapy [9]. The emergence of molecular imaging in the 1970s–1990s, particularly ^{18}F-FDG positron emission tomography (PET), revolutionized cancer diagnosis by enabling visualization of metabolic activity and providing a basis for molecularly guided therapy selection [10]. The introduction of monoclonal antibody technology in 1975 further advanced targeted approaches by enabling antibody-based imaging and drug delivery, contributing to early theranostic concepts [11]. By the 1980s–2010s, receptor-targeted theranostics matured with somatostatin analogs, culminating in ^{177}Lu-DOTATATE peptide receptor radionuclide therapy (PRRT) and its successful phase III validation and regulatory approval for neuroendocrine tumors (NETs) [12]. Expansion into broader oncology occurred in the 2000s–2010s with the approval of ^{223}Ra for metastatic prostate cancer, demonstrating survival benefit and validating targeted radionuclide therapy beyond endocrine tumors [13]. The modern era (2010s–2020s) is defined by PSMA theranostics: ^{68}Ga-PSMA-PET transformed prostate cancer imaging, while ^{177}Lu-PSMA-617 became the first major radioligand therapy (RLT) approved for a common solid tumor [14]. Advances in total-body PET, hybrid PET/MRI, quantitative single-photon emission computed tomography (SPECT), and emerging radionuclides such as ^{225}Ac and ^{161}Tb have further strengthened precision dosing and therapeutic efficacy [10]. Concurrently, novel targets such as fibroblast activation protein (FAP) and gastrin-releasing peptide receptor (GRPR) and the rise of multifunctional nanotheranostic platforms have broadened theranostic applicability across malignancies [15]. Recent integration of artificial intelligence (AI)-driven dosimetry and pharmacokinetic modeling has positioned theranostics as a cornerstone of personalized oncology entering 2025 (Figure 9.2) [16].

9.3 MOLECULAR IMAGING MODALITIES IN THERANOSTICS

Molecular imaging technologies lie at the core of theranostics. These modalities generate anatomical, functional, and molecular information essential for tumor detection, staging, therapy planning, and post-treatment

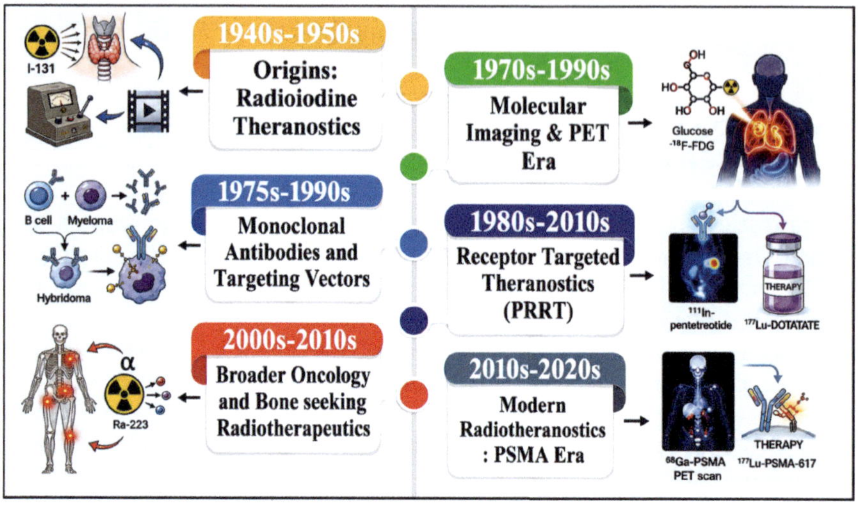

Figure 9.2 Timeline depicting the historical development of theranostics in cancer management, outlining major scientific and technological milestones that advanced the integration of molecular diagnostics, imaging, and targeted treatment strategies.

response assessment. Key techniques include PET, SPECT, MRI, fluorescence imaging, and photoacoustic imaging (PAI).

9.3.1 Positron emission tomography

PET is an exceptionally sensitive nuclear imaging modality that enables the visualization and quantification of biochemical and molecular processes in vivo by using positron-emitting radiotracers. These radiotracers are typically labeled with isotopes such as ^{18}F, ^{68}Ga, ^{64}Cu, ^{89}Zr, and ^{124}I, each offering distinct physical and biochemical properties suitable for diverse oncological applications. PET imaging provides high sensitivity, excellent quantitative accuracy, and the ability to noninvasively assess receptor expression, metabolic activity, tumor perfusion, and drug pharmacokinetics, making it indispensable in modern theranostic workflows. One of its major strengths lies in its capacity for precise patient selection for RLT, exemplified by the use of PSMA-PET imaging to identify prostate cancer patients eligible for ^{177}Lu-PSMA treatment [17,18]. In NETs, somatostatin receptor (SSTR) PET tracers enable optimal assessment prior to PRRT, while fibroblast activation protein inhibitor (FAPI) PET has emerged as a powerful stromal-targeted modality with applications across multiple tumor types, including desmoplastic and treatment-resistant cancers [19,20]. Recent technological advancements, particularly total-body and long-axial field-of-view PET scanners, have further expanded the diagnostic and theranostic utility of

PET by enabling whole-body dynamic imaging, high-resolution pharmacokinetic evaluation, and personalized dosimetry in real time. Collectively, these features position PET as a central pillar in cancer theranostics, guiding target validation, treatment planning, therapy monitoring, and longitudinal assessment of therapeutic response [21,22].

9.3.2 Single-photon emission computed tomography

SPECT is a widely used nuclear imaging modality that employs γ-emitting radionuclides such as ^{99}mTc, ^{111}In, and ^{123}I to generate three-dimensional functional images of biological processes in vivo. Compared with PET, SPECT is generally more accessible, cost-effective, and available in routine clinical settings, making it an important tool in global nuclear medicine practice [23,24]. Hybrid SPECT/CT systems further enhance its capability by providing precise anatomical localization alongside functional information, improving diagnostic accuracy in oncology, cardiology, and endocrine disorders [24]. Within theranostics, SPECT plays a critical role in therapy planning and monitoring. It is especially valuable for post-therapy imaging and patient-specific dosimetry following radioligand treatment, such as ^{177}Lu-DOTATATE PRRT, where SPECT quantifies radiopharmaceutical biodistribution and organ doses to optimize treatment safety and efficacy [25,26]. Additionally, SPECT is used to verify tumor uptake prior to therapy, particularly in cases where pre-printed PET tracers are unavailable, and serves as a reliable modality for routine monitoring during radionuclide therapies [25]. Through its ability to map therapeutic agent distribution, evaluate treatment response, and guide dosimetry calculations, SPECT continues to be an indispensable component of modern theranostic workflows, complementing PET and other imaging techniques in personalized cancer management.

9.3.3 Magnetic resonance imaging

MRI is a powerful, non-ionizing imaging modality that uses strong magnetic fields and radiofrequency pulses to generate high-resolution anatomical and functional images, making it indispensable in oncologic diagnosis and treatment planning. Its exceptional soft-tissue contrast allows precise visualization of tumor margins, infiltration patterns, and microstructural characteristics, which are often difficult to detect with CT or conventional nuclear imaging [27]. In addition to anatomical imaging, advanced MRI techniques, including diffusion-weighted imaging, functional MRI, and dynamic contrast-enhanced perfusion imaging, provide insights into tumor cellularity, vascularity, hypoxia, and treatment-induced physiological changes, thereby contributing critical information for tumor characterization and response assessment [28,29]. MRI is also central to theranostic innovation, as various contrast agents, such as gadolinium chelates and superparamagnetic iron oxide nanoparticles, can be engineered to function as both imaging probes and therapeutic carriers; these materials enable targeted delivery, controlled drug release, and molecular tracking of therapeutic interventions [30]. In nanotheranostics, MRI provides a versatile platform for real-time monitoring of nanoparticle biodistribution, guiding treatment delivery, and evaluating therapeutic outcomes, while its compatibility with photothermal and photodynamic agents allows combined imaging and therapy within a single integrated system [30]. Owing to its multi-parametric capabilities and absence of radiation exposure, MRI continues to play an essential role in the development of precision theranostic strategies for cancer management.

9.3.4 Fluorescence imaging

Fluorescence imaging is a highly sensitive optical modality that employs fluorophores, fluorescent proteins, or near-infrared (NIR) dyes to visualize cellular features, track therapeutic carriers, and enable real-time image-guided interventions. Its sensitivity allows detection of extremely low concentrations of imaging agents, making it valuable for monitoring drug delivery pathways and nanoparticle biodistribution in preclinical and clinical theranostic studies [31]. Technological advances, particularly within the second NIR biological window (NIR-II, 1,000–1,700 nm), have significantly improved tissue penetration depth, reduced autofluorescence, and enhanced signal-to-noise ratios compared with conventional visible or NIR-I imaging [32]. In theranostic applications, fluorescence imaging is widely used to track fluorescently labeled drug-loaded nanoparticles, enabling simultaneous visualization and therapy delivery. Furthermore, fluorescence-guided surgery has become a major clinical application, using agents such as indocyanine green (ICG) or targeted fluorescent probes to delineate tumor margins with high accuracy during oncologic operations [33]. Hybrid fluorescence probes that combine optical imaging with PET or MRI capabilities further expand the theranostic potential of this modality, allowing clinicians to integrate functional, molecular, and anatomical information in a single imaging workflow [34]. These advantages collectively position fluorescence imaging as a versatile and indispensable tool in the implementation of precision theranostics.

9.3.5 Photoacoustic imaging

PAI is an innovative hybrid modality that synergistically combines the high optical contrast of light absorption with the deep tissue penetration and spatial resolution of ultrasound. When short laser pulses are absorbed by endogenous chromophores such as hemoglobin and melanin or by engineered exogenous agents such as gold nanorods and carbon nanotubes, localized thermoelastic expansion generates ultrasonic waves that are detected to form images [35]. Because ultrasound scatters minimally in biological tissues, PAI achieves significantly deeper penetration than conventional fluorescence imaging while maintaining high molecular contrast, making it suitable for visualizing vascular networks, tumor oxygenation, and nanoparticle accumulation [36]. In theranostics, PAI is uniquely suited for integration with photothermal therapy (PTT) since many photothermal agents used for therapy (e.g., gold nanorods, graphene oxide, carbon nanotubes) simultaneously serve as efficient photoacoustic contrast enhancers [37]. This dual functionality enables PAI to guide PTT procedures in real time, providing feedback on treatment zones and monitoring therapeutic efficacy. Additionally, PAI facilitates dynamic tracking of therapeutic nanoparticles, allowing clinicians and researchers to evaluate biodistribution and optimize treatment strategies with exceptional temporal and spatial resolution [38]. As a rapidly evolving modality, PAI continues to demonstrate substantial promise in improving diagnostic accuracy, guiding therapy, and enhancing patient-specific theranostic protocols.

9.4 THERAPEUTIC MODALITIES IN THERANOSTICS

Theranostic systems encompass a broad spectrum of therapeutic strategies ranging from conventional chemotherapy to highly targeted RLT, light-activated phototherapies, and advanced immunomodulatory appro aches. Each therapeutic modality can be paired with an imaging component, creating dual-functional platforms that enable simultaneous diagnosis, real-time treatment monitoring, and precision therapy delivery. Increasingly, multimodal theranostic platforms are being designed to combine multiple therapeutic mechanisms within a single nanocarrier or radiopharmaceutical, offering synergistic tumor killing while minimizing toxicity to healthy tissues.

9.4.1 Chemotherapy-based theranostics

Chemotherapy-based theranostics integrates cytotoxic drugs with imaging agents and targeting ligands to improve drug selectivity, reduce systemic toxicity, and enable real-time visualization of therapeutic distribution. Nanocarriers such as liposomes, polymeric nanoparticles, and iron oxide particles encapsulate chemotherapeutic molecules and exploit the enhanced permeability and retention (EPR) effect, allowing drugs to accumulate preferentially in tumor tissues due to leaky vasculature and impaired lymphatic drainage [39]. Active targeting strategies using peptides, antibodies, or small molecules further enhance tumor-specific delivery and reduce off-target exposure. The addition of imaging moieties such as MRI contrast metals, fluorophores, or radiotracers allows clinicians to track biodistribution and release kinetics noninvasively [40]. Representative examples include doxorubicin-loaded iron oxide nanoparticles that serve as both MRI contrast agents and drug carriers, enabling image-guided chemotherapy, as well as fluorescently labeled paclitaxel liposomes developed to simultaneously visualize and treat tumor tissue [41,42]. These multifunctional platforms illustrate how chemotherapy can be integrated within a theranostic framework to enhance efficacy and personalize treatment strategies.

9.4.2 Radiotherapy and radioligand therapy

RLT represents the backbone of modern theranostics, utilizing radionuclides linked to molecular vectors that selectively bind tumor-associated targets. RLT relies on therapeutic isotopes such as β-emitters (^{177}Lu, ^{90}Y), α-emitters (^{225}Ac, ^{213}Bi), and Auger emitters (^{161}Tb, ^{125}I), each offering distinct radiation penetration ranges and cytotoxic profiles suitable for different tumor sizes and biological contexts [22,43]. These isotopes are conjugated to targeting ligands including peptides, antibodies, and small molecules that direct radiation selectively to malignant cells while sparing surrounding tissues. A key advantage of RLT is its built-in companion imaging: diagnostic PET or SPECT using isotopic analogs (e.g., ^{68}Ga-DOTATATE or ^{68}Ga-PSMA-11) verifies target expression and predicts therapeutic uptake before treatment [17,44]. Clinically validated examples include ^{177}Lu-DOTATATE for SSTR-positive NETs and ^{177}Lu-PSMA-617 for metastatic castration-resistant prostate cancer, both of which demonstrate significant improvements in survival and quality of life [22,43]. Emerging α-therapies, particularly ^{225}Ac-labeled agents, offer potent treatment options for resistant and micrometastatic disease due to their high linear energy transfer and short path length [45]. Thus, RLT exemplifies the precision and personalization central to theranostic medicine.

9.4.3 Photothermal therapy

PTT is a minimally invasive therapeutic modality that employs photothermal agents capable of converting NIR light into heat to induce localized thermal ablation of tumor cells. Common photothermal materials include gold nanorods, graphene oxide, carbon nanotubes, and various conductive polymers, each exhibiting strong NIR absorption and efficient photothermal conversion [46,47]. When functionalized with tumor-targeting ligands or encapsulated in nanocarriers, these agents selectively accumulate within tumors, allowing precise destruction of malignant tissues upon laser irradiation. PTT is frequently integrated within theranostic systems, as many photothermal agents also function as imaging contrast enhancers for PAI, MRI, or fluorescence, enabling real-time visualization and temperature monitoring during treatment [38]. The combination of imaging and therapy improves safety, guides laser placement, and prevents overtreatment. PTT's high spatial precision, minimal invasiveness, and compatibility with multimodal imaging make it an attractive component of advanced cancer theranostics.

9.4.4 Photodynamic therapy

Photodynamic therapy (PDT) utilizes photosensitizers that, when activated by specific wavelengths of light, generate reactive oxygen species (ROS) that exert cytotoxic effects on tumor cells. Photosensitizers naturally possess fluorescent properties, enabling intrinsic imaging capabilities that support tumor visualization, treatment guidance, and therapeutic monitoring within theranostic frameworks [48]. Incorporation of photosensitizers into nanocarriers further enhances tumor selectivity, improves solubility, and enables co-delivery of imaging dyes or contrast agents for multimodal imaging such as fluorescence, MRI, or PET [49]. PDT offers several advantages, including localized action, minimal invasiveness, and reduced systemic toxicity. Moreover, its synergistic potential with PTT and chemotherapy has motivated the development of triple-functional theranostic nanoplatforms that allow simultaneous imaging, ROS generation, and thermal ablation [50]. This modularity has established PDT as a valuable component of integrated imaging–therapy systems for cancer management.

9.4.5 Immunotherapy-based theranostics

Immunotherapy-based theranostics combines immune modulation strategies with advanced imaging to monitor immune activity, predict response, and guide treatment. Molecular imaging tracers targeting PD-1/PD-L1, CTLA-4, or other immune checkpoints have enabled noninvasive assessment of immune status and therapeutic response in real time, overcoming limitations of biopsy-based evaluation [51]. Nanocarriers co-loaded with immunostimulants (e.g., cytokines, adjuvants) and imaging agents allow simultaneous activation and tracking of immune pathways, enhancing therapeutic precision [52]. Emerging approaches include reporter gene imaging for CAR-T cell therapy, enabling visualization of CAR-T trafficking, expansion, and persistence within tumors [53]. By merging immunotherapy with imaging, theranostic systems can better predict responders, monitor early treatment failures, and tailor patient-specific immunotherapeutic regimens.

9.4.6 Combination therapies in theranostics

Combination therapies represent one of the most promising directions in theranostics, leveraging synergistic mechanisms to maximize tumor eradication while minimizing resistance and toxicity. Platforms integrating PTT and PDT combine thermal ablation with ROS-mediated cytotoxicity, demonstrating superior tumor-killing efficiency compared with either therapy alone [50]. Chemotherapy–RLT combinations allow simultaneous eradication of both large tumor masses and micrometastatic disease, enhancing treatment depth and breadth [39]. Radio-immunotherapy platforms pair radiopharmaceuticals with immune checkpoint blockade to enhance antigen presentation and generate systemic antitumor immunity [51]. Nanoplatforms integrating three or more functions such as imaging, chemotherapy, PTT, PDT, and immunomodulation enable multi-mechanistic tumor targeting and real-time visualization of therapeutic delivery [49,50]. Such multimodal strategies exemplify the power of theranostics to tailor cancer treatment with unprecedented precision and efficacy (Figure 9.3).

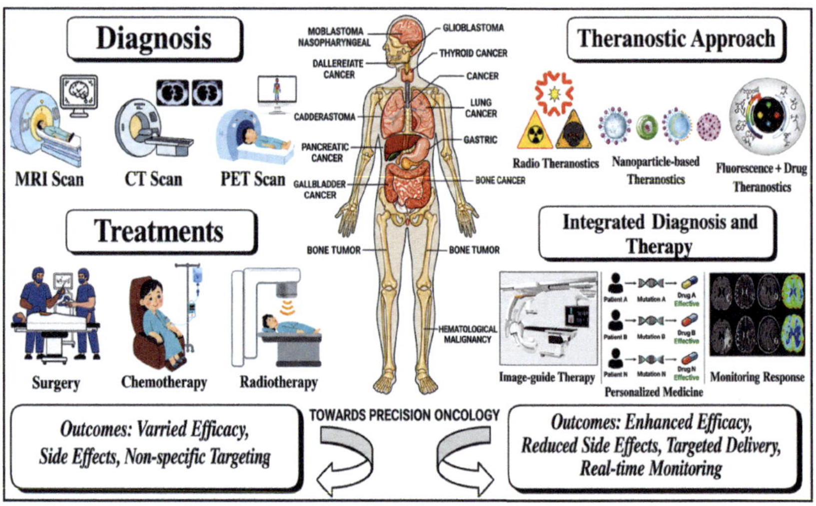

Figure 9.3 Schematic representation of cancer theranostics, highlighting the coordinated use of diagnostic biomarkers, molecular imaging agents, and targeted therapeutic compounds to simultaneously diagnose, monitor, and treat malignant lesions.

9.5 CLASSES OF MEDICINAL COMPOUNDS FOR DUAL IMAGING AND THERAPY

9.5.1 Metal-based agents

Metal-based theranostic agents represent a powerful class of dual-function compounds that combine diagnostic imaging capabilities with therapeutic efficacy through exploitation of their unique physicochemical properties [54]. These metallic nanoparticles interact with magnetic fields, generate heat under external stimulation, emit radiation for both imaging and therapy, and serve as carriers for additional therapeutic payloads [55]. The selection of specific metal elements determines the imaging modality and therapeutic mechanism, enabling personalized treatment strategies tailored to individual cancer types [54].

9.5.1.1 Gadolinium-based theranostic agents

Gadolinium possesses seven unpaired electrons and relatively long electronic relaxation times, making it the most widely used paramagnetic element for positive-contrast MRI [56]. Gadolinium-based contrast agents reduce both T1 and T2 proton relaxation times, causing enhanced signal intensity in injected structures, including vessels, parenchyma, and tumor lesions [56]. Beyond imaging applications, gadolinium exhibits radiosensitization properties related to its high atomic number ($Z=57$), undergoing inner-shell ionization during radiotherapy that generates secondary electrons capable of inducing DNA damage in cancer cells [57,58]. AGuIX® gadolinium-based nanoprobes consist of polysiloxane networks surrounded by approximately 10 gadolinium chelates with ultrasmall size (<5 nm), exhibiting nearly ideal biodistribution for MRI-guided radiotherapy applications [59,60]. Multifunctional theranostic systems combining doxorubicin-loaded PEGylated gadolinium nanoparticles with graphene oxide functionalized by folic acid demonstrate excellent photothermal-chemotherapeutic efficacy with tumor-targeting specificity and diagnostic capability [61].

9.5.1.2 Terbium radionuclides for theranostics

Terbium offers an "isotopic menu" comprising four clinically relevant radionuclides: ^{149}Tb, ^{152}Tb, ^{155}Tb, and ^{161}Tb, each suited for specific diagnostic or therapeutic applications. The same chemical scaffold and targeting vector can be paired with imaging radionuclides (^{152}Tb for PET or ^{155}Tb for SPECT) during treatment planning and dosimetry, then switched to ^{161}Tb for therapy, enabling truly isotopic theranostic pairs [62]. Terbium-161 emits beta particles with longer path lengths (≤12 mm) suitable for heterogeneous tumors, combined with Auger electrons possessing high linear energy transfer (4–26 keV/μm) and limited range (<1 μm) ideal for targeting isolated cancer cells and micrometastases [62]. The cellular absorbed dose of ^{161}Tb exceeds lutetium-177 by more than threefold due to the high emission of electrons below 40 keV, theoretically requiring 40% lower absorbed dose to achieve metastatic control (Table 9.1) [63].

Table 9.1 Clinical trial outcomes comparing ^{161}Tb-PSMA-617 and ^{177}Lu-PSMA-617 therapies in metastatic castration-resistant prostate cancer including PSA response rates, radiographic efficacy, prior treatments, safety, and therapy line

Parameter	*^{161}Tb-PSMA-617 (pilot and registry studies)*	*^{177}Lu-PSMA-617 (VISION Trial)*	*References*
Patient population	mCRPC, refractory to prior PSMA-RLT	mCRPC, treated in an earlier line	[64,65]
Number of patients	6 (pilot), 18 (registry)	831 total patients	[64,65]
PSA50 response rate (≥50% PSA decline)	38.9% (7/18, registry); 50% (3/6, pilot)	46%	[64,65]
Radiographic partial response	44.4% (registry study)	Not reported	[64,65]
Prior PSMA-targeted therapy	Majority previously treated with ^{177}Lu-PSMA or ^{225}Ac-PSMA	Mostly PSMA-naïve in terms of RLT	[64,65]
Safety and tolerability	Favorable; comparable to ^{177}Lu	Well-established and acceptable	[64,65]
Therapy line	Later-line; heavily pretreated, refractory cases	Earlier-line use; less heavily pretreated	[64,65]

This table highlights that ^{161}Tb-PSMA-617 demonstrates comparable or superior efficacy with favorable safety in heavily pretreated populations relative to ^{177}Lu-PSMA-617 in earlier-line settings.

9.5.1.3 Copper radioisotopes for theranostics

Copper-64 ($T_1/_2$=12.7 hours) serves as both a positron emitter for PET imaging and a β-emitter for radiotherapy, while copper-67 ($T_1/_2$=61.8 hours) functions as a β- and γ-emitter appropriate for SPECT imaging and targeted radiotherapy [66]. The longer half-life of cyclotron-produced copper-64 provides logistical advantages over fluorine-18 ($T_1/_2$=109 minutes) and gallium-68 ($T_1/_2$=68 minutes), enabling extended tumor imaging, metastatic detection, and distribution away from cyclotron facilities [66]. Copper-64/67-based radiopharmaceuticals targeting HER2-positive breast cancer show very high tumor uptake on PET imaging with [^{64}Cu] CuSar-trastuzumab, followed by radionuclide therapy using [^{67}Cu] CuSar-trastuzumab for the treatment of immunotherapy-resistant disease [67]. Targeted copper theranostics deliver compelling combinations of high accuracy and precision across diverse cancer types while providing scalability and cost-effectiveness advantages over current theranostic platforms.

9.5.1.4 Manganese oxide nanoparticles

Manganese(II) ions possess five unpaired electrons in 3d orbitals that produce large magnetic moments causing nearby water proton relaxation, making manganese oxide (MnO) nanoparticles ideal candidates for T1-weighted MR contrast agents with bright imaging characteristics [68]. Surface modifications including polymer functionalization, silica coating, and phospholipid modification enhance the relaxation rate and biocompatibility of MnO nanoparticles for clinical applications [68]. Manganese-doped iron oxide theranostic nanoparticles enable dual T1–T2 contrast MRI while simultaneously serving as PTT agents, demonstrating higher contrast imaging effects without local tissue damage under strong MRI magnetic fields [68]. Multifunctional MnO-based nanotheranostics co-loaded with chemotherapeutic agents like doxorubicin or docetaxel provide MRI-guided chemotherapy with real-time monitoring of drug delivery and therapeutic response [68].

9.5.1.5 Gold and platinum nanoparticles

Gold nanoparticles exhibit exceptional photothermal conversion efficiency, converting NIR light to localized heat for tumor ablation while serving as contrast agents for CT and PAI [69]. Platinum nanoparticles function as radiosensitizers and chemotherapeutic agents while providing imaging capabilities, leveraging their high atomic number for enhanced X-ray attenuation in CT imaging [70].

9.5.1.6 Iron oxide nanoparticles

Superparamagnetic iron oxide nanoparticles serve as negative T2-weighted MR contrast agents while enabling magnetic hyperthermia therapy and magnetically guided drug delivery to tumor sites [71]. External magnetic fields precisely direct iron oxide nanoparticles to target tissues and generate localized heating through energy dissipation under alternating magnetic fields, destroying cancer cells through thermal ablation [72]. Polymer-coated iron oxide nanoparticles improve biocompatibility while enabling surface functionalization with targeting ligands, therapeutic drugs, and additional imaging modalities for comprehensive theranostic applications [73]. Hybrid platforms combining iron oxide with MnO create dual T1–T2 contrast agents that overcome the inherent dark imaging limitations of iron oxide alone [74].

9.5.2 Organic/molecular compounds for dual imaging and therapy

9.5.2.1 Polydopamine-based nanoparticles

Polydopamine (PDA) nanoparticles exhibit remarkable versatility as theranostic platforms due to their inherent photothermal conversion properties, metal ion chelation capacity, drug-loading capability, and excellent biocompatibility [75]. PDA nanoparticles serve as multifunctional carriers capable of simultaneously loading ICG, chemotherapeutic drugs like doxorubicin, and paramagnetic ions such as manganese for T1-weighted MRI [75]. Under NIR laser irradiation at 808 nm, PDA-based nanoplatforms demonstrate excellent photothermal conversion efficiency, generating localized hyperthermia that destroys cancer cells while enabling real-time thermal imaging [75]. The abundant catechol groups on PDA surfaces facilitate strong coordination with metal ions and provide reactive sites for surface functionalization with targeting ligands such as hyaluronic acid or RGD peptides, enhancing tumor-specific accumulation [75].

9.5.2.2 *Peptide-based theranostic agents*

Peptide-modified nanovectors represent highly selective theranostic platforms capable of simultaneously carrying therapeutic drugs and imaging probes to target cancer cells with minimal off-target effects [76]. Targeting peptides such as RGD (recognizing $\alpha v\beta 3$ integrins), VEGF-binding sequences, and M2-macrophage-targeting peptides (YEQDPWGVKWWY) [77] enable precise tumor cell recognition and enhanced nanoparticle accumulation at malignant sites [76]. Peptide-based materials hold great promise as immunotherapeutic agents by modulating tumor-associated macrophage (TAM) populations, blocking VEGF-associated signaling pathways, and triggering host immune responses that cause sustained tumor regression [78].

RGD peptide-conjugated doxorubicin nanoparticles (RGD-Dox-NPs) demonstrate superior tumor-targeting ability compared to control peptides, achieving 23% reduction in primary tumor growth and 82% reduction in metastasis in orthotopic pancreatic tumor models with hepatic lymph node metastases [79]. pH-responsive peptide-modified polymeric micelles loaded with doxorubicin (Dox-AP-pH-PMs) exhibit excellent anticancer therapeutic efficacy in human breast tumor-bearing mice, outperforming both free doxorubicin and non-targeted formulations. Peptide-based checkpoint blockade strategies targeting CTLA-4 and PD-1/PD-L1 pathways enhance effector T cell activity while downregulating immunosuppressive Treg capacity, with six FDA-approved antibodies now targeting these pathways across diverse cancer types [80].

9.5.2.3 *Nitroreductase-activable theranostic agents*

Nitroreductase (NTR)-activable agents offer personalized and targeted approaches to cancer theranostics by selectively activating prodrugs within hypoxic tumor microenvironments where NTR expression is significantly elevated [81]. Hypoxia is a common characteristic of solid tumors associated with resistance to radiotherapy and chemotherapy, resulting in poor prognosis and high probability of tumor recurrence. NTR catalyzes the reduction of nitro compounds to corresponding amine derivatives in the presence of NADH as an electron donor, serving as an indicative biomarker of tumor hypoxia whose levels directly correlate with hypoxic degree [82].

The hypoxia-triggered probe IR1048-MZ demonstrates precise tumor imaging through background-free NIR-II fluorescence and deeper-penetrating photoacoustic (PA) imaging, combined with NTR-activated PTT that overcomes cellular resistance during hypoxia. Upon NTR-mediated activation in hypoxic tumors, IR1048-MZ emits strong NIR-II fluorescence and PA signals, enabling accurate tumor localization and deep tissue penetration, while exhibiting significant temperature increases during PTT that effectively ablate tumors without recurrence. In vivo studies demonstrate that untreated hypoxic tumors exhibit strong fluorescence, indicating high NTR levels, whereas NTR inhibitor-treated groups show weak fluorescence, confirming selective hypoxia-triggered activation [83].

9.5.2.4 Porphyrin-based photosensitizers

Porphyrin and porphyrin-related compounds represent the most used photosensitizers for PDT, combining therapeutic efficacy with intrinsic fluorescence imaging capabilities for theranostic applications [49]. Porphyrin-based organic nanoparticles with strong emission in the NIR-IIa window (NIR-IIa, 1,300–1,400 nm) enable deep tissue imaging with minimal autofluorescence background while generating ROS for cancer cell destruction. The conjugation of porphyrins with perylene diamide (PDI) in donor-acceptor (D-A) structures enhances absorption in the NIR region, improves intersystem crossing efficiency, and increases spin–orbit coupling for efficient triplet state generation required for ROS production [84].

Porphyrin photosensitizers achieve therapeutic effects through both Type I mechanisms (direct electron transfer producing superoxide radicals) and Type II mechanisms (energy transfer to molecular oxygen generating singlet oxygen), with computational screening identifying optimal candidates capable of fulfilling both pathways [85]. Porphyrin nano-metal-organic frameworks demonstrate exceptional multifunctionality as theranostic agents, combining imaging modalities (fluorescence, photoacoustic, MRI) with therapeutic functions (PDT, PTT, chemotherapy) within stable crystalline architectures [86]. FDA-approved porphyrin-based photosensitizers including Photofrin, Foscan, and Visudyne have demonstrated clinical success in treating lung, bladder, gastric, and cervical cancers, although skin photosensitivity and limited light penetration depth remain challenges [87].

Next-generation porphyrin photosensitizers with long-wavelength absorptions (>700 nm) increase light penetration depth for treating larger tumors, while activatable porphyrin systems achieve local activation to minimize damage to adjacent healthy tissues. Porphyrin nanocarriers, including polymers, micelles, dendrimers, and liposomes, combined with tumor-targeting strategies improve delivery efficiency and therapeutic outcomes. Computational studies demonstrate that porphyrin–PDI conjugates exhibit sufficient S1-T1 energy gaps (ΔEST) and enhanced spin–orbit coupling, increasing triplet state lifetimes and improving chances of reacting with oxygen to generate ROS [87].

9.5.2.5 Indocyanine green-based systems

ICG is an FDA-approved NIR cyanine dye with over 50 years of clinical use in retinal angiography and cardiovascular/liver function assessment, now extensively investigated for cancer theranostic applications [88]. ICG exhibits strong absorption in the NIR spectrum (700–900 nm), enabling bioimaging with high signal-to-background ratios due to minimal absorption by biological tissues, combined with efficient photothermal conversion for hyperthermia-based cancer therapy [89,90]. ICG-incorporating nanoparticles overcome the limitations of free ICG, including rapid plasma clearance, high protein binding, photobleaching, and concentration-dependent aggregation [89].

Assembly-enhanced ICG nanoparticles achieve high entrapment efficiency (up to 94.5%) through manipulation of electrostatic interactions and noncovalent bonding, maintaining intrinsic fluorescence emission while significantly improving photothermal conversion stability and resistance to complex physiological conditions [91]. Calcium phosphate nanoparticles co-loaded with ICG and gadolinium, then labeled with 99m-technetium-methylene diphosphonate, enable combined optical, magnetic, and nuclear imaging for comprehensive tumor visualization [92]. Polydopamine nanoparticles loaded with ICG, doxorubicin, and manganese ions (PDA-ICG-PEG/DOX(Mn)) demonstrate MRI-guided combined chemotherapy and PTT with remarkable synergistic effects and minimal side effects.

After systemic administration, ICG nanoparticles exhibit preferential accumulation in the liver and steroid-rich organs (adrenal glands, ovaries), with distribution to intestines, kidneys, and lymph nodes, requiring careful formulation design to enhance tumor-specific targeting. Polyethyleneimine coatings protect against ICG leaching from nanocarriers, improving stability and prolonging circulation time. The clinical translation of ICG-based theranostics benefits from its established safety profile and FDA approval status, accelerating development timelines compared to novel compounds.

9.5.3 Nanomaterials for theranostics

Nanotheranostics integrate diagnostic imaging and therapeutic functions within single nanoparticle platforms, enabling simultaneous visualization, drug delivery, and real-time monitoring of treatment response.

These multifunctional nanoscale agents leverage unique physicochemical properties, including high surface area-to-volume ratios, tunable optical characteristics, and capacity, for multivalent surface functionalization to overcome limitations of conventional cancer diagnosis and therapy. Recent advances emphasize the integration of preclinical MRI, CT, PET, SPECT, and PAI with targeted drug delivery systems, providing comprehensive molecular imaging capabilities that enhance diagnostic accuracy and therapeutic precision [93].

Bimetallic hyaluronate-modified gold-platinum (HA-Au@Pt) nanoparticles demonstrate enhanced absorbance in the NIR region and broadened absorption through surface plasmon resonance coupling effects, enabling noninvasive cancer theranostics with simultaneous PTT and imaging in skin cancer models [94]. Polymer-based nanotheranostics incorporating fluorochromes for optical imaging, nanobubbles for ultrasound, and metal chelates for PET/SPECT provide high-sensitivity, multichannel imaging platforms suitable for target confirmation and compound optimization [95]. Integration of nanotheranostics into immunotherapy, gene therapy, and personalized medicine enhances diagnostic and therapeutic effects through data fusion from different imaging modalities [95].

9.5.3.1 Lipid-based nanoparticles for theranostics

Lipid nanoparticles have emerged as versatile carriers for cancer theranostics due to their exceptional biocompatibility, safety profile, high drug-loading capacity, prolonged in vivo circulation half-life, and substantial tumor site accumulation [96]. Different lipidic nanocarriers, including liposomes, nanoemulsions, solid lipid nanoparticles, nanostructured lipid carriers, and lipid nanocapsules, provide diverse structural platforms adaptable to various imaging agents and therapeutic payloads [96]. Liposomes encapsulating MRI contrast agents such as iron oxide or gadolinium-based compounds together with therapeutic agents enable noninvasive real-time monitoring of pharmacokinetics and tumor accumulation [96].

Targeted liposomal theranostics modified with peptides recognizing $\alpha v\beta 3$ or $\alpha v\beta 6$ integrins demonstrate higher tumor accumulation compared to non-targeted liposomes through integrin-mediated endocytosis, resulting in enhanced tumor retardation. pH-sensitive lipid nanoparticles [97] loaded with photosensitizers exhibit superior PDT-induced tumor growth inhibition compared to non-pH-sensitive formulations by facilitating nuclear delivery of photosensitizers rather than lysosomal entrapment [98]. Perfluorohexane-containing nanoemulsions with concurrent laser excitation capability demonstrate tremendous potential for destroying tumor cells through combined acoustic and photothermal effects, emerging as competent theranostic agents [99].

Novel amphiphilic molecules engineered with structure-based fluorescent, photoacoustic, and magnetic resonance properties enable activatable multimodal imaging with prolonged tumor retention time for repeated PDT treatments in melanoma xenograft models [100]. The synergistic combination of anticancer drugs with photothermal effects in liposomal theranostics exhibits superior tumor inhibition compared to standalone monotherapy approaches [101]. Temperature-sensitive liposomes co-loaded with MRI contrast agents and chemotherapeutics enable hyperthermia-triggered drug release with real-time MRI monitoring, demonstrating strong tumor growth suppression when mild heating (44°C for 10 min) is applied at optimal time points post-injection [102].

9.5.3.2 Carbon-based nanomaterials

Carbon dots exhibit high tunable fluorescence, improved aqueous solubility, biocompatibility, bioavailability, chemical inertness, and considerably low photobleaching, making them highly useful for bioimaging, drug delivery, and cancer diagnosis [103]. Sulfur and nitrogen co-doped NIR carbon dots demonstrate strong NIR emission peaking at 720 nm with exceptionally high photothermal conversion efficiency (59.19%) under 655 nm laser irradiation, achieving rapid tumor temperature elevation to 55°C–71°C sufficient for irreversible tissue damage. Carbon dots serve as potential drug carrier vehicles through surface modifications that enable targeted delivery while maintaining intrinsic fluorescence for real-time tracking [103,104].

Single-walled carbon nanotubes incorporated into NIR-II theranostic platforms provide deep tissue penetration capabilities with superior imaging resolution and precision-guided photothermal ablation [105]. Carbon dot-incorporated mesoporous silica nanoparticles (MSNs) synthesized via green approaches demonstrate potent and selective anticancer activity with maximum cell mortality of 66%–72% in breast cancer cells while exhibiting lower toxicity (40% mortality) in normal endothelial cells [106]. Fluorescence imaging capabilities of carbon dot-based nanoplatforms enable visualization of cellular uptake and intracellular distribution, with apoptosis-mediated cell death recognized as the primary anticancer mechanism [103].

The negative surface charge of carbon dots (zeta potential –20.1 mV) arising from abundant carboxyl and hydroxyl functional groups facilitates electrostatic interactions with biomolecules and cellular membranes, enhancing cellular internalization. Compared to organic fluorophores and semiconductor quantum dots, carbon dots exhibit superior photostability, resistance to photobleaching, and lower cytotoxicity, positioning them as next-generation theranostic agents [103]. The mass absorption coefficient of engineered carbon dots in the red to NIR region significantly exceeds that of graphene oxide, enabling efficient light harvesting for both imaging and therapeutic applications [103].

9.5.3.3 Polymeric nanoparticles for theranostics

Organic polymer nanoparticles demonstrate exceptional versatility for cancer theranostics through enhanced targeted drug delivery via passive tumor accumulation through the EPR effect and active targeting using surface-conjugated ligands [107]. Advances in polymer chemistry and fabrication methods including solvent evaporation, nanoprecipitation, emulsion-diffusion, and emulsion polymerization enable precise control over particle size, surface charge, and drug-loading efficiency, optimizing biodistribution and imaging performance. Hybrid polymer-inorganic nanoparticles expand functionality by incorporating magnetic, optical, or radiopaque components, enabling multimodal imaging and stimuli-responsive drug release while maintaining biocompatibility [107].

Biodegradable polymers such as poly(lactic-co-glycolic acid) (PLGA) and chitosan ensure safe metabolism without adverse immune reactions, supporting controlled release of contrast agents for sustained imaging signals during procedures [108]. Passive targeting through the EPR effect allows nanoparticles to preferentially accumulate in tumors due to leaky vasculature and deficient lymphatic drainage, effectively increasing local concentrations of contrast agents and therapeutic agents [109]. Active targeting strategies employing ligand functionalization, biomimetic coatings, and multifunctional nanoparticle designs continue to advance therapeutic specificity and imaging precision [110].

Immunogenic responses to polymeric nanoparticles involve complement system activation through classical, lectin, or alternative pathways leading to opsonization and rapid clearance, as well as Toll-like receptor (TLR) signaling that triggers cytokine release and inflammation [111]. Recent clinical investigations demonstrate that intravenously administered polymeric nanoparticles are well-tolerated even after repeated dosing over several weeks, with extracellular vesicle-based systems emerging as low-immunogenicity alternatives. These pivotal findings have enabled investigational new drug applications, marking major milestones in translating polymeric nanoparticle platforms toward human clinical use .

9.5.3.4 Mitochondria-targeting nanoparticles

Mitochondria-targeted nanoparticles (mitoNANO) represent an innovative strategy for selective, targeted, and safer cancer treatment by delivering therapeutic agents directly to the powerhouse organelles of malignant cells [112]. The selective delivery of nanoparticles to mitochondria provides an elegant approach to overcoming multidrug resistance (MDR) by evading drug efflux mechanisms mediated by P-glycoprotein and other ABC transporters localized at the plasma membrane. MitoNANO exert anticancer effects through multiple mechanisms including electron transport chain inhibition, respiratory blockade, modulation of the mitochondrial permeability transition pore, inhibition of anti-apoptotic Bcl-2 family proteins, and promotion of mitochondria-regulated apoptosis [112].

Acid-triggered nanocarriers for targeted nitric oxide delivery to cancer mitochondria demonstrate enhanced therapy against MDR and metastasis by suppressing ATP production, impairing mitochondrial function, and inhibiting P-glycoprotein activity [113]. Co-loading mitochondria-targeted nanoparticles with lonidamine achieves tenfold higher antitumor effects compared to free lonidamine, demonstrating the therapeutic advantage of subcellular organelle targeting. Selective targeting of mitochondria in the acidic tumor microenvironment leads to superior therapeutic outcomes through enhanced drug accumulation at sites of energy metabolism dysregulation characteristic of cancer cells [114].

Mitochondria-targeted nanoparticles can function as drug carriers or drug-free light-responsive materials capable of inducing mitochondrial impairment via PDT or PTT. The combination of mitochondrial targeting with light-mediated therapies offers considerable reduction in systemic adverse effects while maintaining high local cytotoxicity. This mitochondrial targeting approach holds potential for synergistic combination therapies that simultaneously disrupt energy production, induce oxidative stress, and trigger intrinsic apoptotic pathways, providing innovative solutions to MDR challenges in cancer treatment [112].

9.5.3.5 Mesoporous silica nanoparticles

MSNs have received considerable attention in cancer theranostics as carriers for poorly water-soluble drugs, imaging agents, and multifunctional therapeutic combinations. The large surface area, tunable pore size (2–10 nm), high loading capacity, excellent biocompatibility, and ease of surface functionalization make MSNs ideal platforms for integrating diagnostic and therapeutic modalities. MSN-based formulations demonstrate versatile applications including tissue engineering, wound healing, antibacterial effects, bioimaging, stem cell research, and anticancer tumor therapy [115].

Multifunctional MSN platforms (MSN-Man-DOX@PDA-Gd-PEOz) co-loaded with doxorubicin and gadolinium, then coated with polydopamine and mannose-targeting ligands, demonstrate synergistic chemotherapy-PTT with simultaneous MRI contrast enhancement for real-time disease monitoring in lung cancer xenograft models. Gold nanorod-embedded silica nanoparticles loaded with 5-fluorouracil and ICG (GNR@SiO2–5-FU-ICG) undergo pH-triggered drug release in acidic tumor microenvironments through protonation-induced loosening of electrostatic interactions, enabling combined chemotherapy, PDT, and fluorescence imaging [116].

Transferrin and hydroxypyridinone-modified MSNs loaded with radioactive isotopes (^{225}Ac-HOPO-MSNP and ^{238}Pu-HOPO-MSNP) target transferrin receptors overexpressed on breast cancer cells, increasing cytotoxic effects through targeted alpha particle therapy while demonstrating enhanced clearance in vitro and reduced bone deposition in vivo. Perfluorocarbon-loaded PEGylated mesoporous silica-titania nanophotosensitizers (PFH@P-MSTN) increase ROS generation and induce tumor cell death upon ultrasound exposure, combining sonodynamic therapy with diagnostic imaging. X-ray-stimulated synergistic radiotherapy and PDT using MSN-based platforms loaded with Rose Bengal achieve sixfold higher tumor regression in breast cancer-bearing mice compared to single-modality radiotherapy [117].

9.6 THERANOSTICS IN SPECIFIC CANCER TYPES

9.6.1 Breast cancer

Theranostic approaches in breast cancer target multiple molecular markers including human epidermal growth factor receptor 2 (HER2), FAP, SSTRs in neuroendocrine differentiation, and GRPR [118]. Bhusari et al. demonstrated the feasibility of in-house labeling of trastuzumab with ^{177}Lu via the DOTA chelator, showing 12-hour stability with image confirmation of targeted primary and metastatic lesions at 5 days post-injection. A study of six patients with HER2-positive metastatic breast cancer compared to four patients with HER2-negative disease confirmed selective tumor targeting [119]. Ballal et al. reported compassionate use of ^{177}Lu-DOTA-FAPI in a 31-year-old female with ER-, PR-, HER2+ breast cancer and metastatic brain lesions who had disease progression on conventional therapy, administering 3.2 GBq with target lesion confirmation on dosimetry images resulting in symptomatic relief [120]. Lindner et al. used 2.9 GBq of ^{90}Y-FAPI-04 to treat a patient with metastatic breast cancer, resulting in significant pain reduction without significant side effects, with bremsstrahlung images confirming intended target lesion delivery up to 1 day post-injection [118]. PSMA theranostics shows promise in breast cancer, as PSMA is highly expressed in tumor-associated neovasculature and could potentially serve as a theranostic biomarker.

9.6.2 Prostate cancer

Lutetium-177 PSMA-RLT represents a landmark advancement in metastatic castration-resistant prostate cancer (mCRPC) management [121]. The FDA-approved ^{177}Lu-PSMA-617 in March 2022 based on the VISION phase III trial, which demonstrated improved radiographic progression-free survival and overall survival when combined with standard therapy [122]. A retrospective analysis of 48 mCRPC patients with PSMA-avid diffuse bone metastases treated between May 2018 and January 2023 showed median overall survival of 9.3 months, with any PSA response observed in 75% of patients. Notably, 48% achieved ≥50% PSA reduction, correlating with longer median OS (11.6 vs 8.6 months, p=0.03). Median PSA progression-free survival was 3.2 months, with improved outcomes in patients achieving ≥50% PSA reduction (6.0 vs 1.8 months, $p<0.0001$) [123]. Pain relief was reported in 43% of patients with a median pain score reduction of five points. The 2025 FDA approval expansion now includes chemotherapy-naive patients, significantly broadening treatment eligibility [124].

9.6.3 Hepatocellular carcinoma

Yttrium-90 (^{90}Y) radioembolization has emerged as an effective palliative treatment for hepatocellular carcinoma, delivering high-dose internal radiation via the hepatic artery with preferential concentration in hypervascular tumors [123]. Radioembolization resulted in a significantly longer time to progression (13.3 months) versus transarterial chemoembolization (TACE) (8.4 months, $p=0.046$), though overall survival differences were not statistically significant (TACE: 17.4 months, radioembolization: 20.5 months, $p=0.232$). Downstaging to UNOS T2 was achieved in 31% of TACE patients versus 58% of radioembolization patients ($p=0.023$), with median time to overall progression of 12.8 months for TACE and 33.3 months for radioembolization ($p=0.005$). The short half-life of ^{90}Y (2.67 days) and preferential tumor accumulation via hepatic artery injection enable high radiation doses while limiting radiation-induced liver disease in cirrhotic patients. Some investigators now consider ^{90}Y radioembolization a preferred therapy for advanced Barcelona Clinic Liver Cancer (BCLC)B and early BCLC C patients based on superior time to progression, downstaging capability, and quality of life maintenance compared to TACE [123].

9.6.4 Thyroid cancer

Radioiodine I-131 represents the prototypical theranostic agent with over 50 years of clinical use, exemplifying integration of diagnostic imaging and therapeutic intervention using the same radiolabeled molecule. I-131 treatment is time-tested and effective for papillary and follicular thyroid cancer, the two most common types that efficiently absorb and concentrate iodine. After thyroidectomy, I-131 can find remnant and metastasized thyroid cancer cells throughout the body and destroy them by emitting beta particles that damage cellular DNA. Thyroid cells absorb nearly all iodine in the body, enabling highly selective targeting of thyroid cancer cells without significantly affecting other organs. The radiation is very targeted, mainly affecting the thyroid gland, nearby tissues, and distant thyroid cancer metastases, with excess I-131 eliminated through urine within days. Post-therapy I-131 scans serve dual purposes of confirming treatment delivery and detecting previously unknown metastatic sites, enabling restaging 6–12 months after thyroidectomy and ablation therapy [124].

9.6.5 Neuroendocrine tumors

SSTR theranostics revolutionized NET management through targeting SSTR2 overexpression characteristic of well-differentiated tumors. ^{177}Lu-DOTATATE received FDA approval in January 2018 (Europe in September 2017) for adult patients with SSTR-positive gastroenteropancreatic NETs [125]. The NETTER-1 trial demonstrated that median progression-free survival for the ^{177}Lu-DOTATATE group was not yet reached (estimated at 40 months) compared to 8.4 months for high-dose somatostatin analog control group, yielding a hazard ratio of 0.21 and a 79% reduction in risk of disease progression or death. Pre-treatment ^{68}Ga-DOTATATE PET/CT imaging confirms SSTR expression and predicts therapeutic response, with post-therapy imaging demonstrating excellent responses including significant reduction in size and tracer avidity of multiple liver metastases [126]. Pancreatic NETs often overexpress SSTR2, making them ideal theranostic targets, with ^{177}Lu-DOTATATE significantly extending progression-free survival (22.8 vs 8.4 months). A survey of referring physicians demonstrated substantial impact of ^{68}Ga-DOTATATE PET/CT on intended management of NET patients, fundamentally changing diagnostic and therapeutic paradigms (Table 9.2) [127].

9.7 MOLECULAR THERANOSTICS: PROGRESS IN DUAL-LABELED AND RECEPTOR-TARGETED STRATEGIES

Recent advances in molecular theranostics have focused on the development of dual-labeled and receptor-targeted agents that combine diagnostic imaging and therapeutic capabilities for enhanced cancer management. Among these, GRPR-targeting agents demonstrate high affinity and favorable tumor-to-background ratios in prostate cancer models, with imaging efficacy optimized through careful modulation of fluorophore structure, dose, and pharmacokinetics to balance fluorescence intensity and rapid

Table 9.2 Clinical theranostic advances across major cancer types, highlighting key molecular targets, supporting clinical evidence, and therapeutic outcomes

Cancer type	*Theranostic target(s)*	*FDA-approved*	*Principal findings and clinical validation*	*Clinical outcomes*	*References*
Breast cancer	HER2, FAP, SSTR (neuroendocrine differentiation), GRPR, PSMA (neovasculature)	No FDA approval yet	^{177}Lu-DOTA-Trastuzumab selectively targets HER2+ tumors; compassionate use of ^{177}Lu/^{90}Y-FAPI shows symptomatic benefits; PSMA also emerging.	Effective tumor targeting in HER2+, FAP+, and GRPR tumors with observed pain reduction and symptom relief in advanced cases.	[128]
Prostate cancer	PSMA	Yes; ^{177}Lu-PSMA-617 (2022, expanded 2025)	VISION trial showed 75% PSA response; responders had median OS of 11.6 months; strong radioligand uptake and tolerable toxicity profile.	Shows significant improvements in progression-free and overall survival, with notable pain relief and expanded treatment indications.	[129]
Hepatocellular carcinoma	^{90}Y radioembolization	No FDA approval	Superior to TACE in delaying progression and achieving tumor downstaging; radioligand shows high tumor uptake and selective radiation.	Demonstrates better time to progression, downstaging rates, and quality of life with lower liver toxicity risk.	[130]
Thyroid cancer	Sodium-Iodide Symporter (NIS) — I-131	Yes; established for over 50 years	Highly effective in papillary and follicular thyroid cancers due to selective iodide uptake.	Enables efficient ablation of remnant and metastatic cells with precise restaging and minimal side effects.	[131]
Neuroendocrine tumors	SSTR2	Yes; ^{177}Lu-DOTATATE (2018)	NETTER-1 trial showed major PFS extension (~40 vs 8.4 months); strong receptor targeting enables imaging and therapy alignment.	Extends survival, reduces tumor burden, and has transformed imaging and therapeutic protocols in NET management.	[125]

clearance. Similarly, MMP-2/9-targeted peptides dual-labeled with radiometal chelates and NIR dyes enable effective PET and NIR imaging with combined renal and hepatobiliary clearance, achieving robust tumor selectivity through rational conjugation strategies [132,133].

Theranostic radioligands continue to transform clinical oncology, notably with PSMA-targeted agents such as ^{177}Lu-PSMA-617 now established as a standard of care in metastatic castration-resistant prostate cancer, showing significant PSA reductions and survival benefits despite manageable hematologic toxicities [134]. Advances in SSTR-targeted approaches employ ^{68}Ga- and ^{18}F-labeled analogs to guide therapy in NETs, with ongoing exploration of mTOR inhibitor combinations and alternative radionuclides [135]. Moreover, FAP-targeted radioligands show promising early clinical efficacy by addressing the tumor stroma, a key contributor to therapeutic resistance in epithelial cancers, exemplifying the evolving strategy of combining targeted imaging and treatment to overcome traditional therapeutic barriers [132]. These developments illustrate the paradigm shift toward personalized, image-guided oncologic interventions that leverage molecular specificity for improved diagnostic accuracy and therapeutic outcomes (Table 9.3; Figure 9.4).

9.8 ONGOING CLINICAL TRIALS

Multiple phase 2 and 3 clinical trials are currently recruiting patients for theranostic interventions targeting PSMA, GRPR, and FAP with an anticipated near-term (<5 years) impact on clinical practice. These investigations aim to expand theranostic applications to additional cancer types, optimize dosing regimens, evaluate combination therapies with chemotherapy or immunotherapy, and identify predictive biomarkers for treatment response. Preclinical and early clinical studies continue to explore novel radioligands including α-emitting radionuclides (^{225}Ac, ^{213}Bi) paired with established targeting vectors for enhanced tumoricidal effects. Combination strategies integrating RLT with immune checkpoint inhibitors, targeted therapies, and external beam radiation are under active investigation to achieve synergistic anticancer effects (Table 9.4).

9.9 MULTIFUNCTIONALITY: DUAL IMAGING AND THERAPY WITH ACTIVATABLE AGENTS

Activatable molecular agents represent a major advancement in theranostics by exhibiting higher signal-to-noise ratios, real-time detection of cancer-associated biomarkers, lower normal tissue toxicity, and enhanced therapeutic effects compared to "always-on" systems. These agents remain in a

Table 9.3 Overview of emerging peptide-based and radioligand theranostic platforms, summarizing principal molecular targets, diagnostic and therapeutic performance metrics, and current stages of clinical translation

Specific target	*Key findings*	*Pharmacokinetic*	*Clinical impact*	*References*
GRPR-targeting dual-labeled compounds	Demonstrated strong binding in PC-3 prostate tumor models; Kd = 44–53 nM	Tumor-to-background ratio (TBR): 6.9%–8.5%ID/g; fluorescence–clearance balance required	Lower dose (0.75 nmol) → maximum imaging contrast; metabolic processing enhances tumor specificity	[132]
GRPR probe optimization	Fluorophore structural modifications affect imaging performance	Parent compound retention + tumor-specific metabolism enhances TBR	Optimization needed for clinical translation	[132]
MMP-2/9 dual-labeled peptide (KKAHWGFTLD)	Successful PET/NIR imaging in heterotopic ossification	Efficient renal clearance; TBR = 4.77 ± 0.26%ID/g at 12 hours post-injection	Dual-labeled thiol-yne probes show mixed renal + hepatobiliary excretion	[132]
PSMA-targeted radioligand therapy (^{177}Lu-PSMA)	Phase 2/3 trials underway; major upcoming clinical impact	Acceptable toxicity profile: anemia (27%), neutropenia (27%), thrombocytopenia (21%)	Improves survival & reduces symptoms in Mcrpc	[135]
PSMA therapy response	Significant PSA declines ≥50% in many patients	Higher PSA drop correlates with improved OS and PFS	Strong option for mCRPC with bone metastases	[135]
Somatostatin receptor (SSTR) targeting	^{18}F-labeled analogs emerging as alternatives to ^{68}Ga-DOTATATE	Longer half-life aids distribution; SSTR heterogeneity requires pre-therapy imaging	SSTR2-rich well-differentiated NETs benefit from ^{177}Lu-DOTATATE; combination regimens under study	[136]
SSTR patient selection	^{68}Ga-DOTATATE PET/CT identifies ideal responders	—	Ensures therapy effectiveness by confirming SSTR expression	[136]

(*Continued*)

Table 9.3 (Continued) Overview of emerging peptide-based and radioligand theranostic platforms, summarizing principal molecular targets, diagnostic and therapeutic performance metrics, and current stages of clinical translation

Specific target	*Key findings*	*Pharmacokinetic*	*Clinical impact*	*References*
Fibroblast activation protein (FAP) targeting	FAP highly expressed in >90% epithelial tumors; abundant in pancreatic cancers	FAP+ CAFs drive invasion, metastasis, immunosuppression	Radioligands like ^{90}Y-FAPI-04 and ^{177}Lu-DOTA-FAPi show early therapeutic benefits, promising for stromal-rich tumors	[137]
FAP biology	FAP present in CAFs, CTCs, stromal cells	—	Attractive target for desmoplastic & treatment-resistant malignancies	[137]
GRPR-targeted radioligands (therapy)	First-in-human studies with ^{177}Lu-RM2 show effective tumor dosing	Fast clearance from pancreas (dose-limiting organ)	High tumor uptake with ^{177}Lu-Bombesin; enhanced efficacy with rapamycin	[138]

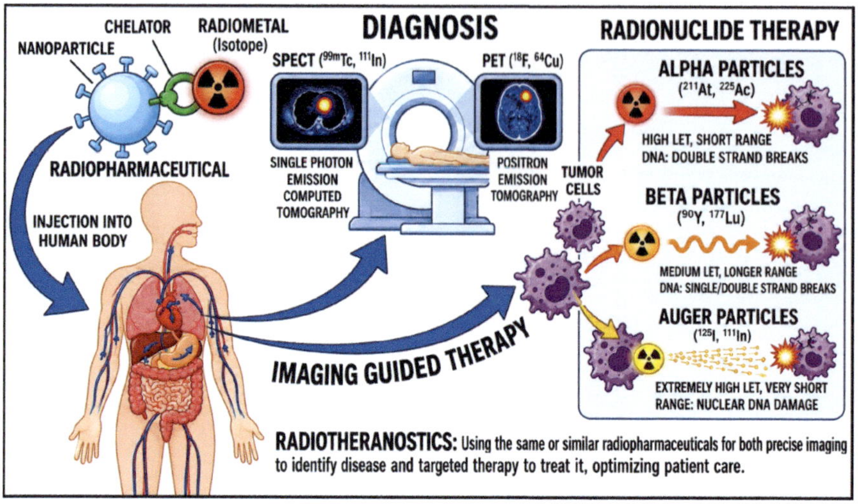

Figure 9.4 Radiotheranostics illustrating the coordinated use of diagnostic radiotracers and therapeutic radionuclides to enable precise tumor detection and targeted treatment.

dormant or "OFF" state during circulation and become activated exclusively within tumor microenvironments in response to specific stimuli, enabling background-free imaging and site-specific therapy. Recent advances in photosensitizer formulations as activatable phototheranostic agents demonstrate promising potential for finely controlled imaging-guided PDT through their propensity to specifically turn on diagnostic signals simultaneously with photodynamic effects in response to tumor-specific stimuli [144].

9.9.1 Enzyme-activated theranostic systems

NAD(P)H:quinone oxidoreductase 1 (NQO1) is frequently overexpressed in various cancer types and has emerged as a significant biomarker for cancer diagnosis and targeted treatment. The NQO1-activated NIR multifunctional theranostic probe I-HCy-Q successfully integrates imaging-guided PDT, demonstrating mitochondria-targeted PDT that produces significant oxidative stress and releases mitochondria-related damage-associated molecular patterns [145]. Enzyme-responsive nanomaterials designed for endogenous enzyme activation exploit cathepsin B, matrix metalloproteinases, histone deacetylases, and other tumor-overexpressed enzymes to trigger drug release and imaging activation. A dual-targeting theranostic agent employing SSTRs for targeting combined with cathepsin B for activation demonstrated effective fluorescence imaging-guided cancer chemotherapy,

Table 9.4 Summary of active and completed clinical trials evaluating molecular theranostic agents targeting PSMA, FAP, GRPR, and SSTR across various cancer types

Trial status	*Target(s)*	*Modality/intervention*	*Cancer types*	*Trial phase and clinical trial ID*	*Trial goals & notes*	*References*
Active recruiting	PSMA	Radioligand therapy (small molecules, peptides), alpha emitters (^{225}Ac, ^{213}Bi), combos with chemo/immunotherapy	Prostate, pancreatic, breast, colorectal, lung, others	(PSMA-617), NCT04071236 (GRPR)	Expand indications, optimize doses, biomarker identification, combos/sequencing studies	[139]
Completed	PSMA	^{177}Lu-PSMA-617 (beta emitter)	Metastatic castration-resistant prostate cancer (mCRPC)	Phase 3; VISION trial (NCT03511664)	Demonstrated survival benefit, symptom control, approved standard	[140]
Completed	FAP	^{90}Y-FAPI-04, ^{177}Lu-DOTA-FAPi	Pancreatic, breast, colorectal, lung cancers	Phase 1/2; NCT04571086, NCT04665445	Early clinical efficacy, symptom relief	[141]
Completed	GRPR	^{177}Lu-RM2, ^{177}Lu-bombesin derivatives	Metastatic castration-resistant prostate cancer	Phase 1/2; NCT04173137, NCT05340374	First-in-human dosimetry, efficacy established; ongoing late phases	[142]
Completed	SSTR	^{177}Lu-DOTATATE, ^{68}Ga-DOTATATE PET/CT	Gastroenteropancreatic neuroendocrine tumors	Phase 3; NETTER-1 (NCT01578239)	Significant PFS improvement	[143,144]

with normal cells expressing low levels of both targets remaining unaffected [146,145].

9.9.2 pH-responsive theranostic platforms

The acidic tumor microenvironment (pH 6.5–6.8) compared to normal physiological pH (7.4) provides a reliable endogenous stimulus for activatable theranostics. pH-sensitive liposomes loaded with gadolinium contrast agents exhibit increased longitudinal relaxivity in acidic environments, enabling enhanced MRI visualization specifically at tumor sites [147]. Imidazole-containing monomers mimicking histidine with pKa around 6 facilitate endosome escape and higher intracellular availability of therapeutic payloads through pH-triggered membrane disruption [148]. Gold nanorod-embedded silica nanoparticles co-loaded with 5-fluorouracil and ICG undergo pH-triggered drug release in acidic tumor microenvironments through protonation-induced loosening of electrostatic interactions, enabling combined chemotherapy, PDT, and fluorescence imaging.

9.9.3 Hypoxia-responsive systems

Tumor hypoxia represents a hallmark of aggressive malignancies associated with resistance to radiotherapy and chemotherapy, providing a selective activation trigger for theranostic agents. Azobenzene linkers undergo reductive cleavage specifically under hypoxic conditions, enabling dissociation of nanoparticle systems for enhanced tumor penetration and increased fluorescence sensitivity. A hypoxia-responsive protein hybrid system cross-linking chlorine 6-conjugated human serum albumin with oxaliplatin via azobenzene formed 100–150 nm particles that dissociated in hypoxic conditions, achieving deep tumor penetration with reduced tumor growth to approximately 3 mm in 4T1 mouse models. pH/hypoxia dual-stimuli-responsive covalent organic framework nanocarriers containing both acidic pH and hypoxia response sites achieve enhanced tumor accumulation through sequential targeting mechanisms [149].

9.9.4 Redox-responsive theranostics

Glutathione (GSH) concentrations in tumor cells reach 2–10 mM, representing 100–1,000-fold higher levels compared to normal tissue extracellular fluid and blood (2–20 μM), enabling selective activation of redox-responsive theranostics. ROS/GSH dual-responsive theranostic agents leveraging multilevel targeting systems accumulate at tumor sites in higher concentrations, facilitating accurate diagnosis and therapy. A TME-targeting theranostic agent employing NIR tracking for multilevel tumor monitoring utilized

ROS/GSH dual-responsiveness combined with biotin receptor targeting to achieve podophyllotoxin delivery with high tumor specificity and reduced systemic toxicity [150–152].

9.9.5 Diagnostic biomarker-driven treatment planning

Biomarker-driven molecular imaging has emerged as an integral part of cancer precision radiotherapy, with molecular imaging probes precisely and noninvasively monitoring spatiotemporal distribution of biomarkers, revealing tumor-killing mechanisms and therapy-induced adverse effects during radiation treatment. Diagnostic biomarker-driven molecular imaging accurately determines gross tumor volume and reduces planning target volume margins to improve tumor control probability or spare organs at risk. Images obtained from biomarker-specific PET tracers such as ^{68}Ga-DOTA-FAPI, ^{68}Ga-DOTATATE, and EGFR-targeted ^{89}Zr-Panitumumab enable offline delineation of gross tumor volume for intracranial meningiomas and other malignancies [153].

Several radiotherapy-related biomarkers have been identified, including tumor-specific biomarkers for targeted radiotherapy and patient stratification. These imaging probes can be tuned for targeting well-established biomarkers with biological significance, offering dynamic and spatiotemporal information for personalized precision therapy to achieve great clinical benefits for cancer patients. Activatable imaging mechanisms enable probes to respond to tumor-specific stimuli before, during, and after radiation-based therapies in a manner of biomarker-driven activatable imaging [153].

9.10 CHALLENGES AND LIMITATIONS

Clinical translation of theranostic radiopharmaceuticals is constrained by multifaceted regulatory, economic, safety, and technical challenges, including stringent approval requirements from agencies such as the FDA and EMA, mandatory GMP compliance, and dual documentation for diagnostic and therapeutic components, which collectively slow bench-to-bedside transfer [154]. High development and production costs driven by complex synthesis, short radionuclide half-lives, specialized infrastructure, and supply chain limitations restrict scalability and equitable global access despite anticipated market growth [155]. Additionally, comprehensive evaluation of safety, toxicity, biodistribution, and pharmacokinetics is essential to address off-target effects, organ-specific toxicity, and interindividual variability, particularly for nanoparticle- and radioligand-based agents. These hurdles are further compounded by technical barriers in imaging sensitivity, tissue penetration, contrast efficiency, and reproducible synthesis, making large-scale clinical implementation dependent on sustained investment, infrastructure expansion, and streamlined, harmonized regulatory frameworks [156].

9.11 FUTURE DIRECTIONS

Alpha-emitting radionuclides are emerging as potent theranostic agents due to their high linear energy transfer and short tissue penetration, enabling precise DNA damage with minimal harm to surrounding tissues [157]. Radium-223 is FDA-approved for metastatic prostate cancer, while other α-emitters like actinium-225, astatine-211, bismuth-213, lead-212, and thorium-227 are actively investigated. Innovations include pan-tumor-targeting probes and tetravalent agents that address tumor heterogeneity by simultaneously targeting two receptors, improving selectivity and efficacy across cancers such as breast and prostate [158]. Concurrently, AI and machine learning enhance theranostics via advanced imaging analysis, tumor characterization, and predictive modeling, facilitating personalized treatment and real-time monitoring. Combining theranostics with immunotherapies leverages immunogenic cell death to overcome tumor immunosuppression, with personalized dosimetry and biomarker-guided approaches advancing individualized cancer care [159].

9.12 CONCLUSION

Theranostics is reshaping how cancer is diagnosed and treated by combining highly accurate imaging with targeted therapeutic agents. Instead of treating cancer based only on where tumors are located, clinicians can now visualize tumor behavior at the molecular level and deliver treatment directly to cancer cells, reducing harm to surrounding healthy tissue. This integrated approach not only makes therapy more precise but also allows treatment strategies to be tailored and adjusted in real time.

The emergence of advanced theranostic platforms, particularly those designed to recognize diverse tumor markers such as PSMA, GRPR, and FAP, highlights considerable progress in overcoming tumor variability and broadening therapeutic options for many cancers, including difficult-to-treat cases. By confirming target expression before treatment and assessing response afterward, molecular imaging helps ensure that patients receive the most suitable and effective care, which is especially valuable when conventional treatments fail.

Looking ahead, the field is rapidly evolving toward even more personalized and effective solutions. Promising developments include novel alpha-emitting radiotherapies, multi-target agents, pan-tumor tracers, and the integration of AI to improve imaging interpretation and individualized dosimetry. Supported by robust clinical trials and translational research, these innovations are expected to improve patient access, enhance therapeutic efficiency, and ensure safer applications. Collectively, these advances signal a future in which theranostics becomes a fundamental element of cancer care, offering genuinely personalized treatment pathways for both adults and children.

REFERENCES

1. Bray F, Laversanne M, Sung H, Ferlay J, Siegel RL, Soerjomataram I, et al. Global cancer statistics 2022: GLOBOCAN estimates of incidence and mortality worldwide for 36 cancers in 185 countries. *CA Cancer J Clin.* 2024 May;74(3):229–63. doi:10.3322/CAAC.21834 PubMed PMID: 38572751.
2. Global cancer burden growing, amidst mounting need for services. [cited 2025 Dec 1]. Available from: https://www.who.int/news/item/01-02-2024-global-cancer-burden-growing--amidst-mounting-need-for-services.
3. Weber WA, Barthel H, Bengel F, Eiber M, Herrmann K, Schafers M. What Is Theranostics? *Journal of Nuclear Medicine.* 2023 May 1;64(5):669–70. doi:10.2967/JNUMED.123.265670 PubMed PMID: 37055220.
4. What is theranostics? | MD Anderson Cancer Center. [cited 2025 Dec 1]. Available from: https://www.mdanderson.org/cancerwise/what-is-theranostics.h00-159701490.html.
5. Haslerud T, Revheim MER. Therapeutics+diagnostics=theranostics. Tidsskrift for Den norske legeforening. 2023 Jan 9;143(1). doi:10.4045/TIDSSKR.22.0776 PubMed PMID: 36655959.
6. What is Nuclear Medicine? – EANM. [cited 2025 Dec 1]. Available from: https://eanm.org/the-eanm-community/about-us/what-is-nuclear-medicine.
7. Yordanova A, Eppard E, Kürpig S, Bundschuh RA, Schönberger S, Gonzalez-Carmona M, et al. Theranostics in nuclear medicine practice. *Onco Targets Ther.* 2017 Oct 3;10:4821. doi:10.2147/OTT.S140671 PubMed PMID: 29042793.
8. Jacob LJ, Deigner HP. Nanoparticles and nanosized structures in diagnostics and therapy. *Precision Medicine: Tools and Quantitative Approaches.* 2018 Jan 1;229–52. doi:10.1016/B978-0-12-805364-5.00010-X
9. Fahey FH, Grant FD, Thrall JH. Saul Hertz, MD, and the birth of radionuclide therapy. *EJNMMI Physics* 2017 4:1. 2017 Apr 27;4(1):15-. doi:10.1186/S40658-017-0182-7
10. Takalkar AM, El-Haddad G, Lilien DL. FDG-PET and PET/CT - Part II. *Indian J Radiol Imaging.* 2008 Feb 1;18(1):17. doi:10.4103/0971-3026.38504
11. Liu JKH. The history of monoclonal antibody development – Progress, remaining challenges and future innovations. *Annals of Medicine and Surgery.* 2014 Dec 1;3(4):113. doi:10.1016/J.AMSU.2014.09.001 PubMed PMID: 25568796.

12. Levine R, Krenning EP. Clinical history of the theranostic radionuclide approach to neuroendocrine tumors and other types of cancer: Historical review based on an interview of Eric P. Krenning by Rachel Levine. *J Nucl Med.* 2017;58:3–9. doi:10.2967/jnumed.116.186502
13. Ehrhardt JD, Güleç S. A review of the history of radioactive iodine theranostics: The origin of nuclear ontology. *Mol Imaging Radionucl Ther.* 2020 Oct 1;29(3):88. doi:10.4274/MIRT.GALENOS.2020.83703 PubMed PMID: 33094571.
14. Hennrich U, Eder M. [177Lu]Lu-PSMA-617 (PluvictoTM): The first FDA-approved radiotherapeutical for treatment of prostate cancer. *Pharmaceuticals.* 2022 Oct 1;15(10):1292. doi:10.3390/PH15101292 PubMed PMID: 36297404.
15. Hennrich U, Kopka K. Lutathera®: The first FDA- and EMA-approved radiopharmaceutical for peptide receptor radionuclide therapy. *Pharmaceuticals.* 2019 Sep 1;12(3):114. doi:10.3390/PH12030114 PubMed PMID: 31362406.
16. Montes AL, Yousefirizi F, Chen Y, Salimi Y, Seifert R, Afshar-Oromieh A, et al. Artificial intelligence for simplified patient-centered dosimetry in radiopharmaceutical therapies. *PET Clin.* 2026 Jan;21(1):73–88. doi:10.1016/J.CPET.2025.09.010 PubMed PMID: 41271260.
17. Strosberg J, El-Haddad G, Wolin E, Hendifar A, Yao J, Chasen B, et al. Phase 3 Trial of 177Lu-dotatate for midgut neuroendocrine tumors. *N Engl J Med.* 2017 Jan 12;376(2):125–35. doi:10.1056/NEJMOA1607427 PubMed PMID: 28076709.
18. Sartor O, de Bono J, Chi KN, Fizazi K, Herrmann K, Rahbar K, et al. Lutetium-177-PSMA-617 for metastatic castration-resistant prostate cancer. *N Engl J Med.* 2021 Sep 16;385(12):1091–103. doi:10.1056/NEJMOA2107322 PubMed PMID: 34161051.
19. Zaknun JJ, Bodei L, Mueller-Brand J, Pavel ME, Baum RP, Hörsch D, et al. The joint IAEA, EANM, and SNMMI practical guidance on peptide receptor radionuclide therapy (PRRNT) in neuroendocrine tumours. *Eur J Nucl Med Mol Imaging.* 2013 May 1;40(5):800–16. doi:10.1007/S00259-012-2330-6 PubMed PMID: 23389427.
20. Hofman MS, Lawrentschuk N, Francis RJ, Tang C, Vela I, Thomas P, et al. Prostate-specific membrane antigen PET-CT in patients with high-risk prostate cancer before curative-intent surgery or radiotherapy (proPSMA): a prospective, randomised, multicentre study. *The Lancet.* 2020 Apr 11;395(10231):1208–16. doi:10.1016/S0140-6736(20)30314-7 PubMed PMID: 32209449.
21. James ML, Gambhir SS. A molecular imaging primer: modalities, imaging agents, and applications. *Physiol Rev.* 2012 Apr 1;92(2):897–965. doi:10.1152/PHYSREV.00049.2010 PubMed PMID: 22535898.
22. Czernin J, Allen-Auerbach M, Nathanson D, Herrmann K. PET/CT in oncology: Current status and perspectives. *Curr Radiol Rep.* 2013 Sep 14;1(3):177. doi:10.1007/S40134-013-0016-X PubMed PMID: 24883234.
23. Physics in Nuclear Medicine - 4th Edition | Elsevier Shop. [cited 2025 Dec 1]. Available from: https://shop.elsevier.com/books/physics-in-nuclear-medicine/cherry/978-1-4160-5198-5
24. Hutton BF. The origins of SPECT and SPECT/CT. *Eur J Nucl Med Mol Imaging.* 2014;41 Suppl 1(SUPPL. 1). doi:10.1007/S00259-013-2606-5 PubMed PMID: 24218098.

25. Dewaraja YK, Frey EC, Sgouros G, Brill AB, Roberson P, Zanzonico PB, et al. MIRD pamphlet No. 23: quantitative SPECT for patient-specific 3-dimensional dosimetry in internal radionuclide therapy. *J Nucl Med.* 2012 Aug 1;53(8):1310–25. doi:10.2967/JNUMED.111.100123 PubMed PMID: 22743252.
26. Vergnaud L, Dewaraja YK, Giraudet AL, Badel JN, Sarrut D. A review of 177Lu dosimetry workflows: how to reduce the imaging workloads? *EJNMMI Phys.* 2024 Dec 1;11(1):65. doi:10.1186/S40658-024-00658-8 PubMed PMID: 39023648.
27. Nussbaum S, Shoukry M, Ashary MA, Kasbi AA, Baksh M, Gabriel E. Advanced tumor imaging approaches in human tumors. *Cancers (Basel).* 2022 Mar 1;14(6):1549. doi:10.3390/CANCERS14061549 PubMed PMID: 35326700.
28. Padhani AR, Koh DM, Collins DJ. Whole-body diffusion-weighted MR imaging in cancer: current status and research directions. *Radiology.* 2011 Dec;261(3): 700–18. doi:10.1148/RADIOL.11110474 PubMed PMID: 22095994.
29. O'Connor JPB, Rose CJ, Waterton JC, Carano RAD, Parker GJM, Jackson A. Imaging intratumor heterogeneity: Role in therapy response, resistance, and clinical outcome. *Clin Cancer Res.* 2014 Jan 15;21(2):249. doi:10.1158/1078-0432.CCR-14-0990 PubMed PMID: 25421725.
30. Oghabian MA, Farahbakhsh NM. Potential use of nanoparticle based contrast agents in MRI: a molecular imaging perspective. *J Biomed Nanotechnol.* 2010 Jun;6(3):203–13. doi:10.1166/JBN.2010.1119 PubMed PMID: 21179937.
31. Hong G, Antaris AL, Dai H. Near-infrared fluorophores for biomedical imaging. *Nature Biomedical Engineering.* 2017 1:1. 2017 Jan 10;1(1):0010-. doi:10.1038/s41551-016-0010
32. Wang T, Chen Y, Wang B, Gao X, Wu M. Recent progress in second near-infrared (NIR-II) fluorescence imaging in cancer. *Biomolecules.* 2022 Aug 1;12(8):1044. doi:10.3390/BIOM12081044 PubMed PMID: 36008937.
33. Vahrmeijer AL, Hutteman M, Van Der Vorst JR, Van De Velde CJH, Frangioni J V. Image-guided cancer surgery using near-infrared fluorescence. *Nat Rev Clin Oncol.* 2013 Sep;10(9):507–18. doi:10.1038/NRCLINONC.2013.123 PubMed PMID: 23881033.
34. Wang C, Dong Y, Shi X, Guo J, Zhang J, Deng L, et al. "Off/on" fluorescence imaging-guided cancer diagnosis and multi-modal therapy. *Biomater Sci.* 2020 Mar 3;8(5):1442–54. doi:10.1039/C9BM01854A PubMed PMID: 31960834.
35. Wang L V., Hu S. Photoacoustic tomography: in vivo imaging from organelles to organs. *Science.* 2012 Mar 23;335(6075):1458–62. doi:10.1126/SCIENCE.1216210 PubMed PMID: 22442475.
36. Mantri Y, Jokerst J V. Engineering plasmonic nanoparticles for enhanced photoacoustic imaging. *ACS Nano.* 2020 Aug 25;14(8):9408. doi:10.1021/ACSNANO.0C05215 PubMed PMID: 32806027.
37. Huang X, El-Sayed MA. Gold nanoparticles: Optical properties and implementations in cancer diagnosis and photothermal therapy. *J Adv Res.* 2010 Jan 1;1(1):13–28. doi:10.1016/J.JARE.2010.02.002
38. De La Zerda A, Zavaleta C, Keren S, Vaithilingam S, Bodapati S, Liu Z, et al. Carbon nanotubes as photoacoustic molecular imaging agents in living mice. *Nat Nanotechnol.* 2008 Sep 5;3(9):557–62. doi:10.1038/NNANO.2008.231 PubMed PMID: 18772918.

39. Maeda H, Wu J, Sawa T, Matsumura Y, Hori K. Tumor vascular permeability and the EPR effect in macromolecular therapeutics: A review. *Journal of Controlled Release.* 2000 Mar 1;65(1–2):271–84. doi:10.1016/S0168-3659(99)00248-5 PubMed PMID: 10699287.
40. Na H Bin, Song IC, Hyeon T. Inorganic nanoparticles for MRI contrast agents. *Advanced Materials.* 2009 Jun 5;21(21):2133–48. doi:10.1002/ADMA.200802366;REQUESTEDJOURNAL:JOURNAL:15214095;WGROUP:STRING:PUBLICATION
41. Veiseh O, Gunn JW, Zhang M. Design and fabrication of magnetic nanoparticles for targeted drug delivery and imaging. *Adv Drug Deliv Rev.* 2010 Mar 8;62(3):284–304. doi:10.1016/j.addr.2009.11.002 PubMed PMID: 19909778.
42. Torchilin VP. Multifunctional nanocarriers. *Adv Drug Deliv Rev.* 2006 Dec 1;58(14):1532–55. doi:10.1016/j.addr.2006.09.009 PubMed PMID: 17092599.
43. Fu H, Du B, Chen Z, Li Y. Radiolabeled peptides for SPECT and pet imaging in the detection of breast cancer: Preclinical and clinical perspectives. *Curr Med Chem.* 2020 Jan 28;27(41):6987–7002. doi:10.2174/0929867327666200128110827 PubMed PMID: 32003658.
44. Kwekkeboom DJ, De Herder WW, Kam BL, Van Eijck CH, Van Essen M, Kooij PP, et al. Treatment with the radiolabeled somatostatin analog [177 Lu-DOTA 0,Tyr3]octreotate: toxicity, efficacy, and survival. *J Clin Oncol.* 2008;26(13):2124–30. doi:10.1200/JCO.2007.15.2553 PubMed PMID: 18445841.
45. Kratochwil C, Bruchertseifer F, Rathke H, Hohenfellner M, Giesel FL, Haberkorn U, et al. Targeted α-therapy of metastatic castration-resistant prostate cancer with 225Ac-PSMA-617: Swimmer-plot analysis suggests efficacy regarding duration of tumor control. *J Nucl Med.* 2018 May 1;59(5):795–802. doi:10.2967/JNUMED.117.203539 PubMed PMID: 29326358.
46. Huang X, Jain PK, El-Sayed IH, El-Sayed MA. Plasmonic photothermal therapy (PPTT) using gold nanoparticles. *Lasers Med Sci.* 2008 Jul;23(3):217–28. doi:10.1007/S10103-007-0470-X PubMed PMID: 17674122.
47. Robinson JT, Tabakman SM, Liang Y, Wang H, Sanchez Casalongue H, Vinh D, et al. Ultrasmall reduced graphene oxide with high near-infrared absorbance for photothermal therapy. *J Am Chem Soc.* 2011 May 4;133(17):6825–31. doi:10.1021/JA2010175 PubMed PMID: 21476500.
48. Dolmans DEJGJ, Fukumura D, Jain RK. Photodynamic therapy for cancer. *Nat Rev Cancer.* 2003 May;3(5):380–7. doi:10.1038/NRC1071;KWRD PubMed PMID: 12724736.
49. Lucky SS, Soo KC, Zhang Y. Nanoparticles in photodynamic therapy. *Chem Rev.* 2015 Feb 25;115(4):1990–2042. doi:10.1021/CR5004198 PubMed PMID: 25602130.
50. Lian H, Wu J, Hu Y, Guo H. Self-assembled albumin nanoparticles for combination therapy in prostate cancer. *Int J Nanomedicine.* 2017 Oct 24;12:7777–87. doi:10.2147/IJN.S144634 PubMed PMID: 29123392.
51. Broos K, Lecocq Q, Raes G, Devoogdt N, Keyaerts M, Breckpot K. Noninvasive imaging of the PD-1:PD-L1 immune checkpoint: Embracing nuclear medicine for the benefit of personalized immunotherapy. *Theranostics.* 2018;8(13):3559. doi:10.7150/THNO.24762 PubMed PMID: 30026866.

52. Xu X, Saw PE, Tao W, Li Y, Ji X, Bhasin S, et al. ROS-responsive polyprodrug nanoparticles for triggered drug delivery and effective cancer therapy. *Adv Mater.* 2017 Sep 6;29(33). doi:10.1002/ADMA.201700141 PubMed PMID: 28681981.
53. Morath V, Fritschle K, Warmuth L, Anneser M, Dötsch S, Živanić M, et al. PET-based tracking of CAR T cells and viral gene transfer using a cell surface reporter that binds to lanthanide complexes. Nature Biomedical Engineering 2025 9:11. 2025 Jun 13;9(11):1886–906. doi:10.1038/s41551-025-01415-7 PubMed PMID: 40514433.
54. Fernandes DA. Review on Metal-based theranostic nanoparticles for cancer therapy and imaging. *Technol Cancer Res Treat.* 2023 Jan 1;22:15330338231191492. doi:10.1177/15330338231191493 PubMed PMID: 37642945.
55. Brito B, Price TW, Gallo J, Bañobre-López M, Stasiuk GJ. Smart magnetic resonance imaging-based theranostics for cancer. *Theranostics.* 2021;11(18):8706–37. doi:10.7150/THNO.57004 PubMed PMID: 34522208.
56. Ibrahim MA, Hazhirkarzar B, Dublin AB. Gadolinium magnetic resonance imaging. *StatPearls.* 2023 Jul 3. PubMed PMID: 29494094.
57. Iyad N, S.Ahmad M, Alkhatib SG, Hjouj M. Gadolinium contrast agents-challenges and opportunities of a multidisciplinary approach: Literature review. *Eur J Radiol Open.* 2023 Dec 1;11:100503. doi:10.1016/J.EJRO.2023.100503
58. Russell E, Mcmahon SJ, Russell B, Mohamud H, Mcgarry CK, Schettino G, et al. Effects of gadolinium MRI contrast agents on DNA damage and cell survival when used in combination with radiation. *Radiat Res.* 2020 Sep 1;194(3):298–309. doi:10.1667/RADE-20-00008.1 PubMed PMID: 32942305.
59. Lux F, Tran VL, Thomas E, Dufort S, Rossetti F, Martini M, et al. AGuIX® from bench to bedside—Transfer of an ultrasmall theranostic gadolinium-based nanoparticle to clinical medicine. *Br J Radiol.* 2018;92(1093):20180365. doi:10.1259/BJR.20180365 PubMed PMID: 30226413.
60. Sancey L, Lux F, Kotb S, Roux S, Dufort S, Bianchi A, et al. The use of theranostic gadolinium-based nanoprobes to improve radiotherapy efficacy. *Br J Radiol.* 2014 Sep 1;87(1041):20140134. doi:10.1259/BJR.20140134 PubMed PMID: 24990037.
61. Shi J, Wang B, Chen Z, Liu W, Pan J, Hou L, et al. A multi-functional tumor theranostic nanoplatform for mri guided photothermal-chemotherapy. *Pharm Res.* 2016 Jun 1;33(6):1472–85. doi:10.1007/S11095-016-1891-7 PubMed PMID: 26984128.
62. Van Laere C, Koole M, Deroose CM, Van de Voorde M, Baete K, Cocolios TE, et al. Terbium radionuclides for theranostic applications in nuclear medicine: from atom to bedside. *Theranostics.* 2024;14(4):1720. doi:10.7150/THNO.92775 PubMed PMID: 38389843.
63. Alcocer-Ávila ME, Ferreira A, Quinto MA, Morgat C, Hindié E, Champion C. Radiation doses from 161Tb and 177Lu in single tumour cells and micrometastases. *EJNMMI Phys.* 2020 Dec 1;7(1):33. doi:10.1186/S40658-020-00301-2 PubMed PMID: 32430671.
64. Fricke J, Westerbergh F, McDougall L, Favaretto C, Christ E, Nicolas GP, et al. First-in-human administration of terbium-161-labelled somatostatin

receptor subtype 2 antagonist ([161Tb]Tb-DOTA-LM3) in a patient with a metastatic neuroendocrine tumour of the ileum. *Eur J Nucl Med Mol Imaging.* 2024 Jul 1;51(8):2517–9. doi:10.1007/S00259-024-06641-W PubMed PMID: 38448550.

65. Chi KN, Yip SM, Bauman G, Probst S, Emmenegger U, Kollmannsberger CK, et al. 177Lu-PSMA-617 in metastatic castration-resistant prostate cancer: A review of the evidence and implications for Canadian clinical practice. *Current Oncology.* 2024 Mar 1;31(3):1400. doi:10.3390/CURRONCOL31030106 PubMed PMID: 38534939.
66. Krasnovskaya OO, Abramchuck D, Erofeev A, Gorelkin P, Kuznetsov A, Shemukhin A, et al. Recent advances in 64Cu/67Cu-based radiopharmaceuticals. *Int J Mol Sci.* 2023 Jun 1;24(11):9154. doi:10.3390/IJMS24119154 PubMed PMID: 37298101.
67. Potential theranostics of breast cancer with copper-64/67 sarcophagine-trastuzumab - Chemical Science (RSC Publishing) DOI:10.1039/D4SC06969B. [cited 2025 Nov 28]. Available from: https://pubs.rsc.org/en/content/articlehtml/2025/sc/d4sc06969b
68. Cai X, Zhu Q, Zeng Y, Zeng Q, Chen X, Zhan Y. Manganese oxide nanoparticles as MRI contrast agents in tumor multimodal imaging and therapy. *Int J Nanomedicine.* 2019;14:8321. doi:10.2147/IJN.S218085 PubMed PMID: 31695370.
69. Vines JB, Yoon JH, Ryu NE, Lim DJ, Park H. Gold nanoparticles for photothermal cancer therapy. *Front Chem.* 2019 Apr 5;7(APR):432792. doi:10.3389/FCHEM.2019.00167/FULL
70. Abed A, Derakhshan M, Karimi M, Shirazinia M, Mahjoubin-Tehran M, Homayonfal M, et al. Platinum nanoparticles in biomedicine: preparation, anti-cancer activity, and drug delivery vehicles. *Front Pharmacol.* 2022 Feb 23;13:797804. doi:10.3389/FPHAR.2022.797804 PubMed PMID: 35281900.
71. Wahajuddin, Arora S. Superparamagnetic iron oxide nanoparticles: magnetic nanoplatforms as drug carriers. *Int J Nanomedicine.* 2012;7:3445. doi:10.2147/IJN.S30320 PubMed PMID: 22848170.
72. Wang YXJ, Xuan S, Port M, Idee JM. Recent advances in superparamagnetic iron oxide nanoparticles for cellular imaging and targeted therapy research. *Curr Pharm Des.* 2013 Sep 27;19(37):6575. doi:10.2174/1381612811319370003 PubMed PMID: 23621536.
73. Zou P, Yu Y, Wang YA, Zhong Y, Welton A, Galbán C, et al. Superparamagnetic iron oxide nanotheranostics for targeted cancer cell imaging and pH-dependent intracellular drug release. Mol Pharm. 2010 Dec 6;7(6):1974–84. doi:10.1021/MP100273T PubMed PMID: 20845930.
74. Im GH, Kim SM, Lee DG, Lee WJ, Lee JH, Lee IS. Fe3O4/MnO hybrid nanocrystals as a dual contrast agent for both T1- and T2-weighted liver MRI. *Biomaterials.* 2013 Mar 1;34(8):2069–76. doi:10.1016/J.BIOMATERIALS.2012.11.054 PubMed PMID: 23246062.
75. Dong Z, Gong H, Gao M, Zhu W, Sun X, Feng L, et al. Polydopamine nanoparticles as a versatile molecular loading platform to enable imaging-guided cancer combination therapy. *Theranostics.* 2016;6(7):1031. doi:10.7150/THNO.14431 PubMed PMID: 27217836.

76. Accardo A, Tesauro D, Morelli G. Peptide-based targeting strategies for simultaneous imaging and therapy with nanovectors. *Polym J.* 2013 May 13;45(5):481–93. doi:10.1038/PJ.2012.215;SUBJMETA
77. Pang L, Pei Y, Uzunalli G, Hyun H, Lyle LT, Yeo Y. Surface modification of polymeric nanoparticles with M2pep peptide for drug delivery to tumor-associated macrophages. *Pharm Res.* 2019 Apr 1;36(4):65. doi:10.1007/S11095-019-2596-5 PubMed PMID: 30859335.
78. Zhang L, Huang Y, Lindstrom AR, Lin TY, Lam KS, Li Y. Peptide-based materials for cancer immunotherapy. *Theranostics.* 2019;9(25):7807. doi:10.7150/THNO.37194 PubMed PMID: 31695802.
79. Murphy EA, Majeti BK, Barnes LA, Makale M, Weis SM, Lutu-Fuga K, et al. Nanoparticle-mediated drug delivery to tumor vasculature suppresses metastasis. *Proc Natl Acad Sci U S A.* 2008 Jul 8;105(27):9343. doi:10.1073/PNAS.0803728105 PubMed PMID: 18607000.
80. Wu XL, Kim JH, Koo H, Bae SM, Shin H, Kim MS, et al. Tumor-targeting peptide conjugated pH-responsive micelles as a potential drug carrier for cancer therapy. *Bioconjug Chem.* 2010 Feb 17;21(2):208–13. doi:10.1021/BC9005283 PubMed PMID: 20073455.
81. Wei L, Wu Y, Zhou Y, Yang M, Li H, Geng WC, et al. A Nitroreductase-activated chemiluminescent prodrug for real-time monitoring of camptothecin release in peritoneal metastasis Theranostics. *J Med Chem.* 2025 Sep 24. doi:10.1021/ACS.JMEDCHEM.5C01674/SUPPL_FILE/JM5C01674_SI_002.CSV PubMed PMID: 40994159.
82. Denny WA. Nitroaromatic Hypoxia-activated prodrugs for cancer therapy. *Pharmaceuticals.* 2022 Feb 1;15(2):187. doi:10.3390/PH15020187 PubMed PMID: 35215299.
83. Meng X, Zhang J, Sun Z, Zhou L, Deng G, Li S, et al. Hypoxia-triggered single molecule probe for high-contrast NIR II/PA tumor imaging and robust photothermal therapy. *Theranostics.* 2018;8(21):6025. doi:10.7150/THNO.26607 PubMed PMID: 30613279.
84. Guo J, Lu M, Zhang X, Sun S, Han C, Zhang Y, et al. Highly stable and efficient light-emitting diodes based on orthorhombic γ-CsPbI3 nanocrystals. *ACS Nano.* 2023 May 23;17(10):9290–301. doi:10.1021/ACSNANO.3C00789 PubMed PMID: 37126487.
85. Figueira F, Tomé JPC, Paz FAA. Porphyrin nanometal-organic frameworks as cancer theranostic agents. *Molecules.* 2022 May 1;27(10):3111. doi:10.3390/MOLECULES27103111 PubMed PMID: 35630585.
86. Agostinis P, Berg K, Cengel KA, Foster TH, Girotti AW, Gollnick SO, et al. Photodynamic therapy of cancer: An update. *CA Cancer J Clin.* 2011 Jul 1;61(4):250–81. doi:10.3322/CAAC.20114; PubMed PMID: 21617154.
87. Yin WH, Li PY, Huang HH, Feng L, Liu SH, Liu X, et al. Porphyrin photosensitizer molecules as effective medicine candidates for photodynamic therapy: electronic structure information aided design. *RSC Adv.* 2024 Sep 16;14(40):29368. doi:10.1039/D4RA05585C PubMed PMID: 39285886.
88. Wang H, Li X, Tse BWC, Yang H, Thorling CA, Liu Y, et al. Indocyanine green-incorporating nanoparticles for cancer theranostics. *Theranostics.* 2018;8(5):1227. doi:10.7150/THNO.22872 PubMed PMID: 29507616.

89. Sevieri M, Silva F, Bonizzi A, Sitia L, Truffi M, Mazzucchelli S, et al. Indocyanine green nanoparticles: Are they compelling for cancer treatment? *Front Chem.* 2020 Jul 16;8:550087. doi:10.3389/FCHEM.2020.00535/BIBTEX
90. Alander JT, Kaartinen I, Laakso A, Pätilä T, Spillmann T, Tuchin V V., et al. A Review of indocyanine green fluorescent imaging in surgery. *Int J Biomed Imaging.* 2012;2012. doi:10.1155/2012/940585 PubMed PMID: 22577366.
91. Assembly-enhanced indocyanine green nanoparticles for fluorescence imaging-guided photothermal therapy. *Journal of Materials Chemistry B* (RSC Publishing). doi:10.1039/D4TB01604A. [cited 2025 Dec 1]. Available from: https://pubs.rsc.org/en/content/articlehtml/2024/tb/d4tb01604a
92. Ashokan A, Gowd GS, Somasundaram VH, Bhupathi A, Peethambaran R, Unni AKK, et al. Multifunctional calcium phosphate nano-contrast agent for combined nuclear, magnetic and near-infrared in vivo imaging. *Biomaterials.* 2013 Sep 1;34(29):7143–57. doi:10.1016/J.BIOMATERIALS.2013.05.077 PubMed PMID: 23791501.
93. Bonlawar J, Setia A, Challa RR, Vallamkonda B, Mehata AK, Vaishali, et al. Targeted Nanotheranostics: Integration of preclinical MRI and CT in the molecular imaging and therapy of advanced diseases. *Nanotheranostics.* 2024;8(3):401. doi:10.7150/NTNO.95791 PubMed PMID: 38751937.
94. Han HH, Kim SJ, Kim J, Park W, Kim C, Kim H, et al. Bimetallic hyaluronate-modified Au@Pt nanoparticles for noninvasive photoacoustic imaging and photothermal therapy of skin cancer. ACS *Appl Mater Interfaces.* 2023 Mar 8;15(9):11609–20. doi:10.1021/ACSAMI.3C01858 PubMed PMID: 36847648.
95. Kim TH, Lee S, Chen X. Nanotheranostics for personalized medicine. *Expert Rev Mol Diagn.* 2013 Apr;13(3):257. doi:10.1586/ERM.13.15 PubMed PMID: 23570404.
96. Luk BT, Fang RH, Zhang L. Lipid- and polymer-based nanostructures for cancer theranostics. *Theranostics.* 2012;2(12):1117. doi:10.7150/THNO.4381 PubMed PMID: 23382770.
97. Pandey P, Arya DK, Deepak P, Ali D, Alarifi S, Srivastava S, et al. αvβ3 integrin and folate-targeted pH-sensitive liposomes with dual ligand modification for metastatic breast cancer treatment. *Bioengineering.* 2024 Aug 1;11(8):800. doi:10.3390/BIOENGINEERING11080800 PubMed PMID: 39199757.
98. Sofias AM, Toner YC, Meerwaldt AE, Van Leent MMT, Soultanidis G, Elschot M, et al. Tumor targeting by αvβ3-integrin-specific lipid nanoparticles occurs via phagocyte hitchhiking. *ACS Nano.* 2020 Jul 28;14(7):7832–46. doi:10.1021/ACSNANO.9B08693/SUPPL_FILE/NN9B08693_SI_016.AVI PubMed PMID: 32413260.
99. Kalidoss M, Basha RY, Doble M, Sampath Kumar TS. Theranostic calcium phosphate nanoparticles with potential for multimodal imaging and drug delivery. *Front Bioeng Biotechnol.* 2019 Jun 4;7(JUN):454228. doi:10.3389/FBIOE.2019.00126/BIBTEX
100. Zhang L, Sheng D, Wang D, Yao Y, Yang K, Wang Z, et al. Bioinspired multifunctional melanin-based nanoliposome for photoacoustic/magnetic resonance imaging-guided efficient photothermal ablation of cancer. *Theranostics.* 2018;8(6):1591–606. doi:10.7150/THNO.22430 PubMed PMID: 29556343.

101. Yu JW, Kim DG, Cha GD. multifunctional liposomes: Smart nanomaterials for enhanced photodynamic therapy. *Biomimetics*. 2025 Oct 1;10(10):689. doi:10.3390/BIOMIMETICS10100689 PubMed PMID: 41149218.
102. Ashrafizadeh M, Delfi M, Zarrabi A, Bigham A, Sharifi E, Rabiee N, et al. Stimuli-responsive liposomal nanoformulations in cancer therapy: Pre-clinical & clinical approaches. *Journal of Controlled Release*. 2022 Nov 1;351:50–80. doi:10.1016/J.JCONREL.2022.08.001 PubMed PMID: 35934254.
103. Nocito G, Calabrese G, Forte S, Petralia S, Puglisi C, Campolo M, et al. Carbon dots as promising tools for cancer diagnosis and therapy. *Cancers (Basel)*. 2021 May 1;13(9):1991. doi:10.3390/CANCERS13091991 PubMed PMID: 33919096.
104. Bartkowski M, Zhou Y, Nabil Amin Mustafa M, Eustace AJ, Giordani S. Carbon dots: Bioimaging and anticancer drug delivery. *Chemistry - A European Journal*. 2024 Apr 2;30(19). doi:10.1002/CHEM.202303982 PubMed PMID: 38205882.
105. Tao W, Farokhzad OC. Theranostic nanomedicine in the NIR-II window: Classification, fabrication, and biomedical applications. *Chem Rev*. 2022 Mar 23;122(6):5405–7. doi:10.1021/ACS.CHEMREV.2C00089 PubMed PMID: 35317560.
106. Kajani AA, Rafiee L, Javanmard SH, Dana N, Jandaghian S. Carbon dot incorporated mesoporous silica nanoparticles for targeted cancer therapy and fluorescence imaging. *RSC Adv*. 2023 Mar 23;13(14):9491. doi:10.1039/D3RA00768E PubMed PMID: 36968033.
107. Shanahan K, Coen D, Nafo W. Polymer-based nanoparticles for cancer theranostics: advances, challenges, and future perspectives. *Open Exploration*. 2019 2: 2025 Jul 30;2:101342-. Doi:10.37349/EBMX.2025.101342
108. Rezvantalab S, Drude NI, Moraveji MK, Güvener N, Koons EK, Shi Y, et al. PLGA-based nanoparticles in cancer treatment. *Front Pharmacol*. 2018 Nov 2;9(NOV):1260. Doi:10.3389/FPHAR.2018.01260 PubMed PMID: 30450050.
109. Khorasani MA, Naghib SM, Jafari T, Takdehghan G. Recent advances in engineering chitosan-based nanoplatforms in biotherapeutic multi-delivery for multi-targeted disease treatments: Promises and outlooks. *Colloid Interface Sci Commun*. 2025 Dec 1;69:100861. Doi:10.1016/J.COLCOM.2025.100861
110. Fan D, Cao Y, Cao M, Wang Y, Cao Y, Gong T. Nanomedicine in cancer therapy. *Signal Transduction and Targeted Therapy*. 2023 8:1. 2023 Aug 7;8(1):293-. doi:10.1038/s41392-023-01536-y
111. Zhang H, Chen H, Hu X, Muhammad W, Liu C, Liu W. Inflammation-modulating polymeric nanoparticles: design strategies, mechanisms, and therapeuticapplications. *EBioMedicine*. 2025 Aug 1;118. doi:10.1016/j.ebiom.2025.105837 PubMed PMID: 40614330.
112. Tabish TA, Hamblin MR. Mitochondria-targeted nanoparticles (mitoNANO): An emerging therapeutic shortcut for cancer. Biomaterials and Biosystems. 2021 Sep 1;3:100023. doi:10.1016/J.BBIOSY.2021.100023 PubMed PMID: 36824307.
113. Deng Y, Jia F, Chen X, Jin Q, Ji J. ATP suppression by pH-activated mitochondria-targeted delivery of nitric oxide nanoplatform for drug resistance reversal and metastasis inhibition. *Small*. 2020 Jun 1;16(23). doi:10.1002/SMLL.202001747 PubMed PMID: 32378343.

114. Wang H, Gao Z, Liu X, Agarwal P, Zhao S, Conroy DW, et al. Targeted production of reactive oxygen species in mitochondria to overcome cancer drug resistance. *Nature Communications*. 2018 9:1. 2018 Feb 8;9(1):562-. doi:10.1038/s41467-018-02915-8 PubMed PMID: 29422620.
115. Xu B, Li S, Shi R, Liu H. Multifunctional mesoporous silica nanoparticles for biomedical applications. *Signal Transduction and Targeted Therapy* 2023 8:1. 2023 Nov 24;8(1):435-. doi:10.1038/s41392-023-01654-7 PubMed PMID: 37996406.
116. Huang L, Zhu J, Xiong W, Feng J, Yang J, Lu X, et al. Tumor-generated reactive oxygen species storm for high-performance ferroptosis therapy. *ACS Nano*. 2023 Jun 27;17(12):11492–506. doi:10.1021/ACSNANO.3C01369 PubMed PMID: 37283506.
117. Koohi Moftakhari Esfahani M, Alavi SE, Cabot PJ, Islam N, Izake EL. Application of mesoporous silica nanoparticles in cancer therapy and delivery of repurposed anthelmintics for cancer therapy. *Pharmaceutics*. 2022 Aug 1;14(8):1579. doi:10.3390/PHARMACEUTICS14081579 PubMed PMID: 36015204.
118. Vorster M, Hadebe BP, Sathekge MM. Theranostics in breast cancer. *Frontiers in Nuclear Medicine*. 2023;3:1236565. doi:10.3389/FNUME.2023.1236565; PubMed PMID: 39355052.
119. Bhusari P, Vatsa R, Singh G, Parmar M, Bal A, Dhawan DK, et al. Development of Lu-177-trastuzumab for radioimmunotherapy of HER2 expressing breast cancer and its feasibility assessment in breast cancer patients. *Int J Cancer*. 2017 Feb 15;140(4):938–47. doi:10.1002/IJC.30500; PubMed PMID: 27813061.
120. Ballal S, Yadav MP, Moon ES, Kramer VS, Roesch F, Kumari S, et al. First-in-human results on the biodistribution, pharmacokinetics, and dosimetry of [177lu]lu-dota.Sa.fapi and [177lu]lu-dotaga.(sa.fapi)2. *Pharmaceuticals*. 2021 Dec 1;14(12):1212. doi:10.3390/PH14121212/S1
121. Ng CWQ, Chua WM, Lam WWC, Tong AKT, Ng DCE, Tang CYL, et al. Safety and efficacy of lutetium-177 PSMA radioligand therapy in metastatic castration-resistant prostate cancer with diffuse bone metastases (Asian population study). *Asia Pac J Clin Oncol*. 2025 Dec 1;21(6):639–46. doi:10.1111/AJCO.14195
122. Fallah J, Agrawal S, Gittleman H, Fiero MH, Subramaniam S, John C, et al. FDA approval summary: Lutetium Lu 177 vipivotide tetraxetan for patients with metastatic castration-resistant prostate cancer. *Clin Cancer Res*. 2023 May 1;29(9):1651. doi:10.1158/1078-0432.CCR-22-2875 PubMed PMID: 36469000.
123. Molvar C, Lewandowski R. Yttrium-90 radioembolization of hepatocellular carcinoma–performance, technical advances, and future concepts. *Semin Intervent Radiol*. 2015 Dec 1;32(4):388. doi:10.1055/S-0035-1564704 PubMed PMID: 26622103.
124. Shengguang Y, Ji-Eun C, Lijuan H. I-131 for remnant ablation in differentiated thyroid cancer after thyroidectomy: A meta-analysis of randomized controlled evidence. *Med Sci Monit*. 2016 Jul 13;22:2439. doi:10.12659/MSM.896535 PubMed PMID: 27406262.

125. Strosberg JR, Caplin ME, Kunz PL, Ruszniewski PB, Bodei L, Hendifar A, et al. 177Lu-Dotatate plus long-acting octreotide versus high-dose long-acting octreotide in patients with midgut neuroendocrine tumours (NETTER-1): final overall survival and long-term safety results from an open-label, randomised, controlled, phase 3 trial. *Lancet Oncol.* 2021 Dec 1;22(12):1752–63. doi:10.1016/S1470-2045(21)00572-6 PubMed PMID: 34793718.
126. Karimi A, Bogdani C, O'Dwyer E, Siolas D. Emerging innovations in theranostics for pancreatic neuroendocrine tumors. *NPJ Precis Oncol.* 2025 Dec 1;9(1):146. doi:10.1038/S41698-025-00938-1 PubMed PMID: 40389624.
127. Herrmann K, Czernin J, Wolin EM, Gupta P, Barrio M, Gutierrez A, et al. Impact of 68Ga-DOTATATE PET/CT on the management of neuroendocrine tumors: the referring physician's perspective. *J Nucl Med.* 2015 Jan 1;56(1):70–5. doi:10.2967/JNUMED.114.148247 PubMed PMID: 25500825.
128. Ray GL, Baidoo KE, Keller LMM, Albert PS, Brechbiel MW, Milenic DE. Pre-clinical assessment of 177Lu-labeled trastuzumab targeting HER2 for treatment and management of cancer patients with disseminated intraperitoneal disease. *Pharmaceuticals.* 2011 Dec 22;5(1):1. doi:10.3390/PH5010001 PubMed PMID: 22229017.
129. Rohith G. VISION trial: 177Lu-PSMA-617 for progressive metastatic castration-resistant prostate cancer. Indian J Urol. 2021 Oct 1;37(4):372. doi:10.4103/IJU.IJU_292_21 PubMed PMID: 34759536.
130. Xing M, Kokabi N, Camacho JC, Kooby DA, El-Rayes BF, Kim HS. 90Y radioembolization versus chemoembolization in the treatment of hepatocellular carcinoma: An analysis of comparative effectiveness. *J Comp Eff Res.* 2013;2(4):435–44. doi:10.2217/CER.13.37; PubMed PMID: 24236684.
131. Yavuz S, Puckett Y. Iodine-131 Uptake Study. *StatPearls.* 2023 Oct 29. PubMed PMID: 32644709.
132. Minges P, Eder M, Eder AC. Dual-labeled small peptides in cancer imaging and fluorescence-guided surgery: Progress and future perspectives. *Pharmaceuticals.* 2025 Feb 1;18(2):143. doi:10.3390/PH18020143 PubMed PMID: 40005958.
133. Baratto L, Jadvar H, Iagaru A. Prostate cancer theranostics targeting gastrin-releasing peptide receptors. *Mol Imaging Biol.* 2018 Aug 1;20(4):501. doi:10.1007/S11307-017-1151-1 PubMed PMID: 29256046.
134. Bozkurt MF, Virgolini I, Balogova S, Beheshti M, Rubello D, Decristoforo C, et al. Guideline for PET/CT imaging of neuroendocrine neoplasms with 68Ga-DOTA-conjugated somatostatin receptor targeting peptides and 18F-DOPA. *Eur J Nucl Med Mol Imaging.* 2017 Aug 1;44(9):1588–601. doi:10.1007/S00259-017-3728-Y PubMed PMID: 28547177.
135. Chi KN, Armstrong AJ, Krause BJ, Herrmann K, Rahbar K, de Bono JS, et al. Safety analyses of the Phase 3 VISION Trial of [177Lu]Lu-PSMA-617 in patients with metastatic castration-resistant prostate cancer. *Eur Urol.* 2024 Apr 1;85(4): 382–91. doi:10.1016/J.EURURO.2023.12.004 PubMed PMID: 38185538.
136. Duan H, Iagaru A, Aparici CM. Radiotheranostics - precision medicine in nuclear medicine and molecular imaging. *Nanotheranostics.* 2022;6(1):103–17. doi:10.7150/NTNO.64141 PubMed PMID: 34976584.
137. Privé BM, Boussihmad MA, Timmermans B, van Gemert WA, Peters SMB, Derks YHW, et al. Fibroblast activation protein-targeted radionuclide therapy: background, opportunities, and challenges of first (pre)clinical studies. *Eur J Nucl*

Med Mol Imaging. 2023 Jun 1;50(7):1906. doi:10.1007/S00259-023-06144-0 PubMed PMID: 36813980.

138. Kurth J, Krause BJ, Schwarzenböck SM, Bergner C, Hakenberg OW, Heuschkel M. First-in-human dosimetry of gastrin-releasing peptide receptor antagonist [177Lu] Lu-RM2: a radiopharmaceutical for the treatment of metastatic castration-resistant prostate cancer. *Eur J Nucl Med Mol Imaging*. 2020 Jan 1;47(1):123–35. doi:10.1007/S00259-019-04504-3 PubMed PMID: 31482426.
139. Study Details | NCT04071236 | Radiation Medication (Radium-223 Dichloride) Versus Radium-223 Dichloride Plus Radiation Enhancing Medication (M3814) Versus Radium-223 Dichloride Plus M3814 Plus Avelumab (a Type of Immunotherapy) for Advanced Prostate Canc. [cited 2025 Nov 30]. Available from: https://clinicaltrials.gov/study/NCT04071236
140. Study Details | NCT03511664 | Study of 177Lu-PSMA-617 In Metastatic Castrate-Resistant Prostate Cancer | ClinicalTrials.gov. [cited 2025 Nov 30]. Available from: https://clinicaltrials.gov/study/NCT03511664
141. Study Details | NCT04571086 | FAPI-PET for Tumor Detection | ClinicalTrials.gov. [cited 2025 Nov 30]. Available from: https://clinicaltrials.gov/study/NCT04571086?term=ga%2068&viewType=Table&checkSpell=&rank=10
142. Study Details | NCT05340374 | Cabazitaxel in Combination With 177Lu-PSMA-617 in Metastatic Castration-resistant Prostate Cancer | ClinicalTrials.gov. [cited 2025 Nov 30]. Available from: https://clinicaltrials.gov/study/NCT05340374
143. Study Details | NCT01578239 | A Study Comparing Treatment With 177Lu-DOTA0-Tyr3-Octreotate to Octreotide LAR in Patients With Inoperable, Progressive, Somatostatin Receptor Positive Midgut Carcinoid Tumours | ClinicalTrials.gov. [cited 2025 Nov 30]. Available from: https://clinicaltrials.gov/study/NCT01578239
144. Zhang J, Ning L, Huang J, Zhang C, Pu K. Activatable molecular agents for cancer theranostics. *Chem Sci*. 2019;11(3):618. doi:10.1039/C9SC05460J PubMed PMID: 34123034.
145. Cao C, Li J, Zhang X, Zhang X, Gong X, Wang S. NQO1-activated multifunctional theranostic probe for imaging-guided mitochondria-targeted photodynamic therapy and boosting immunogenic cell death. *Talanta*. 2024 May 15;272:125786. doi:10.1016/J.TALANTA.2024.125786 PubMed PMID: 38382303.
146. Mu J, Lin J, Huang P, Chen X. Development of endogenous enzyme-responsive nanomaterials for theranostics. *Chem Soc Rev*. 2018 Aug 7;47(15):5554. doi:10.1039/C7CS00663B PubMed PMID: 29856446.
147. Lokling KE, Fossheim SL, Skurtveit R, Bjornerud A, Klaveness J. pH-sensitive paramagnetic liposomes as MRI contrast agents: in vitro feasibility studies. *Magn Reson Imaging*. 2001 Jun 1;19(5):731–8. doi:10.1016/S0730-725X(01)00380-0 PubMed PMID: 11672632.
148. Mapfumo PP, Reichel LS, Leer K, Egger J, Dzierza A, Peneva K, et al. Harnessing guanidinium and imidazole functional groups: A dual-charged polymer strategy for enhanced gene delivery. *ACS Macro Lett*. 2024 Aug 20;13(8):1000. doi:10.1021/ACSMACROLETT.4C00321 PubMed PMID: 39052525.
149. Yang G, Phua SZF, Lim WQ, Zhang R, Feng L, Liu G, et al. A hypoxia-responsive albumin-based nanosystem for deep tumor penetration and excellent therapeutic efficacy. *Advanced Materials*. 2019 Jun 20;31(25):1901513. doi:10.1002/ADMA.201901513; PubMed PMID: 31069885.

150. Chen H, Li F, Yao Y, Wang Z, Zhang Z, Tan N. Redox dual-responsive and O2-evolving theranostic nanosystem for highly selective chemotherapy against hypoxic tumors. *Theranostics.* 2019;9(1):90. doi:10.7150/THNO.30259 PubMed PMID: 30662556.
151. Zhou L, Wang H, Li Y. Stimuli-Responsive Nanomedicines for Overcoming Cancer Multidrug Resistance. *Theranostics.* 2018;8(4):1059–74. doi:10.7150/THNO.22679 PubMed PMID: 29463999.
152. Xu Z, Qian J, Meng C, Liu Y, Ding Q, Wu H, et al. TME-targeting theranostic agent uses NIR tracking for tumor diagnosis and surgical resection and acts as chemotherapeutic showing enhanced efficiency and minimal toxicity. *Theranostics.* 2022;12(6):2535. doi:10.7150/THNO.68074 PubMed PMID: 35401826.
153. Li H, Gong Q, Luo K. Biomarker-driven molecular imaging probes in radiotherapy. *Theranostics.* 2024;14(10):4127. doi:10.7150/THNO.97768 PubMed PMID: 38994026.
154. Kolenc Peitl P, Rangger C, Garnuszek P, Mikolajczak R, Hubalewska-Dydejczyk A, Maina T, et al. Clinical translation of theranostic radiopharmaceuticals: Current regulatory status and recent examples. *J Labelled Comp Radiopharm.* 2019 Aug 1;62(10):673. doi:10.1002/JLCR.3712 PubMed PMID: 30741441.
155. Al-Ibraheem A, Brink A, Lee ST, De Los Reyes A, Paez D, Craviolatti PS, et al. Implementation of radiotheranostics: Challenges, barriers, and IAEA-driven strategies for sustainable access. *Semin Nucl Med.* 2025 Nov 1;55(6):1032–44. doi:10.1053/J.SEMNUCLMED.2025.07.005
156. Badrigilan S, Heydarpanahi F, Choupani J, Jaymand M, Samadian H, Hoseini-Ghahfarokhi M, et al. A review on the biodistribution, pharmacokinetics and toxicity of bismuth-based nanomaterials. *Int J Nanomedicine.* 2020;15:7079. doi:10.2147/IJN.S250001 PubMed PMID: 33061369.
157. Miederer M, Benešová-Schäfer M, Mamat C, Kästner D, Pretze M, Michler E, et al. Alpha-Emitting radionuclides: Current status and future perspectives. *Pharmaceuticals.* 2024 Jan 1;17(1):76. doi:10.3390/PH17010076 PubMed PMID: 38256909.
158. Peng J, Li S, Ti H. Sensitize tumor immunotherapy: Immunogenic cell death inducing nanosystems. *Int J Nanomedicine.* 2024 Jun 13;19:5895–930. doi:10.2147/IJN.S457782 PubMed PMID: 38895146.
159. Satheeskumar R. AI-driven diagnostics and personalized treatment planning in oral oncology: Innovations and future directions. *Oral Oncology Reports.* 2025 Mar 1;13:100704. doi:10.1016/J.OOR.2024.100704

Index

Note: **Bold** page numbers refer to tables and *italic* page numbers refer to figures.